Bushmann

Kurt Bauer, Dorothea Garbe, Horst Surburg

Common Fragrance and Flavor Materials

Preparation, Properties and Uses
Third Completely Revised Edition

 **WILEY-VCH**

Weinheim · New York · Chichester · Brisbane · Singapore · Toronto

Kurt Bauer, Dorothea Garbe, Horst Surburg

Common Fragrance and Flavor Materials

Preparation, Properties and Uses
Third Completely Revised Edition

 **WILEY-VCH**

Weinheim · New York · Chichester · Brisbane · Singapore · Toronto

1st edition 1985
2nd revised edition 1990
3rd completely revised edition 1997
Copyright © 1997 by Wiley-VCH Verlag GmbH, D-69469 Weinheim (Germany)

e-mail (for orders and customer service enquiries): sales-books@wiley-vch.de
Visit our Home Page on http://www.wiley-vch.de

Other Editorial Offices
John Wiley & Sons, Inc., 605 Third Avenue,
New York, NY 10158-0012, USA

John Wiley & Sons Ltd, Baffins Lane,
Chichester, West Sussex, PO19 1UD, England

Jacaranda Wiley Ltd, 33 Park Road, Milton,
Queensland 4064, Australia

John Wiley & Sons (Asia) Pte Ltd, 2 Clementi Loop #02-01,
Jin Xing Distripark, Singapore 0512

John Wiley & Sons (Canada) Ltd, 22 Worcester Road,
Rexdale, Ontario M9W 1L1, Canada

Library of Congress Cataloguing-in-Publication Data
Bauer, Kurt. G.
    Common fragrance and flavor materials: preparation,
  properties, and uses / Kurt Bauer, Dorothea Garbe, Horst Surburg.
  — 3rd completely rev. ed.   p.   cm.
    Includes bibliographical references and indexes.
    ISBN 3-527-28850-3
    1. Flavoring essences.   2. Essences and essential oils.   3. Odors.
I. Garbe, Dorothea.   II. Surburt, Horst.   III. Title.
TP418.B38   1997
664'.52 — dc21   97-22233
                                                            CIP

Deutsche Bibliothek Cataloguing-in-Publication Data:
Die Deutsche Bibliothek – CIP-Einheitsaufnahme
Bauer, Kurt:
Common fragrance and flavor materials: preparation, properties and uses / Kurt Bauer;
Dorothea Garbe; Horst Surburg. – 3., completely rev. ed. – Weinheim; New York;
Chichester; Brisbane; Singapore; Toronto: Wiley-VCH, 1997
    ISBN 3-527-28850-3

British Library Cataloguing in Publication Data

A catalogue record for this book is available from the British Library

ISBN 3-527-28850-3

Typeset by Alden Press
Printed and bound in Great Britain by Bookcraft (Bath) Ltd
This book is printed on acid-free paper responsibly manufactured from sustainable
forestation, for which at least two trees are planted for each one used for paper production.

Preface to the Third Edition

Twelve years after its first publication comes the third edition of "Common Fragrance and Flavor Materials". The content has been updated in many respects while retaining the proven concept.

In the case of the single-component fragrance and flavor materials, those compounds have been included which have become established on the market, as well as those which have attracted considerable interest on account of their outstanding organoleptic properties and have contributed to the composition of new fragrance types. The production processes for all fragrance and flavor materials described in the book have been critically reviewed. New processes have been taken into account, and those that are clearly outdated have been eliminated. A few compounds that have declined in importance or whose use is now restricted for toxicological reasons have been removed from the text, as have several essential oils. The latest publications and standards concerning essential oils and natural raw materials have been included in the new edition, making it an up-to-date reference work. For the first time references are cited for all essential oils, giving newcomers to the field a quick overview of the original literature. The chapters on quality control and product safety have been expanded and brought up to date.

The authors wish to thank all the colleagues whose specialist advice assisted us in revising the book.

Holzminden, February 1997

K. Bauer
D. Garbe
H. Surburg

Preface to the Second Edition

Within three years of publication the first edition of "Common Fragrance and Flavor Materials" was out of print and is now followed by this second edition. As in the case of the first edition this book is based mainly on a chapter on "Flavors and Fragrances" which has since been published in English in Ullmann's Encyclopedia of Industrial Chemistry.

We would like to thank our readers for their suggestions for improvement and further development of the contents which were contained in several book reviews. We have not followed up all the suggestions for the simple reason that we did not wish to change the character of the book, which is expressly aimed at a general audience interested in commonly used fragrance and flavor materials, and not at experts in the field.

The chapter on "Single Fragrance and Flavor Compounds" has been updated to include new developments. production methods have been brought up-to-date and CAS registry numbers have been added to all single compounds described. The former chapters "Essential Oils" and "Animal Secretions" have been grouped together under the heading "Natural Raw Materials in the Flavor and Fragrance Industry" and thoroughly revised to include new literature references.

Holzminden, February 1990

K. Bauer
D. Garbe
H. Surburg

Preface to the First Edition

Fragrance and flavor materials are used in a wide variety of products, such as soaps, cosmetics, toiletries, detergents, alcoholic and non-alcoholic beverages, ice cream, confectioneries, baked goods, convenience foods, tobacco products, and pharmaceutical preparations. This book presents a survey of those natural and synthetic fragrance and flavor materials which are produced commercially on a relatively large scale, or which are important because of their specific organoleptic properties. It provides information concerning their properties, methods employed in their manufacture, and areas of application. Therefore, the book should be of interest to anyone involved or interested in fragrance and flavor, *e.g.*, perfumers, flavorists, individuals active in perfume and flavor application, food technologists, chemists, and even laymen.

The book is, essentially, a translation of the chapter on fragrance and flavor materials in *Ullmanns Encyklopädie der technischen Chemie*, Volume 20, 4th Edition, Verlag Chemie GmbH, Weinheim (Federal Republic of Germany), 1981. The original (German) text has been supplemented by inclusion of recent developments and of relevant information from other sections of the Encylopedia. The present English version will make the information available to a wider circle of interested readers.

The condensed style of presentation of "Ullmann's" has been maintained. A more detailed treatment of various items and aspects was considered but was believed to be outside the scope of this book. Additional information, however, can be obtained from the literature cited.

To improve its usefulness, the book contains
– a formula index, including CAS registry numbers;
– an alphabetical index of single fragrance and flavor compounds, essential oils, and animal secretions.
Starting materials and intermediates are not covered by these indexes.

The authors wish to express their gratitude to:
– Haarmann & Reimer Company, in particular to its General Manager, Dr. C. Skopalik, who suggested the publication of this book in English and who, at an earlier stage, provided time and facilities for writing the chapter on fragrance and flavor materials in *Ullmanns Encyklopädie der technischen Chemie* (1981), and to Dr. Hopp, Vice President Research, for valuable additions to his book;
– all others who provided information and suggestions for the chapter in Ullmann's Encyclopedia and thereby for this book.

The authors are most grateful to Dr. J. J. Kettenes-van den Bosch and Dr. D. K. Kettenes for translating the original German text into English and for their suggestions and help in shaping the present book. Drs. Kettenes thank Mr. W. S. Alexander, Hockessin, Delaware (USA), for critically reviewing the English.

Holzminden, June 1984   K. Bauer
                                                           D. Garbe

Contents

1 Introduction

1.1 History

Since early antiquity, spices and resins from animal and plant sources have been used extensively for perfumery and flavor purposes, and to a lesser extent for their observed or presumed preservative properties. Fragrance and flavor materials vary from highly complex mixtures to single chemicals. Their history began when people discovered that components characteristic of the aroma of natural products could be enriched by simple methods. Recipes for extraction with olive oil and for distillation have survived from pre-Christian times to this day.

Although distillation techniques were improved, particularly in the 9th century A.D. by the Arabs, the production and application of these concoctions remained essentially unchanged for centuries. Systematic development began in the 13th century, when pharmacies started to prepare so-called remedy oils and later recorded the properties and physiological effects of these oils in pharmacopoeias. Many essential oils currently used by perfumers and flavorists were originally prepared by distillation in pharmacies in the 16th and 17th centuries.

Another important step in the history of natural fragrance materials occurred in the first half of the 19th century, when the production of essential oils was industrialized due to the increased demand for these oils as perfume and flavor ingredients. Around 1850, single organic compounds were also used for the same purposes. This development resulted from the isolation of cinnamaldehyde from cinnamon oil by DUMAS and PÉLIGOT in 1834, and the isolation of benzaldehyde from bitter almond oil by LIEBIG and WÖHLER in 1837. The first synthetic 'aroma oils' were introduced between 1845 and 1850. These consisted of lower molecular mass fatty acid esters of several alcohols and were synthesized by the chemical industry for their fruity odor. Methyl salicylate followed in 1859 as 'artificial wintergreen oil' and benzaldehyde in 1870 as 'artificial bitter almond oil.' With the industrial synthesis of vanillin (1874) and coumarin (1878) by Haarmann & Reimer (Holzminden, Federal Republic of Germany), a new branch of the chemical industry was founded.

The number of synthetically produced fragrance and flavor chemicals has since expanded continually as a result of the systematic investigation of essential oils and fragrance complexes for odoriferous compounds. Initially, only major components were isolated from natural products; their structure was then elucidated and processes were developed for their isolation or synthesis. The present trend, however, is to isolate and identify characteristic fragrance and flavor substances that occur in the natural products in only trace amounts. The isolation and structure elucidation of these components requires the use of sophisticated chromatographic and spectroscopic techniques. Interesting products can then be synthesized.

1.2 Definition

Fragrance and flavor substances are comparatively strong-smelling organic compounds with characteristic, usually pleasant odors. They are, therefore, used in perfumes and perfumed products, as well as for the flavoring of foods and beverages. Whether a particular product is called a fragrance or a flavor substance depends on whether it is used as a perfume or a flavor. Fragrances and flavors are, like taste substances, chemical messengers, their receptors being the olfactory cells in the nose, [1], [2].

1.3 Physiological Importance

Chemical signals are indispensable for the survival of many organisms which use chemoreceptors to find their way, to hunt for and inspect food, to detect enemies and harmful objects, and to find members of the opposite sex (pheromones). These functions are no longer vitally important for humans. The importance of flavor and fragrance substances in humans has evolved to become quantitatively and qualitatively different from that in other mammals; this is because humans depend to a greater extent on acoustic and optical signals for orientation. However, humans have retained the ability to detect odors and human behavior can undoubtedly be affected by fragrances and aromas.

Sensory information obtained from the interaction of fragrance and flavor molcules with olfactory and taste receptors is processed in defined cerebral areas, resulting in perception. During the past 10 years much research has been done concerning sensory perception and results have been published in, e.g., [2], [3], [4].

Although food acceptance in humans is determined mainly by appearance and texture, flavor is nevertheless also important. For example, spices are added to food not for their nutritional value, but for their taste and flavor. Furthermore, aromas that develop during frying and baking enhance the enjoyment of food.

Unlike flavoring substances, fragrances are not vitally important for humans. The use of fragrances in perfumery is primarily directed toward invoking pleasurable sensations by shifting the organism's emotional level. Whereas 'naturalness' is preferred in aromas (generally mixtures of many compounds), the talent and imagination of the perfumer is essential for the creation of a perfume.

1.4 Natural, Nature-identical, and Artificial Products

Natural products are obtained directly from plant or animal sources by physical procedures. *Nature-identical* compounds are produced synthetically, but are chemically identical to their natural counterparts. *Artificial* flavor substances are compounds that have not yet been identified in plant or animal products for human consumption. Alcohols, aldehydes, ketones, esters, and lactones are classes of compounds that are represented most frequently in natural and artificial fragrances.

Nature-identical aroma substances are, with very few exceptions, the only synthetic compounds used in flavors besides natural products. The primary functions of the olfactory and taste receptors, as well as their evolutionary development, may explain why artificial flavor substances are far less important. The majority of compounds used in fragrances are those identified as components of natural products, e.g., constituents of essential oils or resins. The fragrance characteristics of artificial compounds nearly always mimic those of natural products.

1.5 Organoleptic Properties and Structure

Similarity between odors arises because dissimilar substances or mixtures of compounds may interact with receptors to create similar sensory impressions in the sensory centers of the brain. The group of musk fragrances (comprising macrocyclic ketones and esters as well as aromatic nitro compounds and polycyclic aromatics) are, for example, compounds with similar odors but totally different structures [5], [6]. Small changes in structure (e.g., the introduction of one or more double bonds in aliphatic alcohols or aldehydes) may, however, alter a sensory impression or intensify an odor by several orders of magnitude. Increasing knowledge of the structure and functioning of olfactory receptors provides a better scientific basis for the correlation of odor and structure in fragrance and flavor substances, and facilitates the more accurate prediction of the odor of still unknown compounds [7].

1.6 Volatility

Fragrances must be volatile to be perceived. Therefore, in addition to the nature of the functional groups and the molecular structure of a compound, the molecular mass is also an important factor. Molecular masses of ca. 200 occur relatively frequently; masses over 300 are an exception.

Since fragrance compounds differ in volatility, the odor of a perfume composition changes during evaporation and is divided into the top note, the middle notes or body, and the end note or dry out, which consists mainly of less volatile compounds. Odor perception also depends largely on odor intensity. Therefore, the typical note is not determined only by the most volatile compounds.

In some cases, substances (fixatives) are added to perfumes to prevent the more volatile components from evaporating too rapidly [8].

1.7 Threshold Concentration

Due to the specificity of olfactory receptors, some compounds can be perceived in extremely low concentrations and significant differences in threshold concentrations are observed. The threshold concentration is defined as the lowest concentration at which a chemical compound can be distinguished with certainty from a blank under standard conditions.

For the compounds described in Chapter 2, threshold concentrations vary by a factor of 10^6–10^7. This explains why some fragrance and flavor compounds are manufactured in quantities of a few kilograms per year, others in quantities of several thousands of tons.

The relative contribution of a particular compound (its odor or flavor value) to the odor impression of a composition can be expressed as the ratio between the actual concentration of the compound and its threshold concentration [9], [9a].

1.8 Odor Description

The odors of single chemical compounds are extremely difficult to describe un-
equivocally. The odors of complex mixtures are often impossible to describe unless
one of the components is so characteristic that it largely determines the odor or
flavor of the composition. Although an objective classification is not possible, an
odor can be described by adjectives such as flowery, fruity, woody, or hay-like,
which relate the fragrances to natural or other known products with similar odors.
    A few terms used to describe odors are listed below:

Aldehydic	odor note of the long-chain fatty aldehydes, e.g., fatty-sweaty, ironed laundry, seawater
Animal(ic)	typical notes from the animal kingdom, e.g., musk, castoreum, skatol, civet, ambergris
Balsamic	heavy, sweet odors, e.g., cocoa, vanilla, cinnamon, Peru balsam
Camphoraceous	reminiscent of camphor
Citrus	fresh, stimulating odor of citrus fruits such as lemon or orange
Earthy	humus-like, reminiscent of humid earth
Fatty	reminiscent of animal fat and tallow
Floral, flowery	generic terms for odors of various flowers
Fruity	generic term for odors of various fruits
Green	typical odor of freshly cut grass and leaves
Herbaceous	noncharacteristic, complex odor of green herbs with, e.g., sage, minty, eucalyptus-like, or earthy nuances
Medicinal	odor reminiscent of disinfectants, e.g., phenol, lysol, methyl salicylate
Metallic	typical odor observed near metal surfaces, e.g., brass or steel
Minty	peppermint-like odor
Mossy	typical note reminiscent of forests and seaweed
Powdery	note associated with toilet powders (talcum), diffusively sweet
Resinous	aromatic odor of tree exudates
Spicy	generic term for odors of various spices
Waxy	odor resembling that of candle wax
Woody	generic term for the odor of wood, e.g., cedarwood, sandalwood

2 Single Fragrance and Flavor Compounds

Fragrance and flavor compounds of commercial interest are arranged according to the Beilstein system of functional groups, not according to their organoleptic properties, since relationships between odor and structure are difficult to establish. However, the Beilstein system has been abandoned in a few cases for practical reasons.

In each class of parent compounds, hydrocarbons and oxygen-containing compounds are described first. Nitrogen- and sulfur-containing compounds are treated at the end of each of these sections under the heading Miscellaneous Compounds. Aliphatic compounds are discussed in Section 2.1, followed by the terpenes. The terpenes constitute a very important group of compounds and are subdivided into acyclic terpenes (Section 2.2) and cyclic terpenes (Section 2.3). Nonterpenoid cycloaliphatics are described in Section 2.4. Aromatic compounds are discussed in Section 2.5. Phenols and phenol derivatives are described under a separate heading (Section 2.6) on account of their biogenetic and odor relationships. Methylenedioxyphenyl derivatives are also described under this heading for the same reason even though, systematically, they belong to the oxygen-containing heterocycles (Section 2.7). Compounds that are only produced in small quantities, but which are important due to their high odor intensity, are mentioned but not described in detail.

When available, trade names are given for individual fragrance and flavor materials. The names of the suppliers are given as follows:

Agan	= Agan Aroma Chemicals Ltd., Israel
Aromor	= Aromor Flavors & Fragrances Ltd., Israel
BASF	= BASF AG, Germany
BBA	= Bush Boake Allen, England
Dragoco	= DRAGOCO Gerberding & Co. AG, Germany
Firmenich	= Firmenich S.A., Switzerland
FR	= Fragrance Resources Inc., USA
Giv.-Roure	= Givaudan-Roure (International) S.A., Switzerland
H&R	= Haarmann & Reimer GmbH, Germany
Henkel	= Henkel KGaA, Germany
Hüls	= Chemische Werke Hüls, Germany
IFF	= International Flavors & Fragrances, USA
Kao	= Kao Corp., Japan
NZ	= Nippon Zeon Co., Ltd., Japan

PFW   = Polak's Frutal Works BV, Netherlands
Polarome   = Polarome Mfg. Co., Inc., USA
Quest   = Quest International, Netherlands
Rhône-Poulenc   = Rhône-Poulenc, France
Soda Aromatic   = Soda Aromatic Co., Japan
Takasago   = Takasago Perfumery Co., Japan

Monographs on fragrance materials and essential oils which have been published by the Research Institute for Fragrance Materials (RIFM) in 'Food and Chemical Toxicology' are cited below the individual compounds as 'FCT' with year, volume and page of publication.

2.1 Aliphatic Compounds

The acyclic terpenes are discussed separately in Section 2.2. Some of the cyclo-aliphatic fragrance and flavor compounds are structurally related to the cyclic terpenes and are, therefore, discussed in Section 2.4 after the cyclic terpenes.

2.1.1 Hydrocarbons

Saturated and unsaturated aliphatic hydrocarbons with straight as well as branched chains occur abundantly in natural foodstuffs, but they contribute to the odor and taste only to a limited extent. The highly unsaturated hydrocarbons 1,3-*trans*-5-*cis*-undecatriene [*51447-08-6*] and 1,3-*trans*-5-*trans*-undecatriene [*19883-29-5*], however, contribute to the odor of galbanum oil [10].

2.1.2 Alcohols

Free and esterified, saturated primary alcohols occur widely in nature, e.g., in fruit. Since their odor is relatively weak, their use as components in fragrance compositions is limited. Their use in aroma compositions, especially for fruit flavors, is by far more important (e.g., straight-chain C_4–C_{10} alcohols, isoamyl alcohol). Unsaturated alcohols are most important (e.g., leaf alcohol with its intensely green odor) and may impart characteristic notes to compositions.

Naturally occurring fatty alcohols used in the fragrance industry are produced principally by reduction of the methyl esters of the corresponding carboxylic acids, which are obtained by transesterification of natural fats and oils with methanol. Industrial reduction processes include catalytic hydrogenation in the presence of copper–chromium oxide catalysts (Adkins catalysts) and reduction with sodium (Bouveault–Blanc reduction). Unsaturated alcohols can also be prepared by the latter method. Numerous alcohols used in flavor compositions are, meantime, produced by biotechnological processes [11]. Alcohols are starting materials for aldehydes and esters.

3-Octanol [*589-98-0*]
$CH_3(CH_2)_4CH(OH)CH_2CH_3$, $C_8H_{18}O$, M_r 130.23, $bp_{97.6 kPa}$ 176–176.5 °C, d_4^{20} 0.8264, n_D^{20} 1.4252, may occur in its optically active form. It is a colorless liquid that has a mushroomy-earthy odor and occurs in mushrooms. 3-Octanol can be obtained by hydrogenation of 3-octanone; it is used in lavender compositions and for imparting mushroom-like odors.
FCT 1979 (**17**) p. 881.

2,6-Dimethyl-2-heptanol [*13254-34-7*]

$C_9H_{20}O$, M_r 144.26, $bp_{101.3 kPa}$ 170–172 °C, d_{20}^{20} 0.8186, n_D^{20} 1.4248, which has not yet been found in nature, is a colorless liquid with a delicate, flowery odor reminiscent of fresias. It is synthesized from 6-methyl-5-hepten-2-one and methylmagnesium chloride by a Grignard reaction, followed by hydrogenation, and is used in flowery perfume compositions.
FCT 1992 (**30**) p. 30.

   Trade Names. Dimetol (Giv.-Roure), Freesiol (H&R), Lolitol (IFF).

trans-**2-Hexen-1-ol** [*928-95-0*]
$CH_3CH_2CH_2CH = CHCH_2OH$, $C_6H_{12}O$, M_r 100.16, $bp_{101.3 kPa}$ 155 °C, d_4^{20} 0.8459, n_D^{20} 1.4382, occurs in many fruits and has a fruity, green odor, which is sweeter than that of the isomeric *cis*-3-hexen-1-ol and is, therefore, preferred in aroma compositions.
FCT 1974 (**12**) p. 911.

cis-**3-Hexen-l-ol** [*928-96-1*], **leaf alcohol**
$CH_3CH_2CH = CHCH_2CH_2OH$, $C_6H_{12}O$, M_r 100.16, $bp_{101.3 kPa}$ 156–157 °C, d_4^{20} 0.8459, n_D^{20} 1.4384, is a colorless liquid with the characteristic odor of freshly cut grass. *Robinia pseudacacia* and mulberry leaf oil contain up to 50% leaf alcohol, and green tea up to 30%. Small quantities occur in the green parts of nearly all plants.
   A stereospecific synthesis for *cis*-3-hexen-1-ol starts with the ethylation of sodium acetylide to 1-butyne, which is reacted with ethylene oxide to give 3-hexyn-1-ol. Selective hydrogenation of the triple bond in the presence of palladium

catalysts yields *cis*-3-hexen-1-ol. Biotechnological processes have been developed for its synthesis as a natural flavor compound, e.g., [12].

Leaf alcohol is used to obtain natural green top notes in perfumes and flavors. In addition, it is the starting material for the synthesis of 2-*trans*-6-*cis*-nonadien-1-ol and 2-*trans*-6-*cis*-nonadien-1-al.
FCT 1974 (12) p. 909.

1-Octen-3-ol [*3391-86-4*]
$CH_3(CH_2)_4CH(OH)CH=CH_2$, $C_8H_{16}O$, M_r 128.21, $bp_{94.6\,kPa}$ 175–175.2 °C, d_4^{20} 0.8383, n_D^{20} 1.4378, may occur in the optically active form. It is found, for example, in lavender oil and is a steam-volatile component of mushrooms. 1-Octen-3-ol is a liquid with an intense mushroom, forest-earthy odor that can be prepared by a Grignard reaction from vinylmagnesium bromide and hexanal. It is used in lavender compositions and in mushroom aromas.
FCT 1976 (14) p. 681.

Trade Name. Matsutake alcohol (Takasago).

9-Decen-1-ol [*13019-22-2*]
$CH_2=CH(CH_2)_7CH_2OH$, $C_{10}H_{20}O$, M_r 157.27, $bp_{0.27\,kPa}$ 85–86 °C, n_D^{20} 1.4480, has been identified as a trace constituent of cognac. It is a colorless liquid with a fresh, dewy, rose note that can be prepared by partial dehydration of 1,10-decanediol. It is used in rosy-floral soap perfumes.
FCT 1974 (12) p. 405.

Trade Name. Rosalva (IFF).

10-Undecen-1-ol [*112-43-6*]
$CH_2=CH(CH_2)_8CH_2OH$, $C_{11}H_{22}O$, M_r 170.29, $bp_{2.1\,kPa}$ 133 °C, d_4^{15} 0.8495, n_D^{20} 1.4500, was tentatively identified, e.g., in the wax gourd (*Benincasa hispida*, Cogn.) [13], and is a colorless liquid with a fatty-green, slightly citrus-like odor. It can be synthesized from 10-undecylenic acid and is used to give flower perfumes a fresh note.
FCT 1973 (11) p. 107.

3,4,5,6,6-Pentamethyl-3(or -4)-hepten-2-ol [81787-06-6] and [81787-07-7] and 3,5,6,6-Tetramethyl-4-methyleneheptan-2-ol [81787-05-5] (mixture)

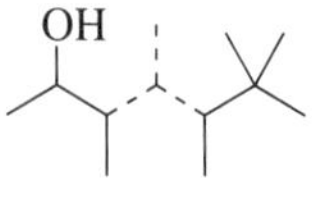

$C_{12}H_{24}O$ M_r 184.32, is a mixture of isomers where one of the dashed lines represents a carbon–carbon double bond, the others a single bond. None of the compounds occur in nature. It is a colorless to slightly yellow liquid, d_4^{20} 0.864–0.872, n_D^{20} 1.454–1.460, with a fine woody, ambra, dry odor with a clean vetivert character. Synthesis starts from 2-methyl-2-butene (isoamylene) which is dimerized and the product acetylated to give the corresponding hepten-2-ones. The hepten-2-ols are obtained by reduction with $NaBH_4$ [14]. The mixture is used in perfume compositions, for example, for detergents.

Trade Name. Kohinool (IFF).

2-*trans*-6-*cis*-Nonadien-1-ol [*28069-72-9*], **violet leaf alcohol**
$CH_3CH_2CH=CHCH_2CH_2CH=CHCH_2OH$, $C_9H_{16}O$, M_r 140.22, $bp_{1.5\,kPa}$ 96 100 °C, d_4^{25} 0.8622, n_D^{25} 1.4631, occurs, for example, in cucumber oil, violet leaf oil, and violet blossom oil. It is a colorless liquid with an intense, heavy-fatty, green odor, reminiscent of violet leaves. The starting material for the synthesis of 2-*trans*-6-*cis*-nonadien-1-ol is *cis*-3-hexen-1-ol, which is converted via its halide into the corresponding Grignard reagent. The Grignard reagent is reacted with acrolein to give 1,6-nonadien-3-ol, which is converted into 2-*trans*-6-*cis*-nonadien-1-ol by allylic rearrangement.

Nonadienol is a powerful fragrance substance. It is used in fine fragrances to create refined violet odors and to impart interesting notes to other blossom compositions. In aroma compositions it is used for fresh-green cucumber notes. FCT 1982 (**20**) p. 771.

2.1.3 Aldehydes and Acetals

Aliphatic aldehydes are among the most important components used in perfumery. Although the lower fatty aldehydes C_2–C_7 occur widely in nature, they are – with the exception of hexanal – seldom used in fragrance compositions. The lower aldehydes (e.g., acetaldehyde, isobutyraldehyde, isovaleraldehyde, and 2-methyl-butyraldehyde) impart fruity and roast characters to flavor compositions. Fatty aldehydes C_8–C_{13}, however, are used, singly or in combination, in nearly all perfume types and also in aromas. Their odor becomes weaker with increasing molecular mass, so that aldehydes $> C_{13}$ are not important as perfume ingredients.

In addition to the straight-chain saturated aldehydes, a number of branched-chain and unsaturated aliphatic aldehydes are important as fragrance and flavoring materials. The double unsaturated 2-*trans*-6-*cis*-nonadienal [*557-48-2*], 'violet leaf aldehyde' (the dominant component of cucumber aroma), is one of the most potent fragrance and flavoring substances; it is, therefore, only used in very small amounts. 2-*trans*,4-*trans*-Decadienal [*25152-84-5*] with its specifically fatty odor character is indispensible in chicken meat flavor compositions.

Acetals derived from aliphatic aldehydes have odor characteristics that resemble those of the aldehydes but are less pronounced. These acetals contribute to the aroma of alcoholic beverages, but can rarely be used in flavoring composi-tions because they are not sufficiently stable. Since they are resistant to alkali, a number of them (e.g., heptanal dimethyl acetal and octanal dimethyl acetal) are occasionally incorporated into soap perfumes.

Fatty aldehydes are generally produced by dehydrogenation of alcohols in the presence of suitable catalysts. The alcohols are often cheap and available in good purity. Aldehyde synthesis via the oxo process is less suitable since the resultant products are often not pure enough for flavor and perfume purposes. Specific syntheses for the branched-chain and unsaturated aldehydes that are important in perfumery and flavoring techniques are described under the individual compounds.

Hexanal [*66-25-1*], **caproaldehyde**
$CH_3(CH_2)_4CHO$, $C_6H_{12}O$, M_r 100.16, $bp_{101.3\,kPa}$ 128 °C, d_4^{20} 0.8139, n_D^{20} 1.4039, occurs, for example, in apple and strawberry aromas as well as in orange and lemon oil. It is a colorless liquid with a fatty-green odor and in low concentration is reminiscent of unripe fruit.
  Hexanal is used in fruit flavors and, when highly diluted, in perfumery for obtaining fruity notes.
FCT 1973 (**11**) p. 111.

Octanal [*124-13-0*], **caprylaldehyde**
$CH_3(CH_2)_6CHO$, $C_8H_{16}O$, M_r 128.21, $bp_{101.3\,kPa}$ 171 °C, d_4^{20} 0.8211, n_D^{20} 1.4217, occurs in several citrus oils, e.g., orange oil. It is a colorless liquid with a pungent odor, which becomes citrus-like on dilution. Octanal is used in perfumery in low concentrations, in eau de cologne, and in artificial citrus oils.
FCT 1973 (**11**) p. 113.

Nonanal [*124-19-6*], **pelargonaldehyde**
$CH_3(CH_2)_7CHO$, $C_9H_{18}O$, M_r 142.24, $bp_{101.3\,kPa}$ 190–192 °C, d_4^{20} 0.8264, n_D^{20} 1.4273, occurs in citrus and rose oils. It is a colorless liquid with a fatty-roselike odor and is used in floral compositions, particularly those with rose characteristics.
FCT 1973 (**11**) p. 115.

Decanal [*112-31-2*], **caprinaldehyde**
$CH_3(CH_2)_8CHO$, $C_{10}H_{20}O$, M_r 156.27, $bp_{101.3\,kPa}$ 208–209 °C, d_4^{15} 0.830, n_D^{20} 1.4287, is a component of many essential oils (e.g., neroli oil) and various citrus peel oils. It is a colorless liquid with a strong odor, reminiscent of orange peel, that changes to a fresh citrus odor when diluted. Decanal is used in low concentrations in blossom fragrances (especially to create citrus nuances) and in the production of artificial citrus oils.
FCT 1973 (**11**) p. 477.

Undecanal [*112-44-7*]
$CH_3(CH_2)_9CHO$, $C_{11}H_{22}O$, M_r 170.29, $bp_{2.4\,kPa}$ 117 °C, d_4^{23} 0.8251, n_D^{20} 1.4325, occurs in citrus oils. It is a colorless liquid with a flowery-waxy odor that has aspects of freshness. Undecanal is the prototype of the perfumery aldehydes and is widely used in perfume compositions for imparting an 'aldehydic note.'
FCT 1973 (**11**) p. 481.

Dodecanal [*112-54-9*], **lauraldehyde, lauric aldehyde**
$CH_3(CH_2)_{10}CHO$, $C_{12}H_{24}O$, M_r 184.32, $bp_{13.3 kPa}$ 185 °C, d_4^{15} 0.8352, n_D^{20} 1.4350, is
a colorless liquid with a waxy odor; in high dilution it is reminiscent of violets.
Dodecanal occurs in several citrus oils and has been found in small amounts in
essential oils obtained from several *Pinus* species. It is used in perfumery in conifer
fragrances with fatty-waxy notes, but also in many other odor types. It is added to
aroma compositions to obtain citrus notes.
FCT 1973 (**11**) p. 483.

Tridecanal [*10486-19-8*]
$CH_3(CH_2)_{11}CHO$, $C_{13}H_{26}O$, M_r 198.34, $bp_{1.3 kPa}$ 128 °C, d_4^{18} 0.8356, n_D^{18} 1.4384,
occurs in lemon oil and has been identified as a volatile constituent of cucumber.
It is a colorless liquid having a fatty-waxy, slightly citrus-like odor. Addition of
tridecanal to fragrance compositions imparts fresh nuances in the top note as well
as in the dry out.

2-Methyldecanal [*19009-56-4*], **methyloctylacetaldehyde**
$CH_3(CH_2)_7CH(CH_3)CHO$, $C_{11}H_{22}O$, M_r 170.29, $bp_{98.8 kPa}$ 119–120 °C, d_4^{23} 0.8948,
n_D^{20} 1.4205, is not reported to have been found in nature. It is a colorless liquid with
an aldehydic, citrus-peel-like, waxy-green odor. 2-Methyldecanal is obtained as
a by-product in the manufacture of 2-methylundecanal by hydroformylation of
1-decene (see 2-methylundecanal). It is used in perfumery to refresh green and
citrus nuances.
FCT 1976 (**14**) p. 609.

2-Methylundecanal [*110-41-8*], **methylnonylacetaldehyde**
$CH_3(CH_2)_8CH(CH_3)CHO$, $C_{12}H_{24}O$, M_r 184.32, $bp_{1.3 kPa}$ 114 °C, d_4^{15} 0.830, n_D^{20}
1.4321, is reported as being found in nature. It is a colorless liquid, with an odor
markedly different from that of the isomeric dodecanal. It has a fatty odor with
incense and ambergris notes.
   2-Methylundecanal is produced by two routes:

1. 2-Undecanone is converted into its glycidate by reaction with an alkyl
    chloroacetate. Saponification of the glycidate, followed by decarboxylation,
    yields 2-methylundecanal

$$H_3C(CH_2)_8 \overset{\overset{\displaystyle CH_3}{|}}{C}=O + ClCH_2COOR \xrightarrow{-HCl} H_3C(CH_2)_8 \overset{\overset{\displaystyle CH_3}{|}}{\underset{\underset{\displaystyle O}{\diagdown \diagup}}{C}}-CHCOOR$$

$$\xrightarrow[\text{2) Decarboxylation}]{\text{1) Hydrolysis}} H_3C(CH_2)_8 \overset{\overset{\displaystyle CH_3}{|}}{C}HCHO$$

2. The second synthesis is based on the conversion of undecanal into
    2-methyleneundecanal by reaction with formaldehyde in the presence of
    catalytic amounts of amines [15]. Hydrogenation of 2-methyleneundecanal

yields methylnonylacetaldehyde. A convenient process starts from l-decene: hydroformylation gives a mixture consisting mainly of undecanal and 2-methyldecanal. Reaction of the crude product with formaldehyde in the presence of dibutylamine yields a mixture containing over 50% 2-methyle-neundecanal. After hydrogenation of the double bond, pure 2-methylunde-canal is separated from by-products by fractional distillation [16].

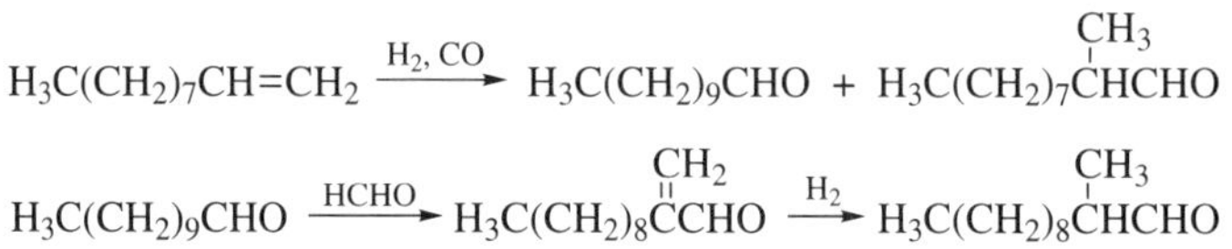

$$H_3C(CH_2)_7CH{=}CH_2 \xrightarrow{H_2, CO} H_3C(CH_2)_9CHO + H_3C(CH_2)_7\overset{\overset{\displaystyle CH_3}{|}}{C}HCHO$$

$$H_3C(CH_2)_9CHO \xrightarrow{HCHO} H_3C(CH_2)_8\overset{\overset{\displaystyle CH_2}{\|}}{C}CHO \xrightarrow{H_2} H_3C(CH_2)_8\overset{\overset{\displaystyle CH_3}{|}}{C}HCHO$$

In comparison with other fatty aldehydes, 2-methylundecanal is used in perfumery in rather large amounts to impart conifer notes, particularly fir impressions, but frequently also in phantasy compositions.
FCT 1973 (**11**) p. 485.

trans-2-Hexenal [6728-26-3], leaf aldehyde

$CH_3CH_2CH_2CH{=}CHCHO$, $C_6H_{10}O$, M_r 98.14, $bp_{101.3\,kPa}$ 146–147 °C, d_4^{20} 0.8491, n_D^{20} 1.4480, is the simplest straight-chain unsaturated aldehyde of interest for perfumes and flavors. It occurs in essential oils obtained from green leaves of many plants.

trans-2-Hexenal is a colorless, sharp, herbal-green smelling liquid with a slight acrolein-like pungency. Upon dilution, however, it smells pleasantly green and apple-like. The aldehyde can be synthesized by reacting butanal with vinyl ethyl ether in the presence of boron trifluoride, followed by hydrolysis of the reaction product with dilute sulfuric acid [17].

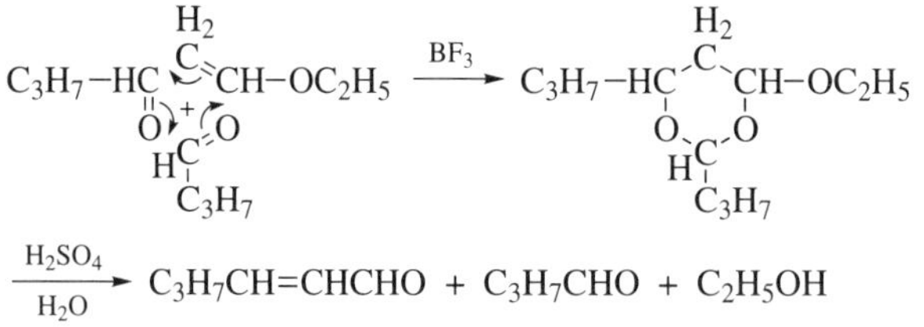

$$\xrightarrow[H_2O]{H_2SO_4} C_3H_7CH{=}CHCHO + C_3H_7CHO + C_2H_5OH$$

Biosynthetic ways for its production as natural flavor compound have been developed [18].

trans-2-Hexenal has an intense odor and is used in perfumes to obtain a green-leaf note, and in fruit flavors for green nuances.
FCT 1975 (**13**) p. 453.

cis-4-Heptenal [6728-31-0]

$CH_3CH_2CH{=}CHCH_2CH_2CHO$, $C_7H_{12}O$, M_r 112.17, $bp_{1.33\,kPa}$ 41 °C, n_D^{20} 1.4343, is a widespread volatile trace constituent of food flavors. It is a colorless, oily liquid with a powerful, fatty, somewhat fishy and, in high dilution, creamy odor. It can be prepared from 1-butyne (via lithium 1-butynide) and acrolein (which is converted into 2-bromopropionaldehyde dimethyl acetal). The resulting 4-heptynal dimethyl acetal is cleaved and the triple bond is hydrogenated catalytically to give *cis*-4-heptenal [19].

cis-4-Heptenal is used in cream, butter, and fat flavors.

2,6-Dimethyl-5-hepten-1-al [*106-72-9*]

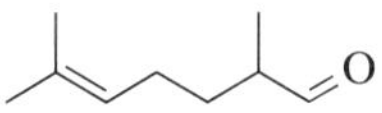

$C_9H_{16}O$, M_r 140.23, bp_{2kPa} 79–80 °C, d_4^{28} 0.848, n_D^{20} 1.4492 was identified in ginger. It is a yellow liquid with a powerful, green, cucumber-like and melon odor. It can be prepared by Darzens reaction of 6-methyl-5-hepten-2-one with ethyl chloroacetate. The intermediate glycidate is saponified and decarboxylated to yield the title compound.

It is used in many fragrance types and is invaluable in the creation of melon and cucumber notes.

Trade Name. Melonal (Giv.-Roure).

10-Undecenal [*112-45-8*]
$CH_2 = CH(CH_2)_8CHO$, $C_{11}H_{20}O$, M_r 168.28, $bp_{0.4kPa}$ 103 °C, d_4^{21} 0.8496, n_D^{21} 1.4464, was identified, e.g., in coriander leaf extract [20]. It is a colorless liquid with a fatty-green, slightly metallic, heavy-flowery odor. The aldehyde can be synthesized from undecylenic acid, for example, by hydrogenation of the acid chloride (Rosenmund reduction) or by reaction with formic acid in the vapor phase in the presence of titanium dioxide. In perfumery, 10-undecenal is one of the aldehydes essential for creating the 'aldehydic note.'
FCT 1973 (**11**) p. 479.

2,6,10-Trimethyl-5,9-undecadienal [*24048-13-3*] and [*54082-68-7*]

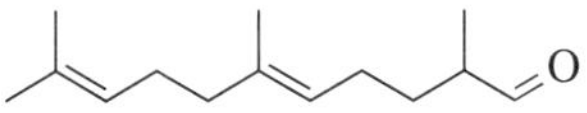

$C_{14}H_{24}O$, M_r 208.35, d_4^{20} 0.870–0.877, n_D^{20} 1.468–1.473, not found in nature, is a clear, colorless to pale yellowish liquid. It has an aldehydic-floral odor reminiscent of nerolidol, with fruity nuances. It can be prepared from geranylacetone (see p. 41) by Darzens reaction with ethyl chloroacetate through the corresponding glycidic ester which is hydrolyzed and decarboxylated.

2,6,10-Trimethyl-5,9-undecadienal is used to modify perfume compositions for soap, detergents and household products.

Trade Names. Oncidal (Dragoco), Profarnesal (H&R).

1,1-Dimethoxy-2,2,5-trimethyl-4-hexene [*67674-46-8*]

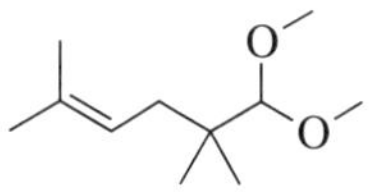

$C_{11}H_{22}O_2$, M_r 186.30, $bp_{1.6\,kPa}$ 82 °C, n_D^{20} 1.441, is a colorless to pale yellow liquid with a fresh fruity citrus grapefruit-peel-like odor. It has not been found in nature. It is prepared by reaction of 2,2,5-trimethyl-4-hexenal (from isobutyraldehyde and prenyl chloride) with methanol in the presence of calcium chloride [21].

Due to its alkali stability it is used in citrus compositions for soaps and detergents.

Trade Name. Methyl Pamplemousse (Giv.-Roure).

2.1.4 Ketones

Aliphatic monoketones are of minor importance as fragrance and aroma substances. 2-Alkanones (C_3–C_{15}) have been found in the volatile fractions of many fruits and foodstuffs, but they do not contribute significantly to their aroma. An exception are the odd-numbered methyl ketones C_7, C_9, C_{11} which possess a characteristic nutty note; they are used, e.g., in cheese flavor compositions. In perfumery, aliphatic ketones are used for accentuation, e.g., 3-octanone [*106-68-3*] for lavender notes. The hydroxyketone acetoin and the diketone 2,3-butanedione are commercially important aroma substances.

3-Hydroxy-2-butanone [*52217-02-4*], **acetoin**
$CH_3COCH(OH)CH_3$, $C_4H_8O_2$, M_r 88.11, $bp_{101.3\,kPa}$ 148 °C, d_{20}^{20} 1.0062, n_D^{20} 1.4171, has a pleasant buttery odor and both of its optical isomers occur widely in nature. It is synthesized by partial oxidation of 2,3-butanediol and is obtained as a by-product in the fermentation of molasses. It is used for flavoring margarine. FCT 1979 (**17**) p. 509.

2,3-Butanedione [*431-03-8*], **diacetyl**
$CH_3COCOCH_3$, $C_4H_6O_2$, M_r 86.09, bp 88 °C, d_4^{20} 0.9831, $n_D^{18.5}$ 1.3933, is a constituent of many fruit and food aromas and well-known as a constituent of butter. Many methods are known for its manufacture, e.g., dehydrogenation of 2,3-butanediol with a copper chromite catalyst [22]. Biotechnological production on an industrial scale is referred [23]. It is used mainly in aromas for butter and roast notes. Large quantities are used for flavoring margarine; small amounts are used in perfumes.
FCT 1979 (**17**) p. 765.

2.1.5 Acids and Esters

Straight-chain, saturated aliphatic acids are found in many essential oils and foods. These acids contribute to aromas, but are not important as fragrance substances. In flavor compositions, aliphatic acids up to C_{10} are used to accentuate certain aroma characteristics (C_3–C_8 for fruity notes; C_4, C_6–C_{12} for cheese flavors). However, straight-chain and some branched-chain aliphatic acids are of considerable importance as starting materials in the manufacture of esters, many of which are valuable fragrance and flavor materials. Aliphatic esters contribute to the aroma of nearly all fruits and many foods. Some are responsible for a particular fruit aroma, or for the smell of a particular flower; however, many of these esters possess a nonspecific fruity odor.

Most of the esters used are acetates and ethanol is the most common alcohol component. In nature, most esters are derived from alcohols and acids with an even number of carbon atoms. In addition to straight-chain saturated compounds, branched-chain compounds such as isoamyl esters, and unsaturated compounds such as hexenyl esters are important.

Although the odor of aliphatic esters with a small number of carbon atoms is strictly fruity, it changes to fatty-soapy and even metallic as the number of carbon atoms increases.

Esters are usually prepared by esterification of carboxylic acids with alcohols. Industrial procedures depend on the physical properties of the esters concerned. Biosynthetic methods may be applied to produce natural esters for flavor purposes [24].

In perfumery, acetates are the most important aliphatic esters; formates do not keep well. Animal and fatty notes become more pronounced in esters of higher fatty acids. Acetates of alcohols up to C_6 are used principally for fruity notes, whereas the acetates of C_8, C_{10}, and C_{12} alcohols are employed for blossom fragrances and for flower notes in general. Lauryl acetate in particular is also used for conifer notes.

In flavor compositions, aliphatic esters are preferred for artificial fruit aromas; as in nature, acetates and ethyl esters prevail.

2-Methyl-2-pentenoic acid [*3142-72-1*]

$C_6H_{10}O_2$, M_r 114.15, was detected to occur, e.g., in capers. It is a yellow liquid, d_4^{20} 0.979–0.987, n_D^{20} 1.457–1.462, with a dry acid note, found in the odor of strawberries. The acid can be prepared from the corresponding saturated one by α-bromination followed by dehydrobromination.

It is used in fragrances to enhance fruity notes and in strawberry flavors.

Trade Name. Strawberriff (IFF).

Ethyl formate [*109-94-4*]
$HCOOCH_2CH_3$, $C_3H_6O_2$, M_r 74.08, $bp_{101.3\,kPa}$ 54.5 °C, d^{20} 0.9168, n_D^{20} 1.3598, occurs widely in fruits. It is a liquid with a slightly pungent, fruity, ethereal odor and is used in fruit flavors.
FCT 1978 (**16**) p. 737.

***cis*-3-Hexenyl formate** [*33467-73-1*]
$HCOO(CH_2)_2CH = CHCH_2CH_3$, $C_7H_{12}O_2$, M_r 128.17, $bp_{101.3\,kPa}$ 155 °C, d_{25}^{25} 0.908, n_D^{20} 1.4270, has been identified in tea. It possesses a green-fruity odor and is used in perfumery and flavor compositions to impart fruity green notes.
FCT 1979 (**17**) p. 797.

Ethyl acetate [*141-78-6*]
$CH_3COOCH_2CH_3$, $C_4H_8O_2$ M_r 88.11, $bp_{101.3\,kPa}$ 77.1 °C, d_4^{20} 0.9003, n_D^{20} 1.3723, is a fruity smelling liquid with a brandy note and is the most common ester in fruits. It is used in fruit and brandy flavors.
FCT 1974 (**12**) p. 711.

Butyl acetate [*123-86-4*]
$CH_3COO(CH_2)_3CH_3$, $C_6H_{12}O_2$, M_r 116.16, $bp_{101.3\,kPa}$ 126.5 °C, d^{20} 0.882, n_D^{20} 1.3942, is a liquid with a strong fruity odor. It occurs in many fruits and is a constituent of apple aromas.
FCT 1979 (**17**) p. 515.

Isoamyl acetate [*123-92-2*]
$CH_3COO(CH_2)_2CH(CH_3)_2$, $C_7H_{14}O_2$, M_r 130.19, $bp_{101.3\,kPa}$ 142.5 °C, d_{25}^{25} 0.868–878, n_D^{18} 1.4017, is a strongly fruity smelling liquid and has been identified in many fruit aromas. It is the main component of banana aroma and is, therefore, also used in banana flavors.
FCT 1975 (**13**) p. 551.

Hexyl acetate [*142-92-7*]
$CH_3COO(CH_2)_5CH_3$, $C_8H_{16}O_2$, M_r 144.21, $bp_{101.3\,kPa}$ 171.5 °C, d_4^{15} 0.8779, n_D^{20} 1.4092, is a liquid with a sweet-fruity, pearlike odor. It is present in a number of fruits and alcoholic beverages, and is used in fruit aroma compositions.
FCT 1974 (**12**) p. 913.

3,5,5-Trimethylhexyl acetate [*58430-94-7*], **isononyl acetate**
$CH_3COO(CH_2)_2CH(CH_3)CH_2C(CH_3)_3$, $C_{11}H_{22}O_2$, M_r 186.29, does not occur in nature. Commercial isononyl acetate contains small amounts of by-products. It is a colorless liquid with a woody-fruity odor and is prepared from diisobutene by the oxo synthesis, followed by hydrogenation to the alcohol and acetylation. It is used in household perfumery.
FCT 1974 (**12**) p. 1009.

Trade Names. Inonyl acetate (Quest), Isononyl acetate (H&R), Vanoris (IFF).

trans-2-Hexenyl acetate [*2497-18-9*]
$CH_3COOCH_2CH = CH(CH_2)_2CH_3$, $C_8H_{14}O_2$, M_r 142.20, $bp_{2.1\,kPa}$ 67–68 °C, d^{20} 0.8980, n_D^{20} 1.4277, occurs in many fruits and in some essential oils, e.g., peppermint. It is a fresh-fruity, slightly green smelling liquid and is used in fruit flavors.
FCT 1979 (**17**) p. 793.

cis-3-Hexenyl acetate [*3681-71-8*]
$CH_3COO(CH_2)_2CH = CHCH_2CH_3$, $C_8H_{14}O_2$, M_r 142.20, $bp_{1.6\,kPa}$ 66 °C, has been identified in many fruit aromas and green tea. It is a prototype for green odors and is often used in combination with *cis*-3-hexenol.
FCT 1975 (**13**) p. 454.

Ethyl propionate [*105-37-3*]
$CH_3CH_2COOCH_2CH_3$, $C_5H_{10}O_2$, M_r 102.13, $bp_{101.3\,kPa}$ 99 °C, d^{20} 0.8917, n_D^{20} 1.3839, is found in many fruits and alcoholic beverages. It has a fruity odor reminiscent of rum and is used in flavor compositions for creating both fruity and rum notes.
FCT 1978 (**16**) p. 749.

Ethyl butyrate [*105-54-4*]
$CH_3(CH_2)_2COOCH_2CH_3$, $C_6H_{12}O_2$, M_r 116.16, $bp_{101.3\,kPa}$ 121–122 °C, d_4^{20} 0.8785, n_D^{20} 1.4000, occurs in fruits and alcoholic beverages, but also in other foods such as cheese. It has a fruity odor, reminiscent of pineapples. Large amounts are used in perfume and in flavor compositions.
FCT 1974 (**12**) p. 719.

Butyl butyrate [*109-21-7*]
$CH_3(CH_2)_2COOCH_2(CH_2)_2CH_3$, $C_8H_{16}O_2$, M_r 144.21, $bp_{101.3\,kPa}$ 166 °C, d_4^{20} 0.8709, n_D^{20} 1.4075, is a liquid with a sweet-fruity odor. It is a volatile constituent of many fruits and honey and is used in fruit flavor compositions.
FCT 1979 (**17**) p. 521.

Isoamyl butyrate [*106-27-4*]
$CH_3(CH_2)_2COO(CH_2)_2CH(CH_3)_2$, $C_9H_{18}O_2$, M_r 158.23, $bp_{101.3\,kPa}$ 178.5 °C, d_4^{20} 0.8651, n_D^{20} 1.4106, is a liquid with strongly fruity odor that occurs, e.g., in banana. It is used mainly in fruit flavors.
FCT 1979 (**17**) p. 823.

Hexyl butyrate [*2639-63-6*]
$CH_3(CH_2)_2COO(CH_2)_5CH_3$, $C_{10}H_{20}O_2$, M_r 172.27, $bp_{101.3\,kPa}$ 208 °C, d_4^{20} 0.8652, n_D^{15} 1.4160, is a liquid with a powerful fruity odor. It has been identified in a number of fruits and berries and is an important constituent of fruit flavor compositions.
FCT 1979 (**17**) p. 815.

cis-3-Hexenyl isobutyrate [*41519-23-7*]
$(CH_3)_2CHCOO(CH_2)_2CH=CHCH_2CH_3$, $C_{10}H_{18}O_2$, M_r 170.25, is found in spearmint oil. It smells fruity-green and is used in perfumery to create freshness in blossom compositions.
FCT 1979 (**17**) p. 799.

Ethyl isovalerate [*108-64-5*]
$(CH_3)_2CHCH_2COOCH_2CH_3$, $C_7H_{14}O_2$, M_r 130.19, $bp_{101.3 kPa}$ 134.7 °C, d_4^{20} 0.8656, n_D^{20} 1.3962, is a colorless liquid with a fruity odor reminiscent of blueberries. It occurs in fruits, vegetables, and alcoholic beverages. It is used in fruity aroma compositions.
FCT 1978 (**16**) p. 743.

Ethyl 2-methylbutyrate [*7452-79-1*]
$CH_3CH_2CH(CH_3)COOCH_2CH_3$, $C_7H_{14}O_2$, M_r 130.19, $bp_{101.3 kPa}$ 131–132 °C, d_4^{25} 0.8689, n_D^{20} 1.3964, is a liquid with a green-fruity odor reminiscent of apples. It is found, for example, in citrus fruits and wild berries and is used in fruit flavor compositions.

Ethyl hexanoate [*123-66-0*], **ethyl caproate**
$CH_3(CH_2)_4COOCH_2CH_3$, $C_8H_{16}O_2$, M_r 144.21, $bp_{101.3 kPa}$ 168 °C, d_4^{20} 0.8710, n_D^{20} 1.4073, is a colorless liquid with a strong fruity odor, reminiscent of pineapples. It occurs in many fruits and is used in small amounts for flowery-fruity notes in perfume compositions and in larger quantities in fruit flavors.
FCT 1976 (**14**) p. 761.

2-Propenyl hexanoate [*123-68-2*], **allyl caproate**
$CH_3(CH_2)_4COOCH_2CH=CH_2$, $C_9H_{16}O_2$, M_r 156.22, $bp_{2 kPa}$ 75–76 °C, d^{20} 0.8869, n_D^{20} 1.4243, has been shown to occur in pineapple. It has a typical pineapple odor and is used in, for example, pineapple flavors.
FCT 1973 (**11**) p. 489.

Ethyl heptanoate [*106-30-9*], **ethyl enanthate**
$CH_3(CH_2)_5COOCH_2CH_3$, $C_9H_{18}O_2$, M_r 158.24, $bp_{101.3 kPa}$ 187–188 °C, d_4^{15} 0.8714, n_D^{15} 1.4144, is a colorless liquid with a fruity odor reminiscent of cognac. It is found in fruits and alcoholic beverages and is used in appropriate aroma compositions.
FCT 1981 (**19**) p. 247.

2-Propenyl heptanoate [*142-19-8*], **allyl enanthate**
$CH_3(CH_2)_5COOCH_2CH=CH_2$, $C_{10}H_{18}O_2$, M_r 170.25, $bp_{101.3 kPa}$ 210 °C, $d^{15.5}$ 0.890, n_D^{20} 1.4290, has been found in, e.g., wild edible mushrooms. It is used in perfume compositions for apple-like (pineapple) notes.
FCT 1977 (**15**) p. 619.

Ethyl octanoate [*106-32-1*], **ethyl caprylate**
$CH_3(CH_2)_6COOCH_2CH_3$, $C_{10}H_{20}O_2$, M_r 172.27, $bp_{101.3\,kPa}$ 208 °C, d_4^{20} 0.8693, n_D^{20} 1.4178, is a liquid with a fruity-flowery odor. It occurs in many fruits and alcoholic beverages and is used in fruit flavors.
FCT 1976 (**14**) p. 763.

Ethyl 2-*trans*-4-*cis*-decadienoate [*3025-30-7*]
$CH_3(CH_2)_4CH=CHCH=CHCOOCH_2CH_3$, $C_{12}H_{20}O_2$, M_r 196.29, $bp_{6\,Pa}$ 70–72 °C, has been identified in pears and has the typical aroma of Williams pears. Synthesis of ethyl 2-*trans*-4-*cis*-decadienoate starts from *cis*-1-heptenyl bromide, which is converted into a l-heptenyllithium cuprate complex with lithium and copper iodide. Reaction with ethyl propiolate yields a mixture of 95% ethyl 2-*trans*-4-*cis*- and 5% ethyl 2-*trans*-4-*trans*-decadienoate. Pure ethyl 2-*trans*-4-*cis*-decadienoate is obtained by fractional distillation [25]. A biotechnological process for its preparation has been developed [26].
FCT 1988 (**26**) p. 317.

Methyl 2-octynoate [*111-12-6*]
$CH_3(CH_2)_4C\equiv CCOOCH_3$, $C_9H_{14}O_2$, M_r 154.21, $bp_{1.3\,kPa}$ 94 °C, d^{20} 0.926, n_D^{20} 1.4464.
FCT 1979 (**17**) p. 375.

Methyl 2-nonynoate [*111-80-8*]
$CH_3(CH_2)_5C\equiv CCOOCH_3$, $C_{10}H_{16}O_2$, M_r 168.24, $bp_{2.7\,kPa}$ 121 °C, d^{20} 0.916–0.918, n_D^{25} 1.4470. Both methyl 2-nonynoate and methyl 2-octynoate have a triple bond and are liquids with a fatty, violet-leaf-like odor. They are used in perfume compositions.
FCT 1975 (**13**) p. 871.

Pentyloxyacetic acid allyl ester [*67634-00-8*], **allyl amylglycolate**

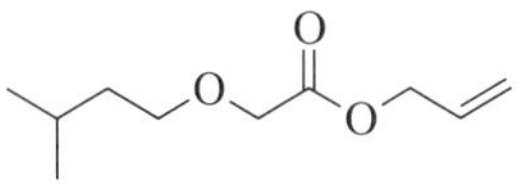

$C_{10}H_{18}O_3$, M_r 186.25, d_4^{20} 0.936–0.944, n_D^{20} 1.428–1.433, is a colorless liquid of strong fruity galbanum odor with pineapple modification. It can be prepared by reaction of chloroacetic acid with isoamyl alcohol in the presence of sodium hydroxide and a phase-transfer catalyst, then treating the resulting sodium amylglycolate with allyl alcohol [27]. Allyl amylglycolate is used in fragrance compositions, e.g., for detergents.

2.1.6 Miscellaneous Compounds

A number of volatile aliphatic compounds that contain nitrogen or sulfur atoms are important aroma constituents. Alkyl thiols, dialkyl sulfides and disulfides, and alkyl thiocyanates belong to this group. They occur widely in foods and spices and determine the odor of, for example, onions, garlic, and mustard. Because of their potent smell, they are used in high dilution and are often produced only in small quantities. The same is true for the following:

1. 3-mercaptohexanol [*51755-83-0*], a tropical fruit aroma component,

2. 3-methylthiohexanol [*51755-66-9*], a volatile constituent of passion fruits,

3. 3-mercapto-2-butanone [*40789-98-8*], a meat aroma constituent, and

4. 4-mercapto-4-methyl-2-pentanone [*19872-52-7*], the so-called 'catty compound' found in blackcurrant flavor and the wine variety 'scheurebe'.

Allyl isothiocyanate, however, is an exception in that it is produced in large quantities. A heptanone oxime and the 2-tridecenenitrile have become important as fragrance materials.

Allyl isothiocyanate [*57-06-7*], **allyl mustard oil**
$CH_2 = CHCH_2N = C = S$, C_4H_5NS, M_r 99.14, $bp_{101.3 \, kPa}$ 152 °C, d_4^{20} 1.0126, is the main component of mustard oil (>95%). It is a colorless oil with a typical mustard odor and can be prepared by reacting allyl chloride with alkaline-earth or alkali rhodanides [28].
FCT 1984 (**22**) p. 623.

5-Methyl-3-heptanone oxime [*22457-23-4*]

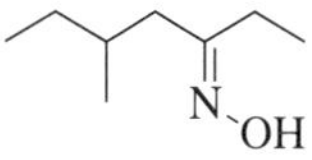

$C_8H_{17}NO$, M_r 143.23, $bp_{0.1 \, kPa}$ 70 °C, n_D^{20} 1.4519, is not reported as being found in nature. It is a colorless to pale yellow liquid with a powerful, green, leafy odor. 5-Methyl-3-heptanone oxime is prepared by oximation of the corresponding ketone. It imparts natural and fresh nuances to fragrance notes and is used in fine perfumery as well as in cosmetics, soaps, and detergents.

   Trade Name. Stemone (Giv.-Roure).

2-Tridecenenitrile [*22629-49-8*]

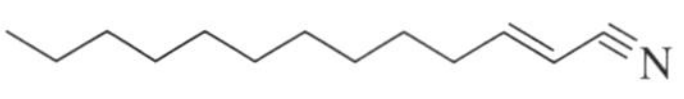

$C_{13}H_{23}N$, M_r 193.34, d_4^{20} 0.833–0.839, n_D^{20} 1.450–1.455, is a clear, colorless to pale yellowish liquid with a very strong, citrus odor reminiscent of nuances in orange and mandarine oil. It can be prepared by Knoevenagel condensation of undecanal with cyano acetic acid and subsequent decarboxylation.

It is used in compositions with agrumen notes for perfuming, e.g., cosmetics, soaps, and detergents.

Trade Names. Ozonil (H&R), 'Tridecen-2-nitril' (Dragoco), 'Tridecene-2-nitrile' (Quest).

2.2 Acyclic Terpenes

2.2.1 Hydrocarbons

Acyclic terpene (C_{10}) and sesquiterpene (C_{15}) hydrocarbons find little use in flavor and fragrance compositions. They are relatively unstable and some have a slightly aggressive odor due to their highly unsaturated structure.

Myrcene, ocimene, and farnesene, are present in many fruits and essential oils, but find only limited use in perfumery.

Myrcene    Ocimene 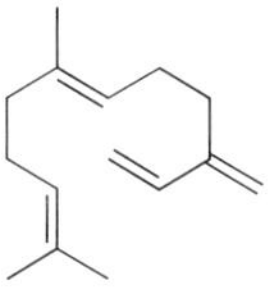   β-Farnesene

2.2.2 Alcohols

Acyclic terpene and sesquiterpene alcohols occur in many essential oils. These alcohols were formerly isolated from oils in which they are major components.

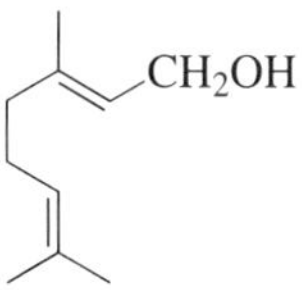

Geraniol 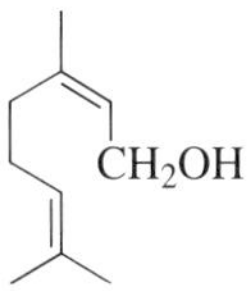   Nerol 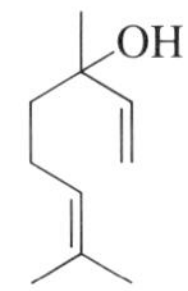   Linalool 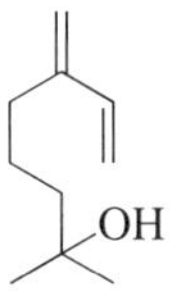   Myrcenol 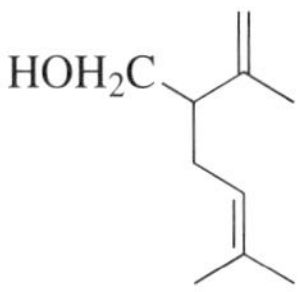   Lavandulol 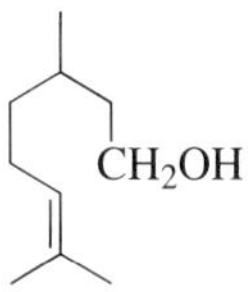   Citronellol

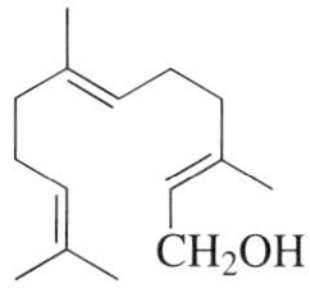

trans-trans-Farnesol 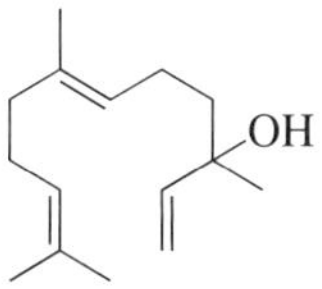   *trans*-Nerolidol

Currently, large-scale synthesis of terpenoids permits production without the uncertainties associated with isolation from natural sources. However, the odor qualities of synthetic products often differ from those of compounds isolated from natural sources, since the desired natural product often is not separated from small amounts of compounds with similar physical properties but different odor quality.

The acyclic terpene alcohols geraniol, linalool, and citronellol are the most important terpene alcohols used as fragrance and flavor substances. Geraniol and linalool are, in addition to nerol and lavandulol, primary products in terpene biosynthesis. The fully saturated alcohols tetrahydrogeraniol and tetrahydrolinalool are also used in large quantities in fragrance compositions. The fragrance materials myrcenol, and its dihydro and tetrahydro derivatives, belong structurally to the terpenes. The sesquiterpene alcohols farnesol and nerolidol are popular materials for perfume compositions.

Geraniol and nerol are *cis–trans* isomers. In the rarely occurring lavandulol, the isoprene units are not coupled in the normal head-to-tail manner.

The farnesols and nerolidols are sesquiterpene analogs of geraniol–nerol and linalool. These compounds are formed by extending one of the methyl groups in the 7-position of the corresponding monoterpene with an isoprene unit. Because these compounds have an extra double bond, they also have an additional possibility for *cis–trans* isomerism. Thus, there are four stereoisomers of farnesol and two of nerolidol.

Geraniol [*106-24-1*], 3,7-dimethyl-*trans*-2,6-octadien-1-ol

$C_{10}H_{18}O$, M_r 154.25, $bp_{101.3\,kPa}$ 230 °C, d^{20} 0.8894, n_D^{20} 1.4777, occurs in nearly all terpene-containing essential oils, frequently as an ester. Palmarosa oil contains 70–85% geraniol; geranium oils and rose oils also contain large quantities. Geraniol is a colorless liquid, with a flowery-roselike odor.

Since geraniol is an acyclic, doubly unsaturated alcohol, it can undergo a number of reactions, such as rearrangement and cyclization. Rearrangement in the presence of copper catalysts yields citronellal. In the presence of mineral acids, it cyclizes to form monocyclic terpene hydrocarbons, cyclogeraniol being obtained if the hydroxyl function is protected. Partial hydrogenation leads to citronellol, and complete hydrogenation of the double bonds yields 3,7-dimethyloctan-1-ol (tetrahydrogeraniol). Citral is obtained from geraniol by oxidation (e.g., with chromic acid), or by catalytic dehydrogenation. Geranyl esters are prepared by esterification.

Production.   Dehydrogenation of geraniol and nerol is a convenient route for synthesizing citral, which is used in large quantities as an intermediate in the synthesis of vitamin A. Large-scale processes have, therefore, been developed for producing geraniol. Currently, these are far more important than isolation from essential oils. Nevertheless, some geraniol is still isolated from essential oils for perfumery purposes.

1. *Isolation from Essential Oils.* Geraniol is isolated from citronella oils and from palmarosa oil. Fractional distillation of, for example, Java citronella oil (if necessary after saponification of the esters present) yields a fraction containing ca. 60% geraniol, as well as citronellol and sesquiterpenes.

    A product with a higher geraniol content and slightly different odor quality for use in fine fragrances is obtained by fractionating palmarosa oil after saponification of the geranyl esters.

2. *Synthesis from β-Pinene.* Pyrolysis of β-pinene yields myrcene, which is converted into a mixture of predominantly geranyl, neryl, and linalyl chloride by addition of hydrogen chloride in the presence of small amounts of catalyst, e.g., copper(I) chloride and an organic quaternary ammonium salt [29]. After removal of the catalyst, the mixture is reacted with sodium acetate in the presence of a nitrogen base (e.g., triethylamine) and converted to geranyl acetate, neryl acetate, and a small amount of linalyl acetate [30].

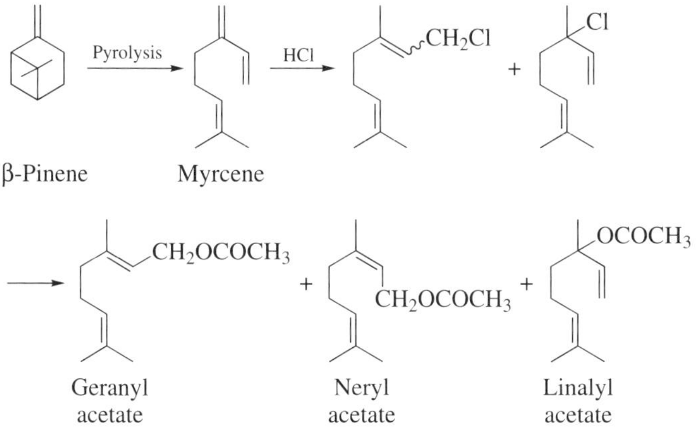

β-Pinene	Myrcene

Geranyl acetate	Neryl acetate	Linalyl acetate

After saponification and fractional distillation of the resulting alcohols, a fraction is obtained that contains ca. 98% geraniol.

3. *Synthesis from Linalool.* A 96% pure synthetic geraniol prepared by isomerization of linalool has become commercially available. Orthovanadates are used as catalysts, to give a >90% yield of a geraniol–nerol mixture [31]. Geraniol of high purity is finally obtained by fractional distillation.

    A considerable portion of commercially available geraniol is produced by a modified process: linalool obtained in a purity of ca. 65% from α-pinene is converted into linalyl borates, which rearrange in the presence of vanadates as catalysts to give geranyl and neryl borates. The alcohols are obtained by hydrolysis of the esters [32].

Uses.   Geraniol is one of the most frequently used terpenoid fragrance materials. It can be used in all flowery-roselike compositions and does not discolor soaps. In flavor compositions, geraniol is used in small quantities to accentuate citrus notes. It is an important intermediate in the manufacture of geranyl esters, citronellol, and citral.
FCT 1974 (**12**) p. 881.

Nerol [*106-25-2*], **3,7-dimethyl-*cis*-2,6-octadien-1-ol**

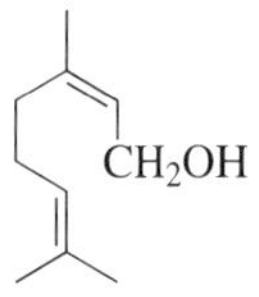

$C_{10}H_{18}O$, M_r 154.25, $bp_{99.3\,kPa}$ 224–225 °C, d_4^{20} 0.8796, n_D^{20} 1.4744, occurs in small quantities in many essential oils where it is always accompanied by geraniol; its name originates from its occurrence in neroli oil. Nerol is a colorless liquid with a pleasant roselike odor which, unlike that of geraniol, has a fresh green note.
   Nerol undergoes the same reactions as geraniol, but cyclizes more readily in the presence of acids.
   Nerol is produced along with geraniol from myrcene in the process described for geraniol. It can be separated from geraniol by fractional distillation.

Uses.   Nerol is used in perfumery not only for the same purposes as geraniol, e.g., in rose compositions, to which it lends a particular freshness, but also in other blossom compositions. In flavor work it is used for bouquetting citrus flavors. Technical-grade nerol, often in a mixture with geraniol, is used as an intermediate in the production of citronellol and citral.
FCT 1976 (**14**) p. 623.

Linalool [*78-70-6*], **3,7-dimethyl-1,6-octadien-3-ol**

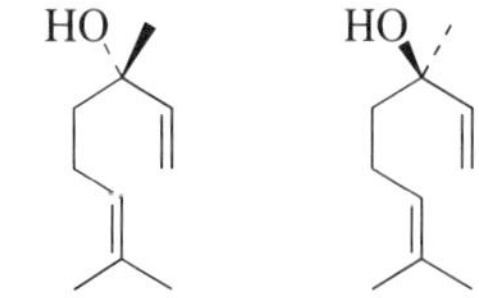

(+)-Linalool   (–)-Linalool

$C_{10}H_{18}O$ M_r 154.25, $bp_{101.3\,kPa}$ 198 °C, d_4^{20} 0.8700, n_D^{20} 1.4616, occurs as one of its enantiomers in many essential oils, where it is often the main component. (−)-Linalool [*126-90-9*], for example, occurs at a concentration of 80–85% in Ho oils from *Cinnamomum camphora*; rosewood oil contains ca. 80%. (+)-Linalool [*126-91-0*] makes up 60–70% of coriander oil.

Properties.   (±)-Linalool [*22564-99-4*] is, like the individual enantiomers, a color-less liquid with a flowery-fresh odor, reminiscent of lily of the valley. However, the enantiomers differ slightly in odor [33]. Together with its esters, linalool is one of the most frequently used fragrance substances and is produced in large quantities.

In the presence of acids, linalool isomerizes readily to geraniol, nerol, and α-terpineol. It is oxidized to citral by chromic acid. Oxidation with peracetic acid yields linalool oxides, which occur in small amounts in essential oils and are also used in perfumery. Hydrogenation of linalool gives tetrahydrolinalool, a stable fragrance compound. Its odor is not as strong as, but fresher than, that of linalool. Linalool can be converted into linalyl acetate by reaction with ketene or an excess of boiling acetic anhydride [34].

Production.   In the 1950s nearly all linalool used in perfumery was isolated from essential oils, particularly from rosewood oil. Currently, this method is used only in countries where oils with a high linalool content are available and where the importation of linalool is restricted.

Since linalool is an important intermediate in the manufacture of vitamin E, several large-scale processes have been developed for its production. Preferred starting materials and/or intermediates are the pinenes and 6-methyl-5-hepten-2-one. Most perfumery-grade linalool is synthetic.

1. *Isolation from Essential Oils.* Linalool can be isolated by fractional distillation of essential oils, for example, rosewood oil, Shiu oil, and coriander oil, of which Brazilian rosewood oil is probably the most important.

2. *Synthesis from α-Pinene.* α-Pinene from turpentine oil is selectively hydro-genated to *cis*-pinane [35], which is oxidized with oxygen in the presence of a

α-Pinene   *cis*-Pinane   *cis-/trans-*
Pinane hydroperoxide

cis-Pinanol   *trans*-Pinanol   (+)-Linalool   (−)-Linalool

radical initiator to give a mixture of ca. 75% *cis*- and 25% *trans*-pinane hydro-peroxide. The mixture is reduced to the corresponding pinanols either with sodium bisulfite ($NaHSO_3$) or a catalyst. The pinanols can be separated by fractional distillation and are pyrolized to linalool: $(-)$-α-pinene yields *cis*-pinanol and $(+)$-linalool, whereas $(-)$-linalool is obtained from *trans*-pinanol [36].

3. *Synthesis from β-Pinene*. For a description of this route, see under Geraniol. Addition of hydrogen chloride to myrcene (obtained from β-pinene) results in a mixture of geranyl, neryl, and linalyl chlorides. Reaction of this mixture with acetic acid–sodium acetate in the presence of copper(I) chloride gives linalyl acetate in 75–80% yield [37]. Linalool is obtained after saponification.

4. *Synthesis from 6-Methyl-5-hepten-2-one*. The total synthesis of linalool starts with 6-methyl-5-hepten-2-one; several large-scale processes have been developed for synthesizing this compound:

(a) Addition of acetylene to acetone results in the formation of 2-methyl-3-butyn-2-ol, which is hydrogenated to 2-methyl-3-buten-2-ol in the presence of a palladium catalyst. This product is converted into its acetoacetate derivative with diketene [38] or with ethyl acetoacetate [39]. The acetoacetate undergoes rearrangement when heated (Carroll reaction) to give 6-methyl-5-hepten-2-one:

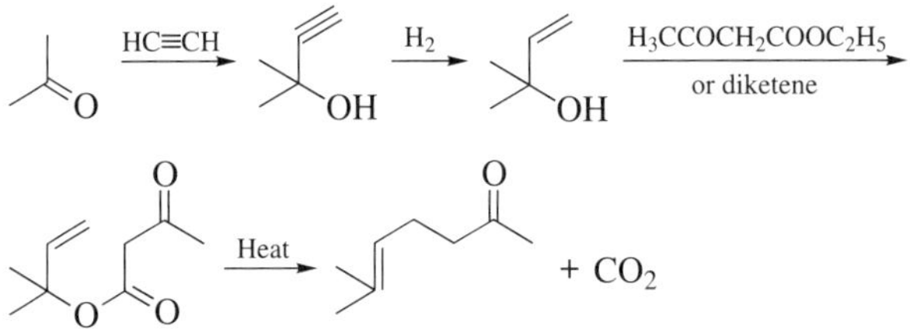

(b) In another process, 6-methyl-5-hepten-2-one is obtained by reaction of 2-methyl-3-buten-2-ol with isopropenyl methyl ether followed by a Claisen rearrangement [40]:

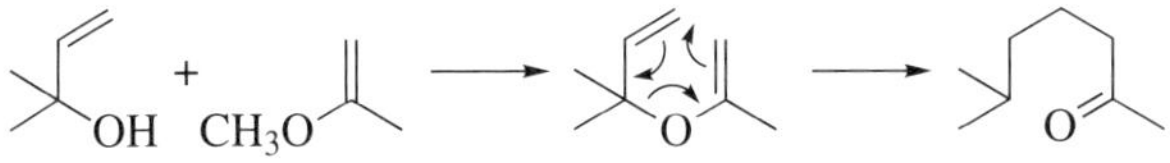

(c) A third synthesis starts from isoprene, which is converted into 3-methyl-2-butenyl chloride by addition of hydrogen chloride. Reaction of the chloride with acetone in the presence of a catalytic amount of an organic base [41] leads to 6-methyl-5-hepten-2-one:

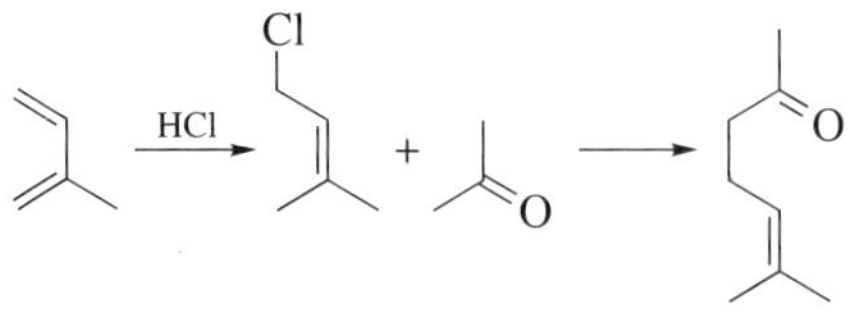

(d) In another process, 6-methyl-5-hepten-2-one is obtained by isomerization of 6-methyl-6-hepten-2-one [42]. The latter can be prepared in two steps from isobutylene and formaldehyde. 3-Methyl-3-buten-1-ol is formed in the first step [43] and is converted into 6-methyl-6-hepten-2-one by reaction with acetone [44].

6-Methyl-5-hepten-2-one is converted into linalool in excellent yield by base-catalyzed ethynylation with acetylene to dehydrolinalool [45]. This is followed by selective hydrogenation of the triple bond to a double bond in the presence of a palladium carbon catalyst.

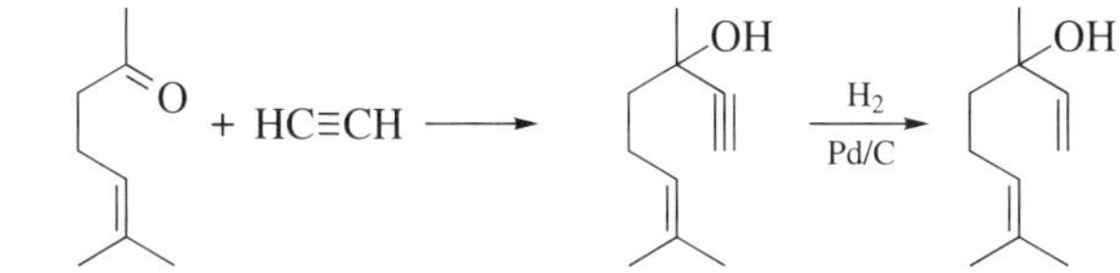

6-Methyl-5-
hepten-2-one   Dehydrolinalool   Linalool

Uses. Linalool is used frequently in perfumery for fruity notes and for many flowery fragrance compositions (lily of the valley, lavender, and neroli). Because of its relatively high volatility, it imparts naturalness to top notes. Since linalool is stable in alkali, it can be used in soaps and detergents. Linalyl esters can be prepared from linalool. Most of the manufactured linalool is used in the production of vitamin E.
FCT 1975 (**13**) p. 827.

Myrcenol [*543-39-5*], **2-methyl-6-methylene-7-octen-2-ol**

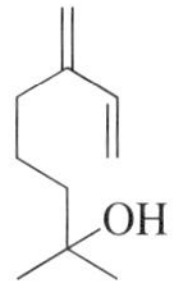

$C_{10}H_{18}O$, M_r 154.25, $bp_{6.7 \, kPa}$ 78 °C, d_{20}^{20} 0.8711, n_D^{20} 1.4731, is an isomer of geraniol and linalool. It has been identified in Chinese lavender oil [46] and some medicinal plants. It is a colorless liquid with a fresh-flowery, slightly limelike odor. Due to its conjugated double bonds, it tends to polymerize; polymerization can be suppressed by adding inhibitors (e.g., antioxidants such as ionol).

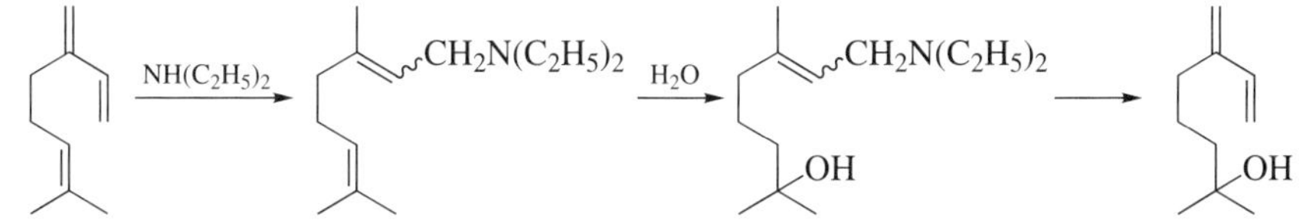

Myrcene   Myrcenol

Myrcenol can be prepared by treating myrcene with diethylamine to give a mixture of geranyl- and neryldiethylamine. These compounds are hydrated with a dilute acid to the corresponding hydroxydiethylamines. Deamination to myrcenol is effected by using a palladium-phosphine-cation complex as a catalyst [47].

Myrcenol is used in perfumery to obtain a lifting top note in citrus and lavender compositions. It is mainly important in the production of 4-(4-hydroxy-4-methylpentyl)-3-cyclohexenecarboxaldehyde (see p. 79).

FCT 1976 (**14**) p. 617.

Citronellol [*26489-01-0*], **3,7-dimethyl-6-octen-1-ol**

$C_{10}H_{20}O$, M_r 156.27, $bp_{101.3\,kPa}$ 224.4 °C, d^{20} 0.8590, n_D^{20} 1.4558, $[\alpha]_D +$ resp. -5 to $6°$, occurs as both (+)-citronellol [*1117-61-9*] and (−)-citronellol [*7540-51-4*] in many essential oils.

(−)-Citronellol isolated from natural sources is often named rhodinol. At present, the name rhodinol is also used for the isopropenyl isomer, α-citronellol; therefore, exclusive use of the systematic name is better.

In many natural products citronellol occurs as a mixture of its two enantiomers; the pure (+) or (−) form is seldom found. (+)-Citronellol dominates in oils from *Boronia citriodora* (total citronellol content ca. 80%) and *Eucalyptus citriodora* (citronellol content 15–20%). (−)-Citronellol is the predominant enantiomer in geranium and rose oils, both of which may contain up to 50% citronellols.

Citronellol is a colorless liquid with a sweet roselike odor. The odor of (−)-citronellol is more delicate than that of (+)-citronellol.

Citronellol undergoes the typical reactions of primary alcohols. Compared with geraniol, which contains one more double bond, citronellol is relatively stable. Citronellol is converted into citronellal by dehydrogenation or oxidation; hydrogenation yields 3,7-dimethyloctan-1-ol. Citronellyl esters are easily prepared by esterification with acid anhydrides.

Production.   (−)-Citronellol is still obtained mainly from geranium oil by saponification followed by fractional distillation. Although of high odor quality, it does not possess the true (−)-citronellol odor due to impurities. Much larger quantities of (+)- and (±)-citronellol are used and are prepared by partial or total synthesis.

1.   *Synthesis of (+)- and (±)-Citronellol from the Citronellal Fraction of Essential Oils.* (+)-Citronellal is obtained by distillation of Java citronella oil and is

hydrogenated to (+)-citronellol in the presence of a catalyst (e.g., Raney nickel). Similarly, (±)-citronellol is prepared from the (±)-citronellal fraction of *Eucalyptus citriodora oil*.

2. *Synthesis of (±)- or Slightly Dextrorotatory Citronellol from Geraniol Fractions of Essential Oils*. This citronellol is produced by catalytic hydrogenation of saponified geraniol fractions (also containing (+)-citronellol) obtained from Java citronella oil, followed by fractional distillation. Selective hydrogenation of the double bond in the 2-position of geraniol in geraniol–citronellol mixtures isolated from essential oils can be achieved by using Raney cobalt as a catalyst; overhydrogenation to 3,7-dimethyloctan-1-ol can be largely avoided by this method [48].

3. *Synthesis of (±)-Citronellol from Synthetic Geraniol–Nerol or Citral*. A considerable amount of commercial synthetic (±)-citronellol is produced by partial hydrogenation of synthetic geraniol and/or nerol. Another starting material is citral, which can be hydrogenated, e.g., in the presence of a catalyst system consisting of palladium, ruthenium, and trimethylamine [49].

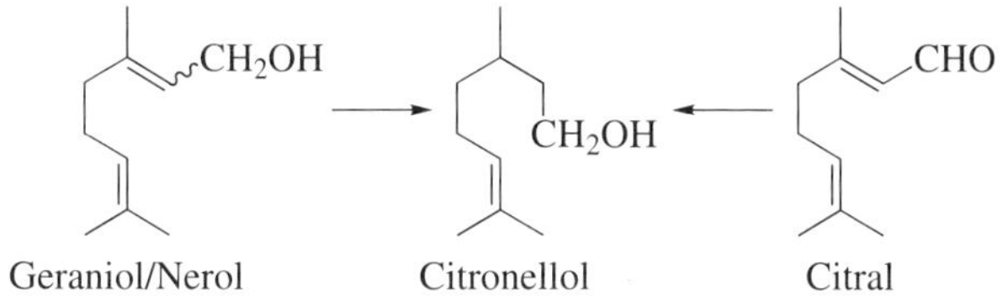

| Geraniol/Nerol | Citronellol | Citral |

4. *Preparation of (−)-Citronellol from Optically Active Pinenes*. (+)-*cis*-Pinane is readily synthesized by hydrogenation of (+)-α-pinene or (+)-β-pinene, and is then pyrolyzed to give (+)-3,7-dimethyl-1,6-octadiene. This compound is converted into (−)-citronellol (97% purity) by reaction with triisobutylaluminum or diisobutylaluminum hydride, followed by air oxidation and hydrolysis of the resulting aluminum alcoholate [50].

Uses.   Citronellol is one of the most widely used fragrance materials, particularly for rose notes and for floral compositions in general. As flavor material, citronellol is added for bouquetting purposes to citrus compositions. It is the starting material for numerous citronellyl esters and for hydroxydihydrocitronellol, an intermediate in the production of hydroxydihydrocitronellal.
FCT 1975 (**13**) p. 757.

Dihydromyrcenol [*18479-58-8*], **2,6-dimethyl-7-octen-2-ol**

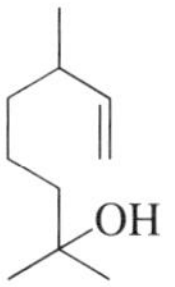

$C_{10}H_{20}O$, M_r 156.27, $bp_{1.3\,kPa}$ 77–79 °C, d_4^{20} 0.841, which was identified in lemon peel oil (*Citrus volkameriana*) is a colorless liquid with a fresh citrus-like odor and a lavender note. It is prepared from 3,7-dimethyl-1,6-octadiene, the pyrolysis product of *cis*-pinane [51], by addition of hydrogen chloride and hydrolysis of the resulting 2,6-dimethyl-2-chloro-7-octene [52].

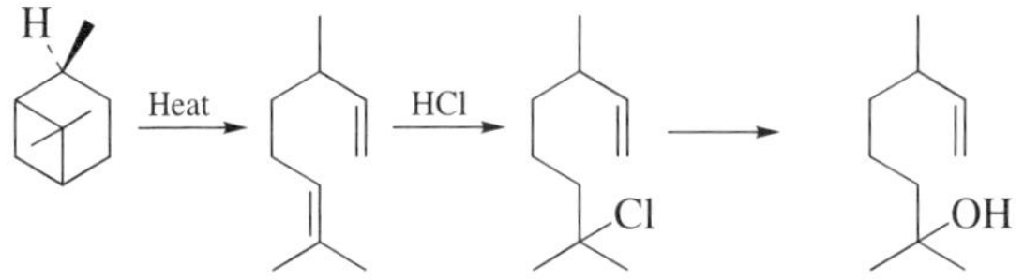

cis-Pinane   Dihydromyrcenol

Dihydromyrcenol is used in soap and detergent perfumes for lime and citrusy-floral notes.
FCT 1974 (**12**) p. 525.

Tetrahydrogeraniol [*106-21-8*], **3,7-dimethyloctan-1-ol**

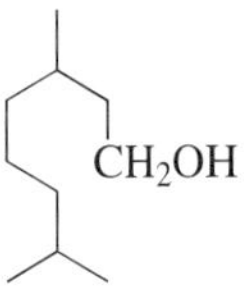

$C_{10}H_{22}O$, M_r 158.28, $bp_{101.3\,kPa}$ 212–213 °C, d_4^{20} 0.8285, n_D^{20} 1.4355, has been iden-tified in citrus oils and is a colorless liquid with a waxy, rose-petal-like odor. It is prepared by hydrogenation of geraniol or citronellol in the presence of a nickel catalyst and is a by-product in the synthesis of citronellol from geraniol or nerol. Because of its stability, it is often used to perfume household products.
FCT 1974 (**12**) p. 535.

Tetrahydrolinalool [*78-69-3*], **3,7-dimethyloctan-3-ol**

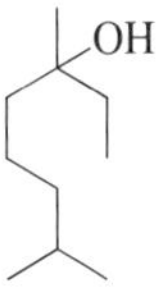

$C_{10}H_{22}O$, M_r 158.28, $bp_{1.3\,kPa}$ 78–79 °C, d_4^{20} 0.8294, n_D^{20} 1.4335, is a constituent of honey aroma. It is a colorless liquid with a linalool-like odor that is slightly fresher but distinctly weaker than that of linalool. Tetrahydrolinalool is prepared by

catalytic hydrogenation of linalool and is used as a substitute for the less stable linalool in perfuming aggressive media.
FCT 1979 (**17**) p. 909.

3,7-Dimethyl-7-methoxyoctan-2-ol [*41890-92-0*]

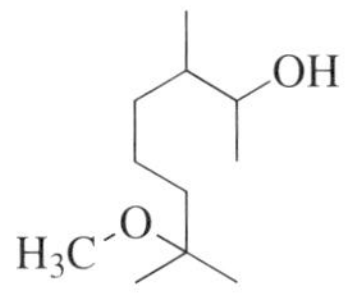

$C_{11}H_{24}O_2$, M_r 188.31, d_4^{20} 0.898–0.900, n_D^{20} 1.446–1.448, is a clear, almost colorless liquid with sandalwood odor, which has a flowery, woody note. It is not known to have been found in nature.

3,7-Dimethyl-7-methoxyoctan-2-ol is prepared by hydrochlorination of dihydromyrcene, methoxylation of the resulting 2-chloro-2,6-dimethyl-7-octene and epoxidation. The alcohol is obtained by hydrogenation of the epoxide in the presence of Raney nickel and triethylamine [53]. It is used in perfumery as a top note in high quality sandalwood compositions for cosmetics, toiletries, and soaps.

Trade Name. Osyrol (BBA).

Farnesol [*4602-84-0*], 3,7,11-trimethyl-2,6,10-dodecatrien-1-ol

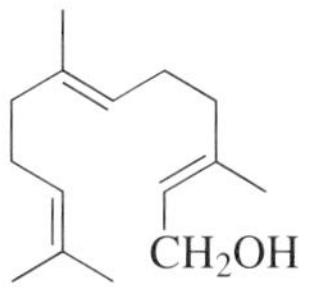

trans-trans-
Farnesol

$C_{15}H_{26}O$, M_r 222.37, $bp_{1.6\,kPa}$ 156 °C, d_4^{20} 0.8846, n_D^{20} 1.4890, is a component of many blossom oils. It is a colorless liquid with a linden blossom odor, which becomes more intense when evaporated, possibly due to oxidation.

Of the four possible isomers (due to the double bonds in the 2- and 6-positions), the trans–trans isomer is the most common in nature and occurs, for example, in ambrette seed oil. 2-*cis*-6-*trans*-Farnesol has been identified in petitgrain oil Bigarade.

Since the odors of the isomers differ very little, natural farnesol in compositions can be replaced by synthetic farnesol, which is a mixture of isomers obtained by isomerization of nerolidol.

Farnesol is particularly suited for use in flower compositions and is valued for its fixative properties.

Nerolidol [*7212-44-4*], **3,7,11-trimethyl-1,6,10-dodecatrien-3-ol**

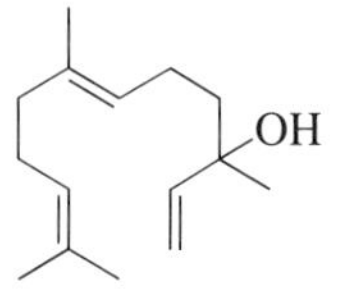

$C_{15}H_{26}O$, M_r 222.37, $bp_{1.6\,kPa}$ 145 °C, d_4^{20} 0.8778, n_D^{20} 1.4898, is the sesquiterpene analogue of linalool. Because of the double bond at the 6-position, it exists as *cis* and *trans* isomers. Each of these isomers can exist as an enantiomeric pair, since the carbon atom in the 3-position is asymmetric.

Nerolidol is a component of many essential oils. (+)-*trans*-Nerolidol occurs in cabreuva oil; (−)-nerolidol has been isolated from *Dalbergia parviflora* wood oils.

Synthetic nerolidol consists of a mixture of (±)-*cis*- and (±)-*trans*-nerolidol and is a colorless liquid with a long-lasting, mild flowery odor.

Industrial synthesis of nerolidol starts with linalool, which is converted into geranylacetone by using diketene, ethyl acetoacetate, or isopropenyl methyl ether, analogous to the synthesis of 6-methyl-5-hepten-2-one from 2-methyl-3-buten-2-ol. Addition of acetylene and partial hydrogenation of the resultant dehydronerolidol produces a mixture of *cis*- and *trans*-nerolidol racemates.

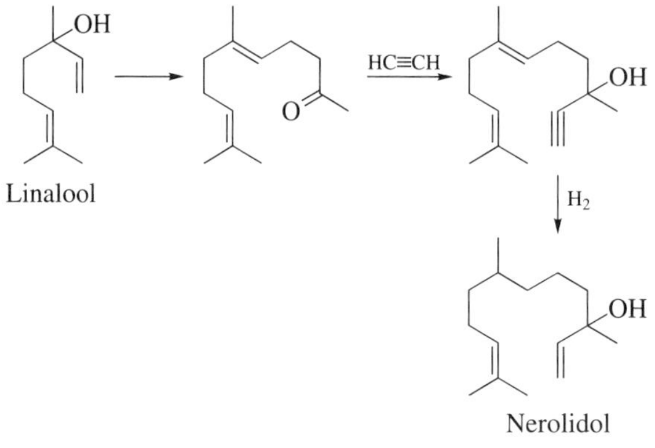

Nerolidol is used as a base note in many delicate flowery odor complexes. It is also an intermediate in the production of vitamins E and K_1.
FCT 1975 (**13**) p. 887.

2.2.3 Aldehydes and Acetals

Among the acyclic terpene aldehydes, citral and citronellal hold key positions as fragrance and flavor chemicals, as well as starting materials for the synthesis of other terpenoids. Hydroxydihydrocitronellal is one of the most important fragrance compounds. Derivatives of these aldehydes, particularly the lower acetals, are also used as fragrance compounds. Acyclic sesquiterpene aldehydes are not very important as such, but they contribute to the characteristic fragrance and aroma of essential oils, for example, in the case of α- and β-sinensal in sweet orange oil.

Citral [*5392-40-5*], **3,7-dimethyl-2,6-octadien-1-al**

Geranial   Neral
(citral a)   (citral b)

$C_{10}H_{16}O$, M_r 152.24, occurs as *cis* and *trans* isomers (citral a and b, respectively) analogous to the corresponding alcohols, geraniol and nerol: citral a [*141-27-5*] (geranial), $bp_{2.7\,kPa}$ 118–119 °C, d^{20} 0.8888, n_D^{20} 1.4898; citral b [*106-26-3*] (neral), $bp_{2.7\,kPa}$ 120 °C, d^{20} 0.8869, n_D^{20} 1.4869.

Natural citral is nearly always a mixture of the two isomers. It occurs in lemongrass oil (up to 85%), in *Litsea cubeba* oil (up to 75%), and in small amounts in many other essential oils. The citrals are colorless to slightly yellowish liquids, with an odor reminiscent of lemon.

Since citral is an α,β-unsaturated aldehyde with an additional double bond, it is highly reactive and may undergo reactions such as cyclization and polymerization. Geraniol, citronellol, and 3,7-dimethyloctan-1-ol can be obtained from citral by stepwise hydrogenation. Citral can be converted into a number of addition compounds; the cis and trans isomers can be separated via the hydrogen sulfite addition compounds. The condensation of citral with active methylene groups is used on an industrial scale in the synthesis of pseudoionones, which are starting materials for ionones and vitamins.

Production. Since citral is used in bulk as a starting material for the synthesis of vitamin A, it is produced industrially on a large scale. Smaller quantities are also isolated from essential oils.

1. *Isolation from Essential Oils.* Citral is isolated by distillation from lemongrass oil and from *Litsea cubeba* oil. It is the main component of these oils.

2. *Synthesis from Geraniol.* Currently, the most important synthetic procedures are vapor-phase dehydrogenation and oxidation of geraniol or geraniol–nerol mixtures. Catalytic dehydrogenation under reduced pressure using copper catalysts is preferred [54].

3. *Synthesis from Dehydrolinalool.* Dehydrolinalool is produced on a large scale from 6-methyl-5-hepten-2-one and acetylene and can be isomerized to citral in high yield by a number of catalysts. Preferred catalysts include organic orthovanadates [55], organic trisilyl oxyvanadates [56], and vanadium catalysts with silanols added to the reaction system [57].

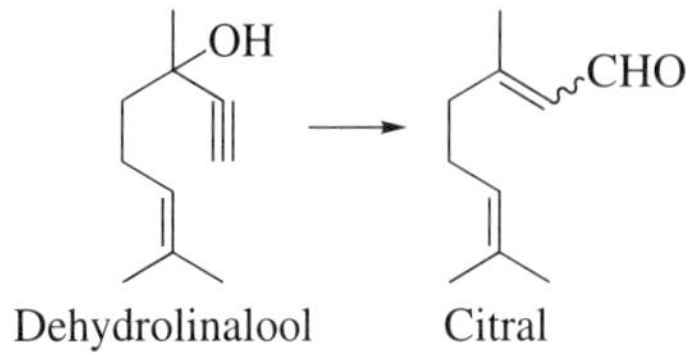

Dehydrolinalool   Citral

4. *Synthesis from Isobutene and Formaldehyde.* 3-Methyl-3-buten-1-ol, obtained from isobutene and formaldehyde [43], isomerizes to form 3-methyl-2-buten-1-ol [58]. However, it is also converted into 3-methyl-2-butenal by dehydrogenation and subsequent isomerization [59], [60]. Under azeotropic conditions in the presence of nitric acid, 3-methyl-2-buten-1-ol and 3-methyl-2-butenal form an acetal (shown below) [61], which eliminates one molecule of 3-methyl-2-buten-1-ol at higher temperatures. The intermediate enol ether undergoes Claisen rearrangement followed by Cope rearrangement to give citral in excellent yield [62]:

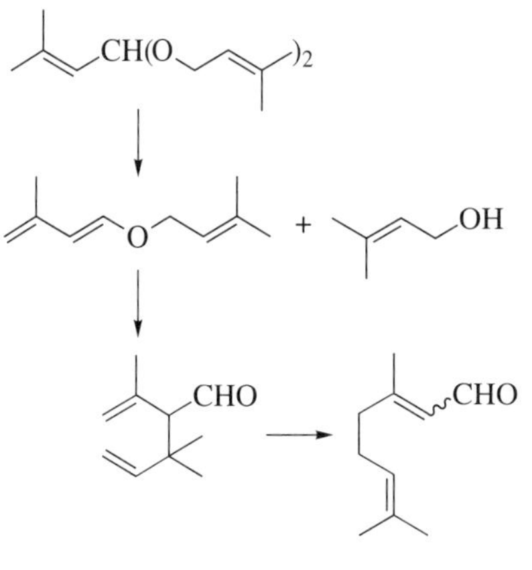

Uses. Because of its strong lemon odor, citral is very important for aroma compositions such as citrus flavors. In perfumery it can be used only in neutral media due to its tendency to undergo discoloration, oxidation, and polymerization. It is used as a starting material in the synthesis of ionones and methylionones, particularly β-ionone, which is an intermediate in vitamin A synthesis.
FCT 1979 (**17**) p. 259.

Citral diethyl acetal [*7492-66-2*], **1,1-diethoxy-3,7-dimethyl-2,6-octadiene**

$C_{14}H_{26}O_2$, M_r 226.36, $bp_{2\,kPa}$ 140–142 °C, d_4^{20} 0.8730, n_D^{20} 1.4503, is a colorless liquid with a flowery, warm-woody citrus odor. It is relatively stable in alkali and can, therefore, be used in soap.
FCT 1983 (**21**) p. 667.

Citronellal [*106-23-0*], **3,7-dimethyl-6-octen-1-al**

$C_{10}H_{18}O$, M_r 154.25, $bp_{101.3\,kPa}$ 207–208 °C, d_4^{20} 0.851, n_D^{20} 1.4477, $[\alpha]_D^{18} + 13.09°$, $[\alpha]_D^{20} - 13.1°$, occurs in essential oils in its (+) and (−) forms, often together with the racemate. (+)-Citronellal [*2385-77-5*] occurs in citronella oil at a concentration of up to 45%; *Backhousia citriodora* oil contains up to 80% (−)-citronellal [*5949-05-3*]. Racemic citronellal [*26489-02-1*] occurs in a number of *Eucalyptus citriodora* oils at a concentration of up to 85%.

Pure citronellal is a colorless liquid with a refreshing odor, reminiscent of balm mint. Upon catalytic hydrogenation, citronellal yields dihydrocitronellal, citronellol, or dihydrocitronellol, depending on the reaction conditions. Protection of the aldehyde group, followed by addition of water to the double bond in the presence of mineral acids or ion-exchange resins results in formation of 3,7-dimethyl-7-hydroxyoctan-1-al (hydroxydihydrocitronellal). Acid-catalyzed cyclization to isopulegol is an important step in the synthesis of (−)-menthol.

Production. Citronellal is still isolated from essential oils in considerable quantities; it is also produced synthetically.

1. *Isolation from Essential Oils.* (+)-Citronellal is obtained from citronella oils by fractional distillation. (±)-Citronellal is isolated from *Eucalyptus citriodora* oil; when necessary, it is purified by using an addition compound, e.g., the bisulfite derivative.

2. *Synthesis from Geraniol or Nerol.* (±)-Citronellal can be obtained by vapor-phase rearrangement of geraniol or nerol in the presence of, e.g., a barium-containing copper–chromium oxide catalyst [63].

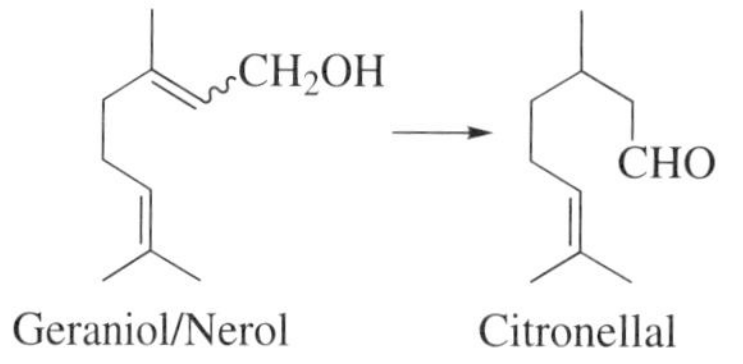

Geraniol/Nerol   Citronellal

3. *Synthesis from Citronellol.* (±)-Citronellal can also be obtained by dehydrogenation of citronellol under reduced pressure with a copper chromite catalyst [64].

4. *Synthesis from Citral.* Selective hydrogenation of citral to citronellal can be accomplished in the presence of a palladium catalyst in an alkaline alcoholic reaction medium [65].

Uses. Citronellal is used to a limited extent for perfuming soaps and detergents. Its main use is as a starting material for the production of isopulegol, citronellol, and hydroxydihydrocitronellal.
FCT 1975 (**13**) p. 755.

7-Hydroxydihydrocitronellal [*107-75-5*], **'hydroxycitronellal,'**
3,7-dimethyl-7-hydroxyoctan-1-al

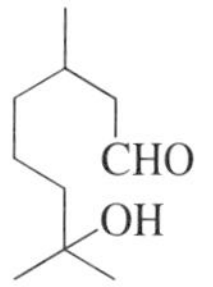

$C_{10}H_{20}O_2$ M_r 172.27, $bp_{0.13\,kPa}$ 85–87 °C, d_4^{20} 0.9220, n_D^{20} 1.4488, has been reported to occur in essential oils [66]. It is a colorless, slightly viscous liquid with a flowery odor reminiscent of linden blossom and lily of the valley. Commercially available 'hydroxycitronellal' is either optically active or racemic, depending on the starting material used. Hydroxydihydrocitronellal prepared from (+)-citronellal, for example, has a specific rotation $[\alpha]_D^{20} + 9$ to $+10°$.

Hydroxydihydrocitronellal is relatively unstable toward acid and alkali and is, therefore, sometimes converted into more alkali-resistant acetals, particularly its dimethyl acetal.

Production. The most important synthetic routes to hydroxydihydrocitronellal are listed below.

1. *Synthesis from Citronellal.* One of the oldest routes to hydroxydihydrocitronellal is the hydration of the citronellal bisulfite adduct (obtained at low temperature) with sulfuric acid, followed by decomposition with sodium carbonate. A more recent development is hydration of citronellal enamines or imines, followed by hydrolysis [67].

2. *Synthesis from Citronellol.* Citronellol is hydrated to 3,7-dimethyloctan-1,7-diol, for example, by reaction with 60% sulfuric acid. The diol is dehydrogenated catalytically in the vapor phase at low pressure to highly pure hydroxydihydrocitronellal in excellent yield. The process is carried out in the presence of, for example, a copper–zinc catalyst [68]; at atmospheric pressure noble metal catalysts can also be used [69].

3. *Synthesis from 7-Hydroxygeranyl/-neryl Dialkylamine.* The starting material can be obtained by treatment of myrcene with a dialkylamine in the presence of an alkali dialkylamide, followed by hydration with sulfuric acid. The 7-hydroxygeranyl/-neryl dialkylamine isomerizes to the corresponding 7-hydroxyaldehyde enamine in the presence of a palladium(II)–phosphine complex as catalyst. Hydrolysis of the enamine gives 7-hydroxydihydrocitronellal [70].

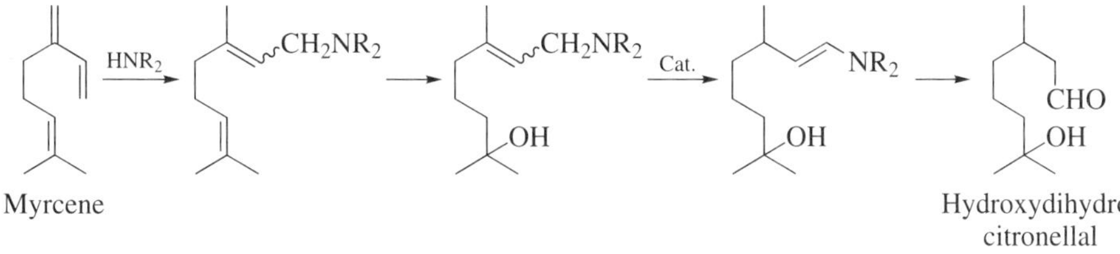

Myrcene   Hydroxydihydro-
                                                                         citronellal

Uses. Because of its fine, flowery odor, hydroxydihydrocitronellal is used in large quantities in many perfume compositions for creating linden blossom and lily of the valley notes. It is also used in other blossom fragrances such as honeysuckle, lily, and cyclamen.
FCT 1974 (**12**) p. 921.

Methoxydihydrocitronellal [*3613-30-7*], **3,7-dimethyl-7-methoxyoctan-1-al**

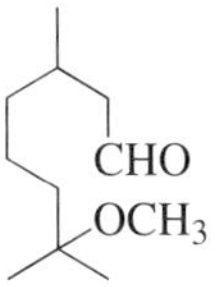

$C_{11}H_{22}O_2$, M_r 186.29, $bp_{0.06\,kPa}$ 60 °C, n_D^{25} 1.4380, is a colorless liquid with a fresh, green, blossom odor and is used in perfumery in floral compositions for fresh-green nuances.
FCT 1976 (**14**) p. 807.

Hydroxydihydrocitronellal dimethyl acetal [*141-92-4*],
8,8-dimethoxy-2,6-dimethyloctan-2-ol

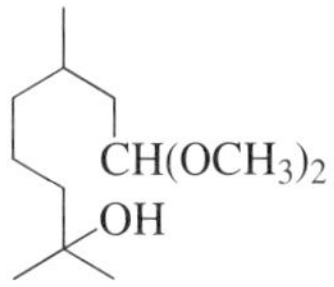

$C_{12}H_{26}O_3$, M_r 218.34, $bp_{1.6\,kPa}$ 131 °C, d_4^{20} 0.931, n_D^{20} 1.4419, is a colorless liquid with a weak, flowery odor. Since the acetal is stable to alkali, it is used occasionally in soap perfumes.
FCT 1975 (**13**) p. 548.

2,6,10-Trimethyl-9-undecenal [*141-13-9*]

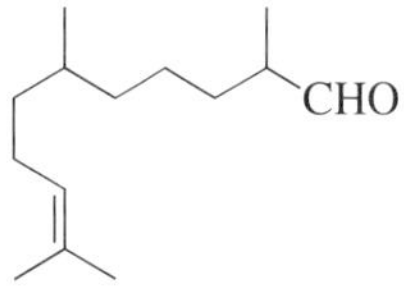

$C_{14}H_{26}O$, M_r 210.36, $bp_{1.2\,kPa}$ 133–135 °C, d_{25}^{25} 0.840–0.853, n_D^{20} 1.447–1.453, is a colorless to slightly yellow liquid with an intense aldehyde-waxy, slightly flowery odor. It is synthesized from a hydrogenated pseudoionone (primarily the tetrahydro compound) and an alkyl chloroacetate by means of a glycidic ester condensation; this is followed by hydrolysis and decarboxylation.

2,6,10-Trimethyl-9-undecenal is a richly fragrant compound that is used in flower compositions to obtain an aldehydic note.
FCT 1992 (**30**) p. 133 S.

Trade Names. Adoxal (Giv.-Roure), Farenal (H&R).

2.2.4 Ketones

Unlike the terpene alcohols, aldehydes, and esters, acyclic terpene ketones are not particularly important as fragrance or flavor substances; thus, they are not discussed here in detail.

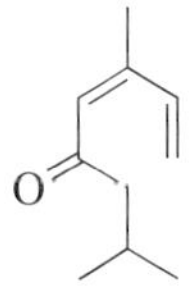

Tagetone 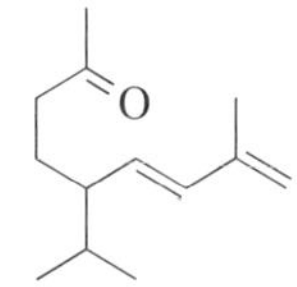   Solanone

6-Methyl-5-hepten-2-one is an important intermediate in the synthesis of terpenoids. Its odor properties are not impressive. It occurs in nature as a decomposition product of terpenes. Tagetone [*6752-80-3*] is a major component of tagetes oil. Solanone [*1937-45-8*] and pseudoionone [*141-10-6*] are acyclic C_{13} ketones with a terpenoid skeleton. Solanone is one of the flavor-determining constituents of tobacco, pseudoionone is an intermediate in the synthesis of ionones.

Geranylacetone [*689-67-8*], **6,10-dimethyl-5,9-undecadien-2-one**

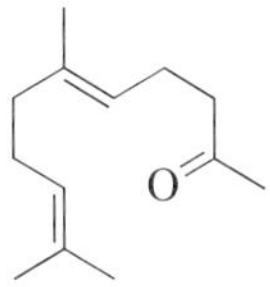

trans-Geranylacetone

$C_{13}H_{22}O$, M_r 194.32, $bp_{1.3\,kPa}$ 124 °C, d_4^{20} 0.8729, n_D^{20} 1.4674, occurs in *cis* as well as *trans* form and has been identified in fruits and in essential oils. It is a colorless liquid with a fresh-green, slightly penetrating, roselike odor.

Geranylacetone is an intermediate in the synthesis of other fragrance compounds. It is used in perfumery in rose compositions, for example, in soap perfumes.
FCT 1979 (**17**) p. 787.

2.2.5 Acids and Esters

Although a small amount of acyclic terpene acids such as geranic acid and citronellic acid occurs in many essential oils, often as esters, they are rarely used in perfume and flavor compositions. Methyl geranate is an intermediate in α-damascone synthesis and is sometimes needed in the reconstitution of essential oils.

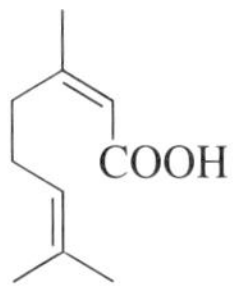

cis-Geranic acid 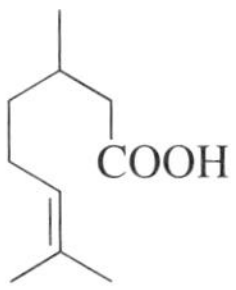   Citronellic acid

However, the lower fatty acid esters (particularly the acetates) of the acyclic terpene alcohols geraniol, linalool, and citronellol are extremely important both as fragrance and as flavor substances. The acetates occur in many essential oils, sometimes in rather high amounts. Formates, propionates, and butyrates occur less frequently. As a result of the development of large-scale production processes for terpenes, the esters of acyclic terpene alcohols are nearly always made synthetically. All acyclic terpene esters that are used as fragrance and flavor materials can be prepared by direct esterification of the appropriate alcohols. However, special precautions are required for the esterification of linalool.

Because the lower fatty acid esters of geraniol, linalool, and citronellol are important contributors to the odor of many essential oils, these esters are widely used in the reconstitution of such oils, as well as in perfume and flavor compositions. The acetates, particularly linalyl acetate, are most widely used. The use of formates is limited by their relative instability. Higher esters are not important in terms of quantity, but are indispensable for creating specific nuances.

In aroma compositions, fatty acid esters of the acyclic terpene alcohols are used for obtaining citrus notes and for rounding off other flavor types.

The most important and most frequently used acyclic terpene esters are described below.

2.2.5.1 Geranyl and Neryl Esters

Geranyl formate [*105-86-2*]
$C_{11}H_{18}O_2$, M_r 182.26, $bp_{101.3\,kPa}$ 229 °C (decomp.), d_4^{25} 0.9086, n_D^{20} 1.4659, is a liquid with a fresh, crisp-herbal-fruity rose odor. It is used as a modifier of, among others, rose, geranium, and neroli compositions.
FCT 1974 (**12**) p. 893.

Geranyl acetate [*105-87-3*]
$C_{12}H_{20}O_2$, M_r 196.29, $bp_{1.5\,kPa}$ 98 °C, d_4^{25} 0.9080, n_D^{20} 1.4624, occurs in varying amounts in many essential oils: up to 60% in oils from *Callitris* and *Eucalyptus* species, and up to 14% in palmarosa oil. A smaller amount occurs in, for example, geranium, citronella, petitgrain, and lavender oils. Geranyl acetate is a liquid with a fruity rose note reminiscent of pear and slightly of lavender. It is used frequently in perfumery to create not only flowery-fruity nuances (e.g., rose), but also for citrus and lavender notes. A small amount is added to fruit aromas for shading.
FCT 1974 (**12**) p. 885.

Geranyl propionate [*105-90-8*]
$C_{13}H_{22}O_2$, M_r 210.32, $bp_{101.3\,kPa}$ 253 °C, d^{15} 0.902, n_D^{20} 1.459, has a fruity rose odor and is used in perfumery in heavy blossom fragrances with a secondary fruity note.
FCT 1974 (**12**) p. 897.

Geranyl isobutyrate [*2345-26-8*]
$C_{14}H_{24}O_2$, M_r 224.34, $bp_{101.3\,kPa}$ 265 °C, d^{15} 0.8997, n_D^{20} 1.4576, is a liquid with a fruity rose odor. It is used in floral perfume compositions and in fruit flavors.
FCT 1975 (**13**) p. 451.

Geranyl isovalerate [*109-20-6*]
$C_{15}H_{26}O_2$, M_r 238.37, $bp_{101.3\,kPa}$ 279 °C, $d^{15.5}$ 0.890, n_D^{20} 1.4640, is a liquid with a strongly fruity rose odor. It is used in perfume and flavor compositions.
FCT 1976 (**14**) p. 785.

Neryl acetate [*141-12-8*]
$C_{12}H_{20}O_2$, M_r 196.29, $bp_{3.4\,kPa}$ 134 °C, d_{15}^{15} 0.903–0.907, n_D^{20} 1.4624, is the *cis* isomer of geranyl acetate. It is present in helichrysum oil and has also been identified in, among others, neroli oil and petitgrain oil Bigarade. It is a colorless, flowery-sweet-smelling liquid and is used in perfumery for blossom compositions (e.g., orange blossom and jasmin).
FCT 1976 (**14**) p. 625.

2.2.5.2 Linalyl and Lavandulyl Esters

Among the linalyl esters, the acetate is by far the most important fragrance and flavor substance. The formate, propionate, and butyrates are used in small amounts.

Linalyl formate [*115-99-1*]
$C_{11}H_{18}O_2$, M_r 182.26, $bp_{1.3\,kPa}$ 100–103 °C, d_4^{25} 0.915, n_D^{20} 1.4530, is a liquid with a fruity odor, reminiscent of bergamot. Linalyl formate is moderately stable and is used in lavender fragrances and eau de cologne.
FCT 1975 (**13**) p. 833.

Linalyl acetate [*115-95-7*]
$C_{12}H_{20}O_2$ M_r 196.29, $bp_{101.3\,kPa}$ 220 °C, d_4^{20} 0.8951, n_D^{25} 1.4480, occurs as its (−)-isomer [*16509-46-9*] as the main component of lavender oil (30–60%, depending on the origin of the oil), of lavandin oil (25–50%, depending on the species), and of bergamot oil (30–45%). It has also been found in clary sage oil (up to 75%) and in a small amount in many other essential oils. (±)-Linalyl acetate [*40135-38-4*] is a colorless liquid with a distinct bergamot–lavender odor.

Production.   Linalyl acetate is synthesized by two methods:

1. Esterification of linalool requires special reaction conditions since it tends to undergo dehydration and cyclization because it is an unsaturated tertiary alcohol. These reactions can be avoided as follows: esterification with ketene in the presence of an acidic esterification catalyst below 30 °C results in formation of linalyl acetate without any byproducts [71]. Esterification can be achieved in good yield, with boiling acetic anhydride, whereby the acetic acid is distilled off as it is formed; a large excess of acetic anhydride must be

maintained by continuous addition of anhydride to the still vessel [34]. Highly pure linalyl acetate can be obtained by transesterification of *tert*-butyl acetate with linalool in the presence of sodium methylate and by continuous removal of the *tert*-butanol formed in the process [72].

2. Dehydrolinalool, obtained by ethynylation of 6-methyl-5-hepten-2-one, can be converted into dehydrolinalyl acetate with acetic anhydride in the presence of an acidic esterification catalyst. Partial hydrogenation of the triple bond to linalyl acetate can be carried out with, for example, palladium catalysts deactivated with lead [73].

Uses.   Linalyl acetate is used extensively in perfumery. It is an excellent fragrance material for, among others, bergamot, lilac, lavender, linden, neroli, ylang-ylang, and phantasy notes (particularly chypre). Smaller amounts are used in other citrus products. Since linalyl acetate is fairly stable toward alkali, it can also be employed in soaps and detergents.

Linalyl propionate [144-39-8]
$C_{13}H_{22}O_2$, M_r 210.31, $bp_{101.3\,kPa}$ 226 °C, d^{15} 0.9000, n_D^{20} 1.4505, is a liquid with a fresh bergamot note, reminiscent of lily of the valley. It is used in perfumery in, for example, bergamot, lavender, and lily of the valley compositions.
FCT 1975 (**13**) p.839.

Linalyl butyrate [78-36-4]
$C_{14}H_{24}O_2$, M_r 224.34, $bp_{101.3\,kPa}$ 232 °C, d^{15} 0.8977, n_D^{20} 1.4523, is a liquid with a fruity bergamot note and a subdued animalic tone. It is used in lavender perfumes and in many blossom compositions.
FCT 1976 (**14**) p.805.

Linalyl isobutyrate [78-35-3]
$C_{14}H_{24}O_2$, M_r 224.34, $bp_{0.02\,kPa}$ 63–65 °C, d^{15} 0.8926, n_D^{25} 1.4450, has a fresh-fruity lavender odor, which is more refined than that of the butyrate. It is used in lavender compositions and in several flowery notes.
FCT 1975 (**13**) p.835.

Lavandulyl acetate [25905-14-0]
$C_{12}H_{20}O_2$, M_r 196.29, $bp_{1.7\,kPa}$ 106–107 °C, d_4^{17} 0.9122, n_D^{17} 1.4561, occurs in its (−)-form at a concentration of ca. 1% in French lavender oil and lavandin oil. It is a liquid with a fresh-herbal rose odor and is used in perfumery for lavender and lavandin oil reconstitutions.
    One synthetic route to lavandulyl acetate starts with prenyl acetate, which dimerizes in the presence of a Friedel Crafts catalyst, such as boron trifluoride–diacetic acid [74].
FCT 1978 (**16**) p.805.

2.2.5.3 Citronellyl and Dihydromyrcenyl Esters

The following esters are used in relatively large amounts as fragrance and flavor materials:

Citronellyl formate [*105-85-1*]
$C_{11}H_{20}O_2$, M_r 184.28, bp_{2kPa} 97–98 °C, d_4^{15} 0.8919, n_D^{20} 1.4556, is a liquid with a strongly fruity, roselike odor, which is suitable for fresh top notes in rose and lily of the valley fragrances.
FCT 1973 (**11**) p. 1073.

Citronellyl acetate [*67650-82-2*]
$C_{12}H_{22}O_2$, M_r 198.30, $bp_{101.3kPa}$ 240 °C, d^{20} 0.8901, n_D^{20} 1.4515, occurs in many essential oils either as one of its optical isomers or as the racemate. The odor of racemic citronellyl acetate differs little from that of the optical isomers. (±)-Citronellyl acetate is a liquid with a fresh-fruity rose odor. It is often used as a fragrance, for example, for rose, lavender, and geranium notes as well as for eau de cologne with citrus nuances. Since it is relatively stable to alkali, it can be used in soaps and detergents. Citrus flavors acquire specific character through the addition of citronellyl acetate; it is also used to round off other fruit flavors.
FCT 1973 (**11**) p. 1069.

Citronellyl propionate [*141-14-0*]
$C_{13}H_{24}O_2$, M_r 212.33, bp_{2kPa} 120–124 °C, d_{15}^{15} 0.8950, n_D^{20} 1.4452, is a fresh-fruity, roselike smelling liquid. It is used in perfume and flavor compositions in the same way as the acetate.
FCT 1975 (**13**) p. 759.

Citronellyl isobutyrate [*97-89-2*]
$C_{14}H_{26}O_2$, M_r 226.36, $bp_{1.6kPa}$ 131–132 °C, d^{15} 0.8816, n_D^{20} 1.4418, is a liquid with a sweet-fruity note and is used in perfumery for fruity-floral nuances.
FCT 1978 (**16**) p. 693.

Citronellyl isovalerate [*68922-10-1*]
$C_{15}H_{28}O_2$, M_r 240.39, bp_{4kPa} 194–196 °C, has a heavy, rosy-herbal odor and is used in oriental perfume compositions among others.

Citronellyl tiglate [*24717-85-9*]
(with *trans*-$CH_3CH=C(CH_3)COOH$ as the acid component), $C_{15}H_{26}O_2$, M_r 238.37, $bp_{0.9kPa}$ 144–145 °C, d_{15}^{15} 0.9090, is a liquid with a flowery-rosy, fruity, mushroom-like odor. It is used in geranium oil reconstitutions.

Dihydromyrcenyl acetate [*53767-93-4*], **2,6-dimethyl-7-octen-2-yl acetate**
$C_{12}H_{22}O_2$, M_r 198.31, is a colorless liquid, d_4^{20} 0.870–0.878, n_D^{20} 1.429–1.434 with a fresh, clean, citrus, floral odor (dihydromyrcenol see p. 31). It can be prepared by

esterification of dihydromyrcenol with acetic acid in the presence of magnesium oxide as a catalyst [75]. It is used for flowery-citric topnotes, especially in soaps. FCT 1983 (**21**) p. 847.

2.2.6 Miscellaneous Compounds

The number of nitrogen- and sulfur-containing derivatives of acyclic terpenoids that are known to be important fragrance and flavor substances is smaller than in the nonterpenoid aliphatic series discussed in Section 2.1.6. However, a few nitriles are used in rather large amounts in soap perfumes because of their relatively high stability toward alkali.

Geranic acid nitrile [*5146-66-7*], **geranonitrile**

cis-Geranic acid nitrile

$C_{10}H_{15}N$, M_r 149.24, $bp_{1.3\,kPa}$ 110 °C, d^{20} 0.8709, n_D^{20} 1.4759, occurs as a mixture of its *cis* and *trans* isomers. It is a liquid with a crisp-fresh, lemon-like, green odor. The nitrile can be prepared from citral by reaction with hydroxylamine and subsequent dehydration with acetic anhydride.
FCT 1976 (**14**) p. 787.

Trade Name. Citralva (IFF).

Citronellic acid nitrile [*51566-62-2*]

Citronellic acid nitrile

$C_{10}H_{17}N$, M_r 151.25, $bp_{2\,kPa}$ 110–111 °C, d_4^{20} 0.845–0.846, n_D^{20} 1.4485–1.4500, is a colorless liquid with a strong, lemon-like odor. The nitrile can be prepared from citronellal oxime in the same way as geranic acid nitrile.
FCT 1979 (**17**) p. 525.

Trade Name. Agrunitril (Dragoco).

2.3 Cyclic Terpenes

2.3.1 Hydrocarbons

Cyclic terpene hydrocarbons occur in essential oils, sometimes in large amounts. They often serve as starting materials for the synthesis of fragrance and flavor compounds. By themselves they generally contribute relatively little to fragrance and aroma. They are used mainly in household perfumery and for reconstitution of essential oils.

Of the various types of monocyclic terpene hydrocarbons, those with the *p*-menthadiene structure are the most important. Examples are as follows:

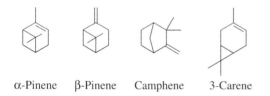

Limonene   α-Terpinene   γ-Terpinene   Terpinolene   α-Phellandrene   β-Phellandrene

Of the bicyclic terpene hydrocarbons, the pinenes are by far the most important industrially. Camphene and 3-carene are used as starting materials for fragrance compounds.

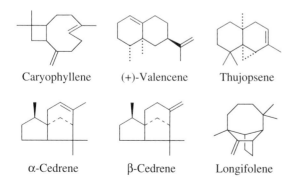

α-Pinene   β-Pinene   Camphene   3-Carene

Many cyclic sesquiterpenes of various structural types have been isolated from essential oils. Typical examples are as follows:

Caryophyllene   (+)-Valencene   Thujopsene

α-Cedrene   β-Cedrene   Longifolene

As in the case of the cyclic monoterpene hydrocarbons, a number of the cyclic sesquiterpenes are used as starting materials in the synthesis of fragrance and flavor compounds or for the reconstitution of essential oils.

Limonene [*138-86-3*], **1,8-*p*-menthadiene**

$C_{10}H_{16}$, M_r 136.24, $bp_{101.3\,kPa}$ 178 °C, d_4^{20} 0.8411, n_D^{20} 1.4726, $[\alpha]_D^{20}$ + or − 126.3°; (+)-limonene [*5989-27-5*] and (−)-limonene [*5989-54-8*] as well as the racemate (dipentene) [*7705-14-8*] occur abundantly in many essential oils. The (+)-isomer is present in citrus peel oils at a concentration of over 90%; a low concentration of the (−)-isomer is found in oils from the *Mentha* species and conifers.

Limonene is a liquid with lemon-like odor. It is a reactive compound; oxidation often yields more than one product. Dehydrogenation leads to *p*-cymene. Limonene can be converted into cyclic terpene alcohols by hydrohalogenation, followed by hydrolysis. Nitrosyl chloride adds selectively to the endocyclic double bond; this reaction is utilized in the manufacture of (−)-carvone from (+)-limonene (see page 59).

(+)-Limonene is obtained in large amounts as a byproduct in the production of orange juice; (−)-limonene is isolated in relatively small quantities from essential oils. Racemic limonenes, which are commercially available under the name dipentene, are formed as by-products in many acid-catalyzed isomerizations of α- and β-pinene. Distillation of the so-called dipentene fraction yields limonenes in varying degrees of purity.

The limonenes are used as fragrance materials for perfuming household products and as components of artificial essential oils.
FCT 1975 (**13**) p. 825: (+)-limonene.
    1978 (**16**) p. 809: (−)-limonene.
    1974 (**12**) p. 703: (±)-limonene.

γ-Terpinene [*99-85-4*], **1,4-*p*-menthadiene**

$C_{10}H_{16}$, M_r 136.24, $bp_{101.3\,kPa}$ 183 °C, d_4^{20} 0.8493, n_D^{20} 1.4747, is a colorless liquid with an herbaceous citrus odor and can be prepared by isomerization of limonene. FCT 1976 (**14**) p. 875.

(−)-α-Phellandrene [*4221-98-1*], 1,5-*p*-menthadiene

$C_{10}H_{16}$, M_r 136.24, $bp_{99\,kPa}$ 172 °C, d_4^{20} 0.8410, n_D^{20} 1.4708, $[\alpha]_D^{20}$ − 183°, is a color-less liquid with a citrus odor and a slight peppery note. It is isolated, for example, from *Eucalyptus dives* oil.
FCT 1978 (**16**) p. 843.

Pinenes are widespread, naturally occurring terpene hydrocarbons. The α- and β-forms occur in varying ratios in essential oils.

α-Pinene [*80-56-8*], 2-pinene

α-Pinene

$C_{10}H_{16}$, M_r 136.24, $bp_{101.3\,kPa}$ 156 °C, d_4^{20} 0.8553, n_D^{20} 1.4662, $[\alpha]_D^{20}$ + or − 51.9° is the most widespread pinene isomer. (+)-α-Pinene [*7785-70-8*] occurs, for example, in oil from *Pinus palustris* Mill. at a concentration up to 65%; oil from *Pinus pinaster* Soland. and American oil from *Pinus caribaea* contain 70% and 70–80%, respectively of the (−)-isomer [*7785-26-4*].

α-Pinene undergoes many reactions, of which the following are used in the fragrance industry: upon hydrogenation α-pinene is converted to pinane, which has become an important starting material in the industrial processes used in the fragrance and flavor industry. α-Pinene can be isomerized to β-pinene with high selectivity for β-pinene formation [76]. Hydration with simultaneous ring opening yields terpineol and *cis*-terpin hydrate. Pyrolysis of α-pinene yields a mixture of ocimene and alloocimene.

Pure α-pinene is obtained by distillation of turpentine oils. As a fragrance substance it is used to improve the odor of industrial products. However, it is far more important as a starting material in industrial syntheses, for example, terpineols, borneol, and camphor.
FCT 1978 (**16**) p. 853.

β-Pinene [*127-91-3*], **2(10)-pinene**

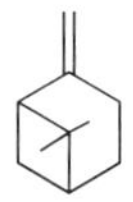

β-Pinene

$C_{10}H_{16}$, M_r 136.24, $bp_{101.3\,kPa}$ 164 °C, d_4^{20} 0.8712, n_D^{20} 1.4763, $[\alpha]_D^{20}$ + or − 22.6°, occurs in many essential oils. Optically active and racemic β-pinenes are present in turpentine oils, although in smaller quantities than α-pinene.

β-Pinene is similar to α-pinene in its reactions. Pyrolytic cleavage to myrcene, the starting material for acyclic terpenes, is used on an industrial scale. Addition of formaldehyde results in the formation of nopol; nopyl acetate is used as a fragrance material. β-Pinene is produced in large quantities by distillation of turpentine oils. It is used as a fragrance material in household perfumery. However, most β-pinene is used in the production of myrcene.
FCT 1978 (**16**) p. 859.

2.3.2 Alcohols and Ethers

Although cyclic terpene alcohols occur widely in nature, few have the physiological properties that make them important fragrance or flavor compounds. Exceptions are α-terpineol and (−)-menthol, the latter because of its cooling/refreshing effect. Of the bicyclic monoterpene alcohols, borneol deserves mention.

Many cyclic sesquiterpene alcohols are key odor components in essential oils, for example, cedrol in cedarwood oil, the vetiverols in vetiver oil, and the santalols in sandalwood oil. Since these alcohols have not yet been synthesized on an industrial scale, they are described under the oil in which they occur (Chapter 3). Some of their derivatives, however, are discussed in this section.

Menthol, *p*-menthan-3-ol

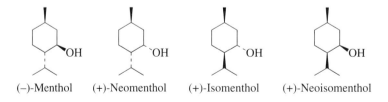

(−)-Menthol   (+)-Neomenthol   (+)-Isomenthol   (+)-Neoisomenthol

$C_{10}H_{20}O$, M_r 156.27, has three asymmetric carbon atoms in its cyclohexane ring and, therefore, occurs as four pairs of optical isomers. The configuration of four of these isomers is given above; the other four are their mirror images.

(−)-Menthol is the isomer that occurs most widely in nature. It is the main component of peppermint and cornmint oils obtained from the *Mentha piperita* and *Mentha arvensis* species. Esterified menthol also occurs in these oils (e.g., as the acetate and isovalerate). Other menthol stereoisomers may be present in these oils as well.

Physical Properties.   The eight optically active menthols differ in their organoleptic properties [77]. (−)-Menthol has a characteristic peppermint odor and also exerts a cooling effect. The other isomers do not possess this cooling effect and are, therefore, not considered to be 'refreshing.' (±)-Menthol occupies an intermediate position; the cooling effect of the (−)-menthol present is distinctly perceptible.

The enantiomeric menthols have identical physical properties (apart from their specific rotation), but the racemates differ from the optically active forms in, for example, their melting points. Although the differences between the boiling points of the stereoisomers are small, the racemates can be separated by fractional distillation. Boiling points (in °C at 101.3 kPa) are as follows:

neomenthol	211.7
neoisomenthol	214.6
menthol	216.5
isomenthol	218.6

Other physical constants of commercially available levorotatory and racemic menthols are: (−)-menthol [*2216-51-5*], mp 43 °C, n_D^{20} 1.4600, $[\alpha]_D^{20} - 50°$; (±)-menthol [*15356-70-4*], mp 38 °C, n_D^{20} 1.4615.

Chemical Properties.   Hydrogenation of menthols yields *p*-menthane; oxidation with chromic acid or catalytic dehydrogenation yields menthones. Dehydration under mild conditions yields 3-*p*-menthene as the main product. Reaction with carboxylic acids or their derivatives yields menthyl esters, which are used mainly as aroma substances and in pharmaceutical preparations and formulations. The esterification of menthols with benzoic acid is used on an industrial scale in the resolution of racemic menthol.

Production.   Many industrial processes exist for the production of menthols. For (−)-menthol, isolation from peppermint oil (see Mint Oils) competes with partial and total syntheses. When an optically active compound is used as a starting material, optical activity must be retained throughout the synthesis, which generally consists of several steps. Total syntheses or syntheses starting from optically inactive materials require either resolution of racemic mixtures or asymmetric synthesis of an intermediate. Recently used processes are the following:

1.   (−)-*Menthol from Cornmint Oil. Mentha arvensis* oils, which may contain 70–80% free (−)-menthol, are cooled and the crystals separated by centrifugation. Since the crystalline product contains traces of cornmint oil, this

menthol has a slightly herbaceous-minty note. Pure (−)-menthol is obtained by recrystallization from solvents with low boiling points.

2. (−)-*Menthol from Dementholized Cornmint Oil.* Dementholized cornmint oil, from which (−)-menthol has been removed by crystallization and which still contains 40–50% free menthol, can be reused for producing (−)-menthol. The fairly large quantity of (−)-menthone in the oil (30–50%) is hydrogenated to form a mixture of mainly (−)-menthol and (+)-neomenthol; the (−)-menthyl esters present (chiefly (−)-menthyl acetate) are saponified. Additional (−)-menthol is then separated from other components by crystallization, distillation, or via the boric acid esters.

3. (−)-*Menthol from* (+)-*Citronellal.* This process uses the readily occurring cyclization of citronellal to isopulegol. (+)-Citronellal can be isolated with an optical purity of ca. 80% from citronella oil. Alternatively, it can be synthesized with a purity of 98% from dialkylgeranylamine (obtained from myrcene and a dialkylamine) by enantioselective isomerization to (+)-citronellaldialkylenamine followed by hydrolytic cleavage to (+)-citronellal. Isomerization is effected in the presence of a chiral rhodium–phosphine complex as a catalyst [78]. (+)-Citronellal is cyclized in the presence of acidic catalysts (e.g., silica gel) to give a mixture of optically active isopulegol isomers containing ca. 20% of the corresponding racemates:

(+)-Citronellal   (−)-Isopulegol (+)-Neoisopulegol (+)-Isoisopulegol (+)-Neoisoisopulegol

(−)-Isopulegol can be isolated from this mixture and hydrogenated to (−)-menthol. The remaining isopulegol stereoisomers can be partly reconverted into (+)-citronellal by pyrolytic cleavage and reused in the cyclization procedure [79].

However, the isopulegol mixture can also be hydrogenated to produce a mixture of menthols; the individual stereoisomers are then separated by distillation. To obtain optically pure (−)-menthol, a resolution step involving a suitable crystalline derivative (such as the benzoate) is required. The undesired stereoisomeric menthols mainly (+)-neomenthol and (+)-isomenthol, are epimerized to an equilibrium mixture (e.g., by heating in the presence of sodium menthylate). (−)-Menthol is then again separated from the mixture.

4. (−)-*Menthol from* (−)-*Piperitone or Piperitol.* (−)-Menthol can also be prepared from (−)-piperitone, the main component of *Eucalyptus dives* Typus oils.

Hydrogenation in the presence of Raney nickel yields a mixture of menthols, from which (−)-menthol can be separated by crystallization and saponification of its chloroacetate.

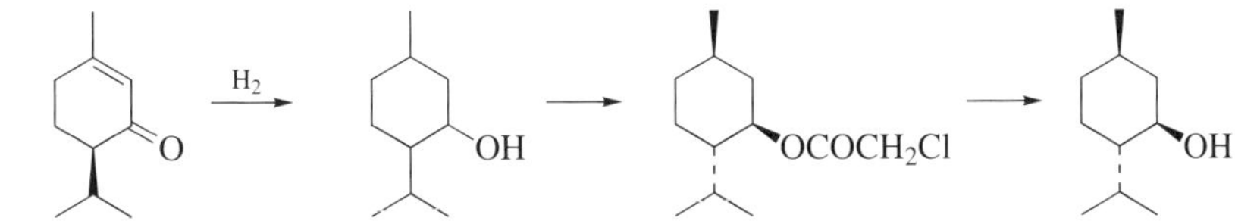

| (−)-Piperitone | Isomeric menthol mixture | (−)-Menthyl chloroacetate | (−)-Menthol |

Analogously, (+)-*trans*-piperitol (obtained from α- or β-phellandrene via piperityl chloride [80]) can be hydrogenated to give a mixture of 97% (+)-isomenthol and 3% (+)-menthol. Pure (+)-isomenthol is obtained by crystallization and undergoes rearrangement to give an equilibrium mixture of (+)-neomenthol and (−)-menthol; the latter is separated by distillation.

5. (−)-*Menthol from* (+)-*3-Carene.* An Indian manufacturing process for (−)-menthol starts from 3-carene, the major component of Indian turpentine oil (55–65%). (+)-3-Carene isomerizes to (+)-2-carene, which can be pyrolyzed to (+)-*trans*-2,8-*p*-menthadiene. Isomerization of the latter yields (+)-isoterpinolene, which is hydrogenated to give >50% (+)-3-*p*-menthene. Epoxidation and subsequent rearrangement lead to a menthone–isomenthone mixture, which gives a mixture of menthols when it is catalytically hydrogenated. Fractional distillation and crystallization yield commercially acceptable (−)-menthol [81].

6. (−)-*Menthol from* (±)-*Menthol.* (±)-Menthol can be prepared via several routes and subsequently resolved into the optical isomers:

(a) Racemic menthol can be synthesized by hydrogenation of thymol. This yields a mixture containing the four stereoisomeric menthols in various proportions. (±)-Menthol is separated from the other isomers by distillation.

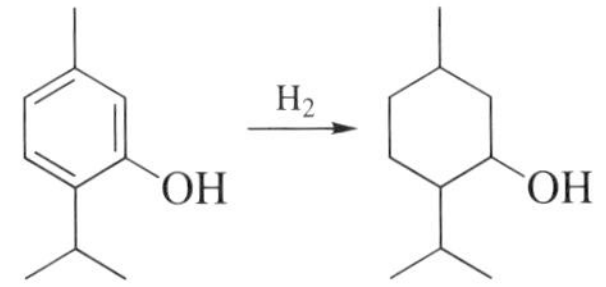

Thymol   Menthol

The remaining isomeric menthols, neomenthol, isomenthol, and a trace of neoisomenthol, can be epimerized, under the conditions used for the thymol hydrogenation, to give ca. 6 : 3 : 1 equilibrium mixture of (±)-menthol, (±)-neomenthol, and (±)-isomenthol, respectively. (±)-Menthol can, again, be distilled from the equilibrium mixture.

(b) (±)-Menthol can be resolved into its optical antipodes by several routes. A large-scale industrial process utilizes selective crystallization of either (+)- or (−)-menthyl benzoate by seeding saturated solutions or supercooled melts of (±)-menthyl benzoates with crystals of (+)- or (−)-menthyl benzoate. Pure (+)- or (−)-menthol is obtained following hydrolysis of the esters [82]. The undesired (+)-menthol can be reconverted into the racemate.
Biochemical resolution methods have also been developed.

Uses.   Because of its cooling and refreshing effect, (−)-menthol is used in large quantities in cigarettes, cosmetics, toothpastes, chewing gum, sweets, and medicines. (±)-Menthol can be used in medicines and liniments.
FCT 1976 (**14**) p. 471: (−)-menthol.
        1976 (**14**) p. 473: (±)-menthol.

Isopulegol, 8-*p*-menthen-3-ol

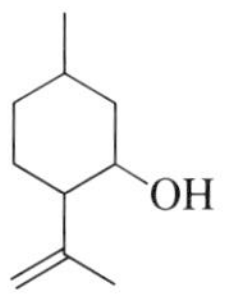

$C_{10}H_{18}O$, M_r 154.25, pure (−)-isopulegol [89-79-2], $bp_{1 kPa}$ 74 °C, d_4^{26} 0.9062, n_D^{26} 1.4690, $[\alpha]_D^{26} - 23.6°$. Like menthol, isopulegol has three asymmetric carbon atoms and, therefore, four stereoisomers, each occurring as a pair of optically active antipodes.

The isopulegols occur in a large number of essential oils, often in optically active or partly racemic form. Since citronellal readily cyclizes to isopulegol, the latter occurs frequently in citronellal-containing essential oils, in which it is formed during the recovery of the oil.

Isopulegol produced industrially from (+)-citronellal is a mixture of isomers containing a high percentage of (−)-isopulegol. The isopulegols are colorless liquids with a minty-herbaceous odor. They are converted into the corresponding menthols by means of hydrogenation. Cyclization of citronellal, in the presence of acidic catalysts, yields a mixture of isomeric isopulegols; (+)-citronellal obtained synthetically or from citronella oil is most frequently used as the starting material.

Isopulegol is used in perfumery in various blossom compositions, as well as for geranium notes. It is an important intermediate in (−)-menthol production.
FCT 1975 (**13**) p. 823.

Terpineols are unsaturated monocyclic terpene alcohols and are formed by acid-catalyzed hydration of pinenes; α-, β-, γ- and δ-isomers exist:

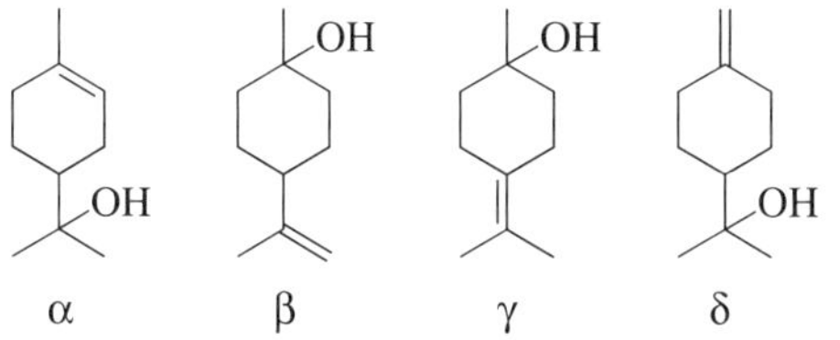

α β γ δ

α- and β-Terpineol occur in optically active forms and as racemates. α-Terpineol is an important commercial product. It occurs in a large number of essential oils primarily as $(-)$-α-terpineol (for example, in conifer and lavandin oils). Small quantities of $(+)$- and $(\pm)$-α-terpineol are found in many other essential oils, β-, γ-, and δ-terpineol do not occur widely in nature.

α-Terpineol [98-55-5], **1-*p*-menthen-8-ol**
$C_{10}H_{18}O$, M_r 154.25, *mp* (enantiomers) 40–41 °C, *mp* (racemate) 35 °C, $bp_{101.3\,kPa}$ 218–219 °C, d_4^{20} 0.9357, n_D^{20} 1.479, $[\alpha]_D + 106.4°$ (solution in ether, 4%) is a colorless, crystalline solid, smelling of lilac. The most important commercial grade of terpineol consists of a liquid mixture of isomers, that contains mainly α-terpineol and a considerable amount of γ-terpineol. This mixture has a stronger lilac odor than does pure crystalline α-terpineol.

Hydrogenation of α-terpineol yields *p*-menthan-8-ol. Terpineol is readily dehydrated by acids, yielding a mixture of unsaturated cyclic terpene hydrocarbons. Under mildly acidic conditions, terpin hydrate is formed. The most important reaction for the fragrance industry is esterification, particularly acetylation to terpinyl acetate.

Production. Although α-terpineol occurs in many cssential oils, only small quantities are isolated, e.g., by fractional distillation of pine oils.

A common industrial method of α-terpineol synthesis consists of the hydration of α-pinene or turpentine oil with aqueous mineral acids to give crystalline *cis*-terpin hydrate (*mp* 117 °C), followed by partial dehydration to α-terpineol. Suitable catalysts are weak acids or acid-activated silica gel [83].

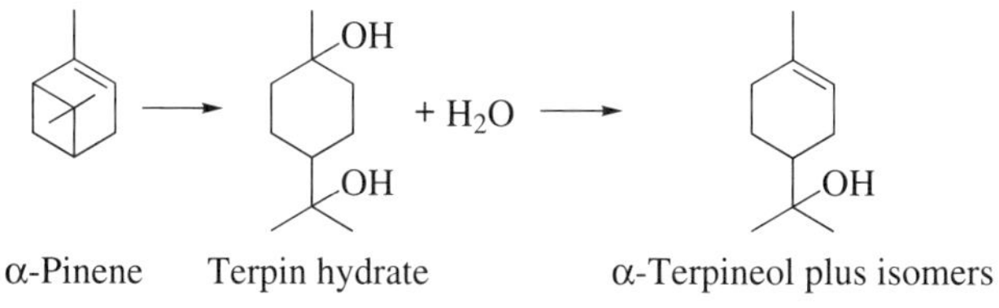

α-Pinene   Terpin hydrate α-Terpineol plus isomers

Selective conversion of pinene, 3-carene, and limonene or dipentene to terpineol, without terpin hydrate formation, is also used. Addition of organic acids (weak acids require catalytic amounts of mineral acids) produces terpinyl esters, which are subsequently hydrolyzed to terpineol, sometimes *in situ*.

Uses. Terpineol with its typical lilac odor is one of the most frequently used fragrance compounds. It is stable and inexpensive, and is used in soaps and cosmetics.
FCT 1974 (**12**) p. 997.

1-Terpinen-4-ol [*562-74-3*], **1-*p*-menthen-4-ol**

$C_{10}H_{18}O$, M_r 154.25, $bp_{101.3\,kPa}$ 212 °C, $bp_{0.5\,kPa}$, 73.5 °C, d^{20} 0.9315, n_D^{20} 1.4799, occurs as (+)-, (−)-, and racemic 1-terpinen-4-ol in many essential oils, e.g., from *Pinus* and *Eucalyptus* species, and in lavender oil. It is a colorless liquid with a spicy, nutmeg-like, woody-earthy, and also lilac-like odor.

1-Terpinen-4-ol is a byproduct in the synthesis of terpineol from terpin hydrate, and occurs in commercial terpineol. Pure 1-terpinen-4-ol can be prepared from terpinolene by photosensitized oxidation, reduction of the resulting 1-methyl-4-isopropenyl-l-cyclohexene-4-hydroperoxide, and selective hydrogenation of the corresponding alcohol [84].

Terpinolene   1-Terpinen-4-ol

It is used, for example, in artificial geranium and pepper oils and in perfumery for creating herbaceous and lavender notes.
FCT 1982 (**13**) p. 833.

cis-**Hexahydrocuminyl alcohol** [*13828-37-0*], ***p*-menthan-7-ol**

$C_{10}H_{20}O$, M_r 156.27, d_{20}^{20} 0.912–0.920, n_D^{20} 1.466–1.471 is a colorless liquid with a fresh soft and clean floral odor which recalls the fragrance associated with the white petals and blossoms of many flowers.

The title compound can be obtained by epoxidizing β-pinene (see p. 50) with peracetic acid, cleaving the oxirane ring by treating the epoxide with diatomaceous earth and reducing the resulting mixture over Raney nickel. The *cis*- and *trans*-isomer mixture is separated by distillation [85]. Because of its excellent stability the compound can be used in a wide range of products, e.g., soaps, detergents, cosmetics.

Trade Name. Mayol (Firmenich).

Borneol, 2-bornanol

$C_{10}H_{18}O$, M_r 154.25, *mp* (enantiomers) 208 °C, *mp* (racemate) 210.5 °C, d_4^{20} 1.011, $[\alpha]_D^{20}$ + or − 37.7 °C, is a bicyclic terpene alcohol. Borneol is an endo isomer; the corresponding exo isomer is isoborneol [*124-76-5*]:

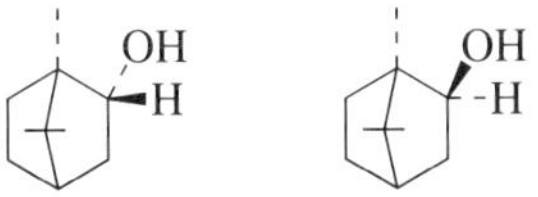

(−)-Borneol   (+)-Isoborneol

Borneol occurs abundantly in nature as a single enantiomer or, less frequently, as the racemate. (−)-Borneol [*464-45-9*] occurs particularly in oils from *Pinaceae* species and in citronella oil. (+)-Borneol [*464-43-7*] is found, for example, in camphor oil (Hon-Sho oil), in rosemary, lavender, and olibanum oils.

Borneol is a colorless, crystalline solid. (+)-Borneol has a camphoraceous odor, with a slightly sharp, earthy-peppery note, which is less evident in (−)-borneol. Commercial borneol is often levorotatory ($[\alpha]_D^{20}$ − 18 to − 28° in ethanol), and contains (−)-borneol and up to 40% isoborneol.

Borneol is oxidized to camphor with chromic or nitric acid; dehydration with dilute acids yields camphene. Borneol is readily esterified with acids, but on an industrial scale bornyl esters are prepared by other routes. For example, levorotatory borneol is synthesized industrially from levorotatory pinenes by Wagner–Meerwein rearrangement with dilute acid, followed by hydrolysis of the resulting esters [86].

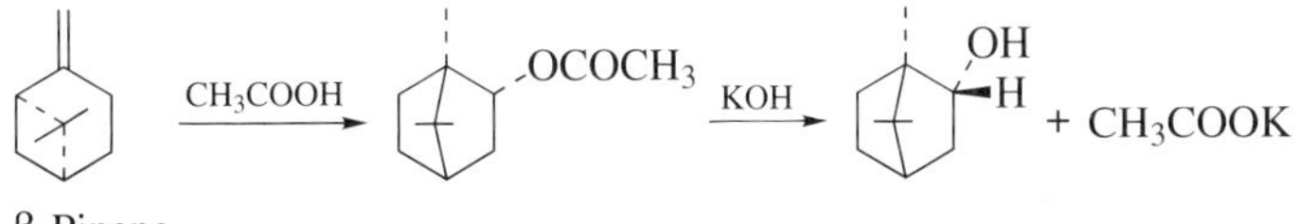

(−)-β-Pinene

Borneol is used in the reconstitution of the essential oils in which it occurs naturally.
FCT 1978 (**16**) p. 655: (−)-borneol.

Cedryl methyl ether [*19870-74-7*] and [*67874-81-1*]

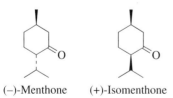

$C_{16}H_{28}O$, M_r 236.40, d_{25}^{25} 0.974–0.979, n_D^{20} 1.494–1.498, is a colorless liquid with a fine cedarwood odor and a distinct amber nuance. It is prepared by methylation of cedrol and is used in perfumes, soaps, and cosmetics.
FCT 1979 (**17**) p. 747.

Trade Name. Cedramber (IFF).

2.3.3 Aldehydes and Ketones

Cyclic terpene aldehydes occur in essential oils only in low concentration. These aldehydes are seldom used as single fragrance compounds. A few of the cyclic terpene ketones are commercially important as fragrance and flavor compounds, for example, menthone and carvone, which have the *p*-menthane skeleton, and the ionones, which have a (trimethylcyclohexenyl)alkenone skeleton. The ionones and their methyl-substituted homologues are some of the most valuable fragrance materials. Some cyclic terpene ketones are the main components of essential oils (e.g., camphor in camphor oil); others, although not main components, may be essential for a fragrance (e.g., β-damascenone, which is an important component of Bulgarian rose oil). The cyclic sesquiterpene ketone, nootkatone, is one of the characteristic components of grapefruit aroma.

Menthone, *p*-menthan-3-one

(−)-Menthone   (+)-Isomenthone

$C_{10}H_{18}O$, M_r 154.25, exists as two stereoisomers, menthone and isomenthone, each of which occurs as a pair of enantiomers, due to the two asymmetric centers present in the molecule.

Table 1. Physical properties of industrially important menthone isomers

Compound	CAS registry number	bp, °C	d_t^{20}	n_D^{20}	$[\alpha]_D$
(−)-Menthone	[14073-97-3]	210	0.896*	1.450	−28.5°
(±)-Menthone	[1074-95-9]	210	0.896*	1.450	
(+)-Isomenthone	[1196-31-2]	212	0.900**	1.453	+95.0°
(±)-Isomenthone	[36977-92-1]	212	0.900**	1.453	

* $t = 20\,°C$, ** $t = 4\,°C$

Both stereoisomers occur in many essential oils, often as a single enantiomer species. A particularly high concentration (sometimes >50%) is found in oils from *Mentha* species. The menthones are colorless liquids that possess a typically minty odor; the odor of isomenthone is slightly musty. They have a strong tendency to interconvert and are, therefore, difficult to obtain in high purity. Industrial products are mixtures of varying composition. Physical constants of industrially important menthone isomers are listed in Table 1.

The menthones are converted into the corresponding menthols by means of hydrogenation; for example, (−)-menthone yields (+)-neomenthol and (−)-menthol.

(−)-Menthone can be obtained by distillation of dementholized cornmint oil or by oxidation of (−)-menthol (e.g., with chromic acid). Dehydrogenation of (−)- menthol (e.g., with copper chromite) yields a mixture of (−)-menthone and (+)-isomenthone.

(±)-Menthone is prepared analogously from (±)-menthol. However, it can also be synthesized by hydrogenation of thymol in the presence of palladium–carbon catalysts [87].

Menthone and isomenthone are used for synthetic peppermint oils and bases. FCT 1976 (**14**) p. 475: (±)-menthone.

Carvone, 1,8-*p*-menthadien-6-one

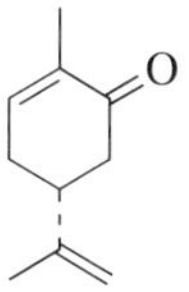

$C_{10}H_{14}O$, M_r 150.22, $bp_{101.3\,kPa}$ 230 °C, d_4^{20} 0.960, n_D^{20} 1.499, $[\alpha]_D^{18}$ (+)-carvone + 64.3°, $[\alpha]_D^{20}$ (−)-carvone − 62.5 °, occurs as (+)-carvone [2244-16-8], (−)-carvone [6485-40-1], or racemic carvone [22327-39-5]. The optical isomers differ considerably in their organoleptic properties. They occur in high percentages in a number of essential oils. (+)-Carvone is the main component of caraway oil (ca. 60%) and dill oil; (−)-carvone occurs in spearmint oil at a concentration of 70–80%.

Properties.   The carvones are colorless to slightly yellow liquids. (+)-Carvone has a herbaceous odor reminiscent of caraway and dill seeds, whereas (−)-carvone has a herbaceous odor reminiscent of spearmint. Depending on the reaction conditions, hydrogenation of carvone yields either carveols or dihydrocarvone, which are also used as flavor compounds. When treated with strong acids, carvone isomerizes to carvacrol.

Production.   In the past, (+)- and (−)-carvones were isolated by fractional distillation of caraway oil and spearmint oil, respectively. However, these carvones are now prepared synthetically, the preferred starting material being (+)- and (−)-limonenes, which are converted into the corresponding optically active carvones. Since optical rotation is reversed in the process, (+)-limonene is the starting material for (−)-carvone.

   The preferred industrial method of carvone synthesis utilizes the selective addition of nitrosyl chloride to the endocyclic double bond of limonene. If a lower aliphatic alcohol is used as solvent, limonene nitrosochloride is obtained in high yield. It is converted into carvone oxime by elimination of hydrogen chloride in the presence of a weak base. Acid hydrolysis in the presence of a hydroxylamine acceptor, such as acetone, yields carvone [88].

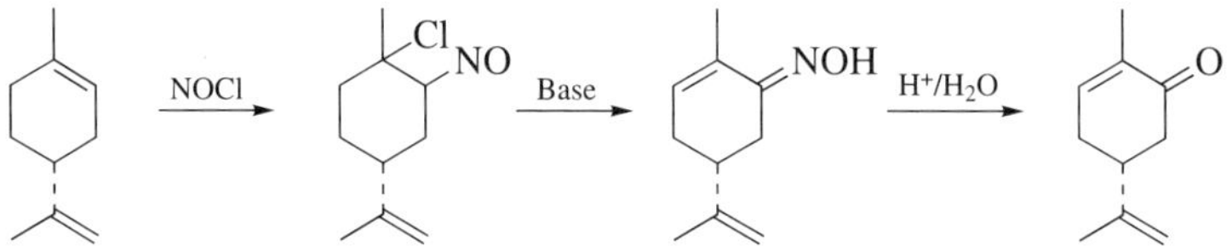

| (+)-Limonene | (+)-Limonene nitrosochloride | (−)-Carvone oxime | (−)-Carvone |

Uses.   Both (+)- and (−)-carvone are used to flavor a number of foods and beverages. (−)-Carvone is produced in much larger quantities and is also used in oral hygiene products.

FCT 1978 (**16**) p. 673: (+)-carvone.
      1973 (**11**) p. 1057: (−)-carvone.

Camphor, 1,7,7-trimethylbicyclo[2.2.1]heptan-2-one

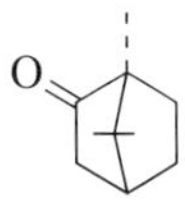

(+)-Camphor

$C_{10}H_{16}O$, M_r 152.24, (+)-camphor: $bp_{101.3\,kPa}$ 204 °C, mp 178.8 °C, $[\alpha]_D^{20}$ + 44.3 °; both optical isomers are found widely in nature, (+)-camphor [464-49-3] being the more abundant. It is, for example, the main component of oils obtained from the camphor tree *Cinnamomum camphora*.

Camphor is produced by fractional distillation and crystallization of camphor oil or, synthetically, by dehydrogenation of isoborneol (from isobornyl acetate, see page 70) over a copper catalyst.

Due to its characteristic penetrating, slightly minty odor, camphor is only used in perfuming industrial products. It is far more important as a plasticizer.
FCT 1978 (**16**) p. 665.

Fenchone, 1,3,3-trimethylbicyclo[2.2.1]heptan-2-one

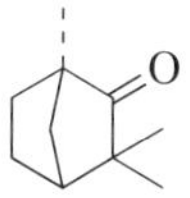

(+)-Fenchone

$C_{10}H_{16}O$ M_r 152.24, $bp_{101.3\,kPa}$ 193 °C, d_4^{20} 0.9484, n_D^{20} 1.4628, $[\alpha]_D^{20}$ + or − 66.8 °, occurs as its (−)-isomer in a number of fennel oils. It is a colorless, slightly viscous liquid with a camphoraceous odor.

(+)-Fenchone [*7787-20-4*] containing a small amount of the (−)-isomer [*4695-62-9*] is prepared by dehydrogenation of (−)-fenchol. (−)-Fenchyl esters are obtained, along with other compounds, by addition of carboxylic acids to α-pinene. Hydrolysis of the esters yields (−)-fenchol.

Fenchone is used to prepare artificial fennel oils and to perfume household products.
FCT 1976 (**14**) p. 769.

Ionones and Homologous Compounds

The C_{13} ketones α- and β-ionone are cyclic terpenoids that occur in many essential oils. However, being metabolites of the corresponding carotenoids [89], they occur in only small amounts. A third isomer, γ-ionone, has not yet been observed in nature.

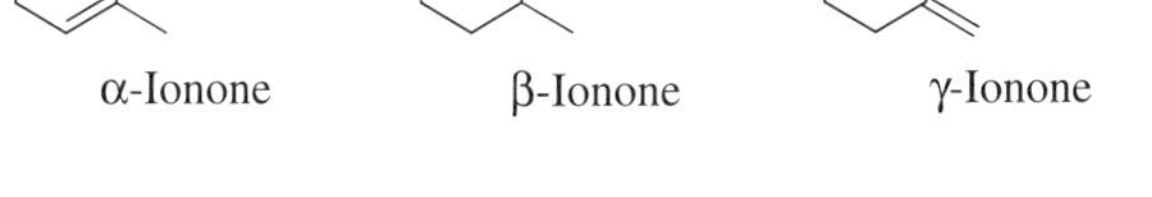

α-Ionone   β-Ionone   γ-Ionone

Both optical isomers of α-ionone are found in nature. Generally, ionones have a *trans* configuration. *trans-α*-Ionone can be converted into the *cis* isomer by exposure to ultraviolet light. Under the same conditions, *trans-β*-ionone rearranges to the retro compound.

The *irones* are ionone homologues that have an additional methyl group adjacent to the twin methyl groups in the cyclohexane ring. The number of possible irone isomers is larger than that of the ionones due to the additional methyl group on the ring. Some of these irone isomers occur in essential oils from the roots of *Orris* species (see Orris Root Oil).

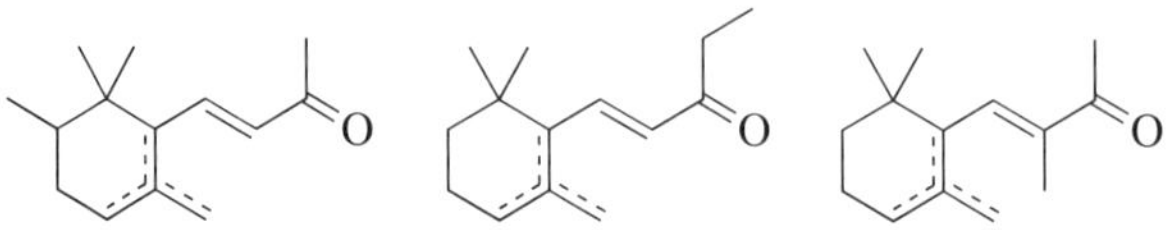

Irone, α-, β-, γ-   *n*-Methylionone,   Isomethylionone,
                              α-, β-, γ-   α-, β-, γ-

Other ionone homologues are the *methylionones*, in which the oxoalkenyl group carries an extra methyl substituent. The methylionones also exist as α-, β, and γ-isomers, each of which can occur in the *cis* or *trans* form; the isomers may also be optically active. Their natural occurrence is debated [90].

All ionones, irones, and methylionones, as well as the corresponding pseudo-compounds (their synthetic acyclic precursors) are slightly viscous yellowish liquids. Commercial irones and methylionones are mixtures of isomers that are named according to their main component. Their composition varies with the method used to prepare and cyclize the pseudocompound and fluctuates considerably between different manufacturers.

Physical and Chemical Properties.   Physical and odor properties of the best known ionones are listed in Table 2.

β-Ionone is converted into intermediates for vitamin A synthesis. The hydrogenation of ionones and methylionones is of some importance. Dihydro or tetrahydro derivatives or ionols can be obtained depending on reaction conditions.

With Raney nickel–copper chromite catalysts, methylionones are converted into tetrahydromethylionols, which are also used as fragrance materials [91].

Production.   Ionones, irones, and methylionones, as well as allylionone, are all produced by analogous routes. Special procedures must be used to obtain a partic-ular isomer, either pure or as the main component. These are described where appropriate.

In all processes an acyclic precursor, called a pseudoionone, pseudoirone, etc., is prepared by base-catalyzed condensation of citral or 6-methylcitral with acetone, methyl ethyl ketone, or allylacetone, as appropriate.

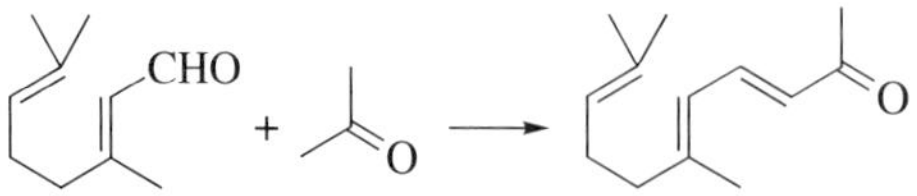

Citral   Acetone   Pseudoionone

Table 2. Physical properties of ionones

Name	Formula	CAS registry number	M_r	bp °C (p, kPa)	d_4^{20}	n_D^{20}	Odor
α-Ionone	$C_{13}H_{20}O$	[127-41-3]	192.30	121-122 (1.3)	0.9319	1.4982	sweet-floral, reminiscent of violets
β-Ionone	$C_{13}H_{20}O$	[79-77-6]	192.30	121.5 (0.93)	0.9461	1.5202	reminiscent of cedarwood, violet-like upon dilution
γ-Ionone	$C_{13}H_{20}O$	[76-76-5]	192.30	82 (0.16)	0.9317	1.4985	violet-like with woody-resinous tonality (intermediate in the synthesis of γ-dihydro-ionone, a component of ambergris)
α-Irone	$C_{14}H_{22}O$	[79-69-6]	206.33	109 (0.36)	0.9340	1.4998	responsible for the fragrance of natural orris oil
β-Irone	$C_{14}H_{22}O$	[79-70-9]	206.33	108–109 (0.21)	0.9465	1.5183	reminiscent of β-ionone, but slightly more intense
α-n-Methylionone	$C_{14}H_{22}O$	[127-42-4]	206.33	97 (0.35)	0.9210^a	1.4938^b	reminiscent of α-ionone, but milder and more delicate
β-n-Methylionone	$C_{14}H_{22}O$	[127-43-5]	206.33	102 (0.35)	0.9370	1.5155	β-ionone-like, but with a distinct leather note
α-Isomethylionone	$C_{14}H_{22}O$	[127-51-5]	206.33	130–131 (1.3)	0.9345	1.5019	reminiscent of orris and violets, possesses the finest odor of all ionones
β-Isomethylionone	$C_{14}H_{22}O$	[79-89-0]	206.33	94 (0.4)	0.9376	1.5033	interesting, powdery, orris-like odor with slightly woody aspects
Allylionone	$C_{16}H_{24}O$	[79-78-7]	232.35	102–104 (0.02)	0.9289	1.5040	floral violet odor with a fruity pineapple note and high tenacity

$^a d_4^{25}$. $^b n_D^{25}$.

In the synthesis of vitamin A, the dependence on natural sources as well as steadily increasing production via β-ionone as an intermediate have led to the development of a method for synthesizing citral from dehydrolinalool (see p. 36). More recent routes employ dehydrolinalool as the starting material for pseudoionone. Dehydrolinalool is converted into pseudoionone by using either diketene [92] or a suitably substituted acetoacetate (Carroll reaction) [93]:

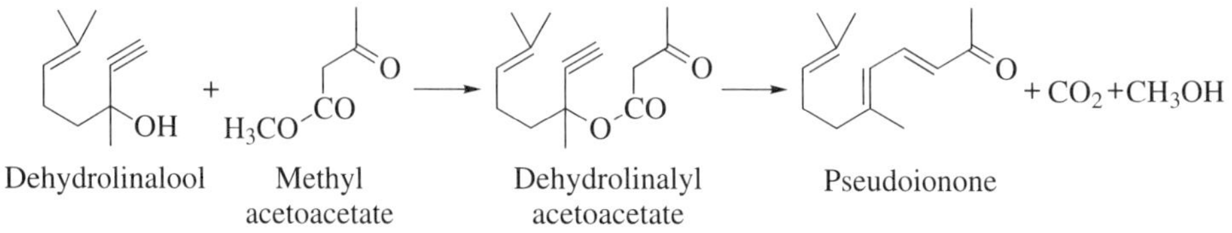

Dehydrolinalool + Methyl acetoacetate → Dehydrolinalyl acetoacetate → Pseudoionone $+ CO_2 + CH_3OH$

A milder reaction for synthesizing pseudoionone from dehydrolinalool is trans-etherification with an alkoxyalkene [94]:

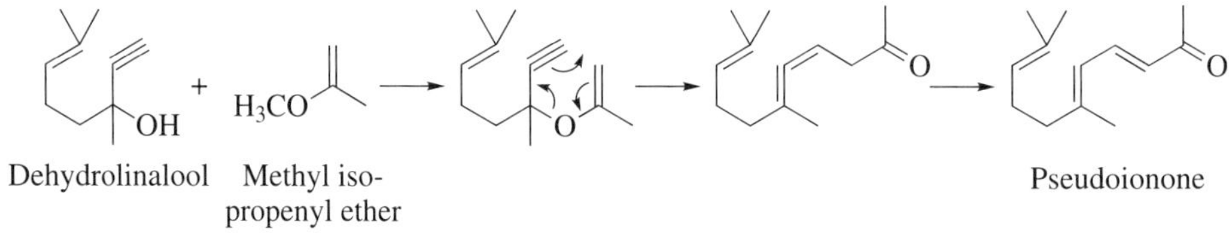

Dehydrolinalool + Methyl iso-propenyl ether → → → Pseudoionone

In the methylionone synthesis, condensation of citral with methyl ethyl ketone results in a mixture of *n*-methyl- and isomethylpseudoionone, each of which may occur as one of four possible *cis–trans* isomers.

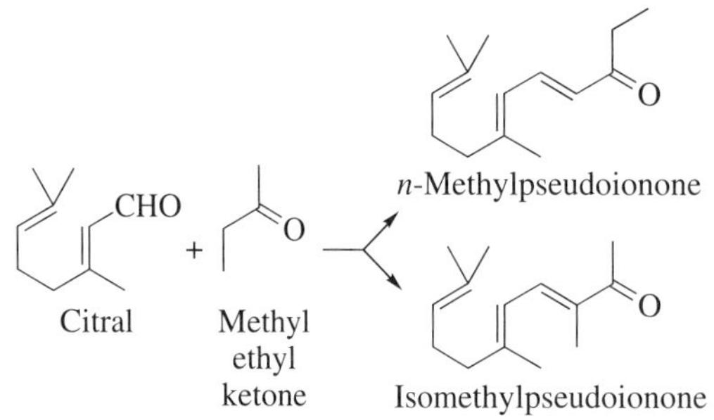

n-Methylpseudoionone

Citral + Methyl ethyl ketone →

Isomethylpseudoionone

The ratio of the major isomers in the mixture depends on the condensation catalyst and the reaction conditions. In the presence of common alkaline catalysts (e.g., sodium hydroxide), straight-chain isomers are formed preferentially. Strongly alkaline catalysts, such as quaternary ammonium bases, favor the formation of isomethylpseudoionone [95]. This compound is a precursor for the highly valued fragrance substance α-isomethylionone and can be obtained as the main component by reacting dehydrolinalool with the enol ether of methyl ethyl ketone and methanol [96]. Acidic and Lewis catalysts are employed in the cyclization of the pseudocompounds to the cyclic ketones. The primary cyclization products are α-ionone or its homologues, which are isomerized to the β-compounds by strong acids. Concentrated sulfuric acid converts pseudoionone almost exclusively into β-ionone; 85% phosphoric acid yields α-ionone in ca. 80% purity. γ-Ionone can be obtained together with a small amount of α- and β-ionone when boron trifluoride etherate is used as the catalyst and dimethylformamide as the solvent [97]; γ-ionone is of little commercial importance. Since α- and β-ionone can be separated on an industrial scale by fractional distillation in high-performance columns, other methods of separation are seldom used.

Uses. The volume of the production of β-ionone, which serves as an intermediate in vitamin A synthesis, shows that it is by far the most important. The ionones and their homologues are components of blossom and phantasy perfume compositions. The ionones and irones are used in aroma compositions as well, although on a much smaller scale. α-Ionone is a highly valued fragrance material. The methylionones are among the most important fragrance substances, α-isomethylionone being the most important. The irones, isomers of the methylionones, are produced in limited quantities, mainly due to their high cost. Likewise, allylionone is manufactured in small amounts.
FCT 1975 (**13**) p. 549: Ionone,
p. 551: α-Irone,
p. 863: Methylionone.

Damascones, 1-(2,6,6-trimethylcyclohexenyl)-2-buten-1-ones
$C_{13}H_{20}O$, M_r 192.30, are ionone isomers. Damascone exists in α-, β-, γ-, δ-, and ε-forms, depending on the position of the double bond in the cyclohexane ring. Commercially important are α-, β-, γ- and δ-damascone as mixtures of their *cis* and *trans* isomers.

The α- and β-damascones have been identified as components of tea aroma. They are liquids with a fruity, roselike odor. The δ-isomer, not reported to have been found in nature has a leathery, somewhat laurel-like odor. One synthetic route to the damascones starts with an appropriate cyclogeranic acid derivative (halide, ester, etc.). This is reacted with an allyl magnesium halide to give 2,6,6-trimethylcyclo-hexenyl diallyl carbinol, which on pyrolysis yields the desired 1-(2,6,6-trimethyl-cyclohexenyl)-3-buten-1-one. Damascone is obtained by rearrangement of the double bond in the side-chain [98].

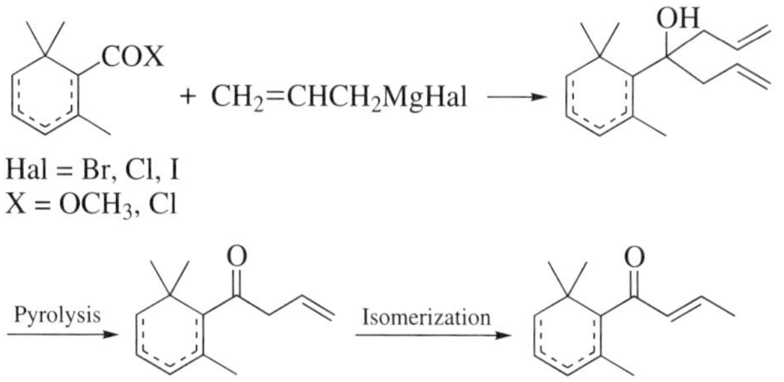

Hal = Br, Cl, I
X = OCH$_3$, Cl

The α- and β-damascones are used in perfume compositions, especially rose perfumes, and in flavor compositions, to which they impart naturalness and body. δ-Damascone is used in perfumes to create masculine notes.

γ-Damascone [*35087-49-1*], **1-(2,2-dimethyl-6-methylenecyclohexyl)-2-buten-1-one**

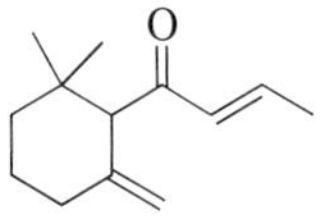

C$_{13}$H$_{20}$O, M_r 192.30, $bp_{0.013\,kPa}$ 58–60 °C, d^{20} 0.9335, n_D^{20} 1.4940, is not reported as being found in nature. It is a liquid with very powerful floral, rosy and fruity note, with a complex odor of apple, green and slight pine effect.

It is prepared from methyl β-cyclogeranate by deprotonation with butyl lithium, treating the enolate with allyl magnesium chloride and isomerizing the resulting 1-(2-methylene-6,6-dimethylcyclohexyl)-3-buten-1-one to γ-damascone [99]. It is used in fine fragrances.

β-Damascenone [*23726-93-4*], **1-(2,6,6-trimethyl-1,3-cyclohexadienyl)-2-buten-1-one**

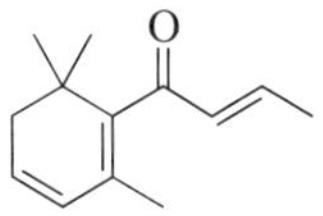

C$_{13}$H$_{18}$O, M_r 190.28, is a constituent of Bulgarian rose oil, to the flavor of which it contributes an important role although it is only present at a concentration of 0.05%.

It is used in small quantities in flavor and perfume compositions to impart natu-ralness and brilliance.

1-(2,4,4-Trimethyl-2-cyclohexen-1-yl)-2-buten-1-one [*39872-57-6*]

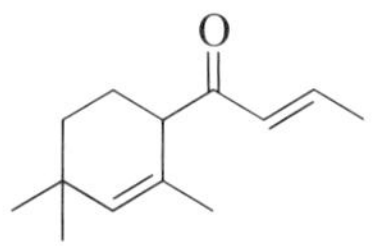

$C_{13}H_{20}O$, M_r 192.30, d_4^{20} 0.917–0.925, n_D^{20} 1.485–1.493, is a colorless to slightly yellow liquid with a highly diffusive fruity-flowery odor. It is prepared from 1,5,5-trimethyl-1-cyclohexene by reaction with crotonic anhydride in the presence of catalysts [100].

It is used in perfumery for soaps, cosmetics, and toiletries to give naturalness and freshness.

Trade Name. Isodamascon (Dragoco).

Nootkatone [*4674-50-4*], 5,6-dimethyl-8-isopropenylbicyclo[4.4.0]-1-decen-3-one

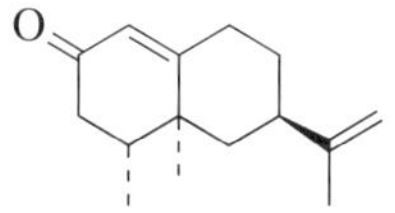

$C_{15}H_{22}O$, M_r 218.33, *mp* 35 °C, has been isolated from grapefruit peel and juice and identified in other citrus oils as well. The commercially available product is a colorless to yellowish liquid with a typical grapefruit odor.

Nootkatone can be prepared by oxidation of valencene, a sesquiterpene hydrocarbon isolated from orange oils.

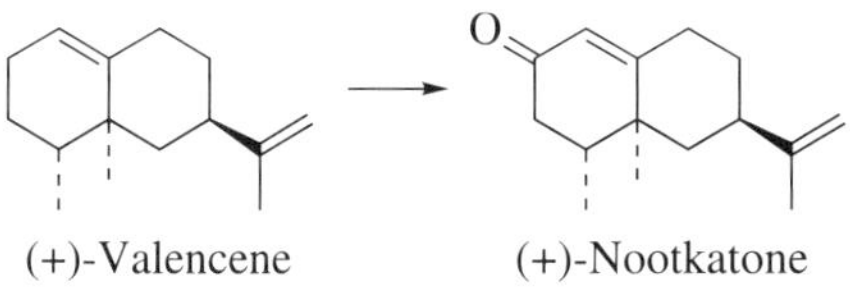

(+)-Valencene   (+)-Nootkatone

Nootkatone is used for flavoring beverages.

Cedryl methyl ketone
$C_{17}H_{26}O$, M_r 246.39, is a long-lasting wood fragrance which is prepared by acetyl-ation of cedarwood oil fractions that contain sesquiterpene hydrocarbons, mainly α-cedrene and thujopsene. Acetylation is carried out in the presence of an acidic

catalyst (e.g., polyphosphoric acid). Commercially available cedryl methyl ketone is a multicomponent mixture. One of its odor-determining compounds is 4-acetyl-1,1,6-trimethyl-6,8a-ethano-1,2,3,5,6,7,8,8a-octahydronaphthalene [*32388-56-0*]:

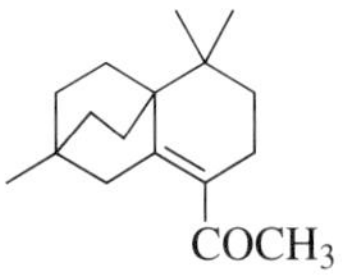

FCT 1978 (**16**) p. 639.

Trade Names. Lignofix (Dragoco), Lixetone (Quest), Vertofix (IFF).

2.3.4 Esters

Esters derived from cyclic terpene alcohols, especially the acetates, are common fragrance and flavor components. Menthanyl, menthenyl, bicyclic bornyl acetates and a few acetates of sesquiterpene alcohols are extensively used in perfume and aroma compositions.

(−)-Menthyl acetate [*2623-23-6*], **(−)-*p*-menthan-3-yl acetate**

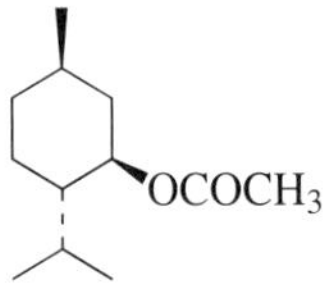

$C_{12}H_{22}O_2$, M_r 198.30, $bp_{3\,kPa}$ 116 °C, d_4^{20} 0.9253, n_D^{20} 1.4467–1.4468, $[\alpha]_D^{20} - 81.1°$, occurs in peppermint oils. It is a colorless liquid with a fresh-fruity, peppermint odor.

(−)-Menthyl acetate is prepared by acetylation of (−)-menthol (e.g., with acetic anhydride). It is used mainly in peppermint flavors and reconstituted peppermint oils, but also to a small extent in perfumery.
FCT 1976 (**14**) p. 477.

(±)-Menthyl acetate [*29066-34-0*]
Occurs in essential oils. It is synthesized by esterification of racemic menthol. Its odor is crisper and less fruity than that of (−)-menthyl acetate. It is used for essential oil compositions and occasionally in household perfumery.
FCT 1976 (**14**) p. 479.

p-Menthanyl acetate

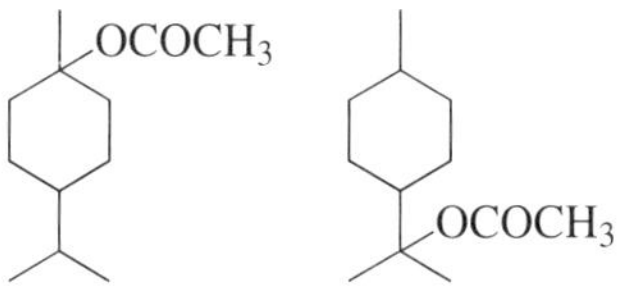

1   2

$C_{12}H_{22}O_2$, M_r 198.30, $bp_{0.2 kPa}$ 67–70 °C, d_{25}^{25} 0.931–0.937, n_D^{20} 1.446–1.451, is commercially available as a *cis–trans* mixture of *p*-menthan-l-yl (**1**) and *p*-menthan-8-yl (**2**) acetates. It is a colorless liquid with a citrus-fresh, pine-needle odor and a secondary, slightly herbaceous note. Menthanyl acetate is produced by hydrogenation of terpinyl acetates (mixtures of isomers) (e.g., in the presence of Raney nickel [101]) or by esterification of a mixture of isomeric *p*-menthanols.

The ester mixture is highly stable and is, therefore, used in perfumery for detergents and other household products.

α-Terpinyl acetate [*80-26-2*], **1-*p*-menthen-8-yl acetate**

$C_{12}H_{20}O_2$, M_r 196.29, $bp_{5.3 kPa}$ 140 °C, d_4^{20} 0.9659, n_D^{21} 1.4689, $[\alpha]_D^{20}$ of the enantiomers + or− 79°; the enantiomers and the racemate occur in many essential oils (e.g., Siberian pine-needle oil and cypress oil), but generally not as the main component. Pure *α*-terpinyl acetates are colorless liquids with a fresh bergamot-lavender odor. Commercially available terpinyl acetate consists mainly of *α*-terpinyl acetate, but also contains a number of other isomeric compounds. It can be prepared by acetylating the terpineol mixture obtained from terpin hydrate, using a customary procedure for tertiary alcohols.

Because of its odor properties, stability, and low price, large quantities of terpinyl acetate are used in perfumery for lavender and bergamot types, as well as in essential oil reconstitutions.
FCT 1974 (**12**) p. 999.

Nopyl acetate [*35836-72-7*], **(−)-2-(6,6-dimethylbicyclo[3.1.1]hept-2-en-2-yl)-ethyl acetate**

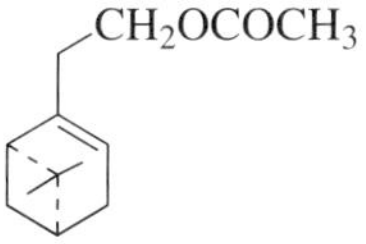

$C_{13}H_{20}O_2$ M_r 208.30, $bp_{1.5 kPa}$ 122 °C, d_4^{20} 0.9811, n_D^{20} 1.4733, $[\alpha]_D$ − 30.9°, is not found in nature and has a fresh-fruity-woody odor. It is prepared by acetylation of (−)-nopol with acetic anhydride. (−)-Nopol is obtained from (−)-β-pinene and paraformaldehyde in a Prins reaction.

Nopyl acetate is used in perfumes for soap and household products.
FCT 1974 (**12**) p. 943.

Bornyl acetate, 2-*endo*-bornanyl acetate

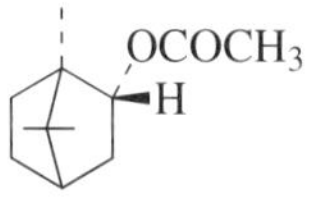

(−)-Bornyl acetate

$C_{12}H_{20}O_2$, M_r 196.29, $bp_{101.3 kPa}$ 223–224 °C, d_4^{20} 0.9838, n_D^{20} 1.4630, $[\alpha]_D^{20}$ + or − 44.4°, *mp* (+) and (−) form 29 °C, occurs in its optically active forms and as a racemate in many essential oils.

(−)-Bornyl acetate [5655-61-8] is a characteristic component of most conifer oils. It has a camphoraceous, pine-needle-like odor. Both (+)-bornyl acetate [20347-65-3] and (−)-bornyl acetate form colorless crystals; the racemate [36386-52-4] is a colorless liquid. Bornyl acetate is prepared by esterification of borneol with acetic anhydride or via the process described under borneol (see page 57).

Due to its characteristic pine-needle odor, bornyl acetate is frequently used in conifer needle compositions, soap, bath products, room sprays, and pharmaceutical products.

Isobornyl acetate [125-12-2], 2-*exo*-bornanyl acetate

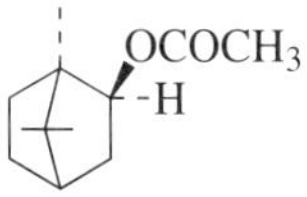

$C_{12}H_{20}O_2$, M_r 196.29, $bp_{1.6-1.7 kPa}$ 102–103 °C, d_4^{20} 0.9841, n_D^{20} 1.4640, has been identified in a number of essential oils. It is a colorless liquid with a pleasant, pine-needle odor. Isobornyl acetate is prepared from camphene and acetic acid in the presence of acidic catalysts (e.g., sulfuric acid) [102], or on a styrene-divinylbenzene acid ion-exchanger [103].

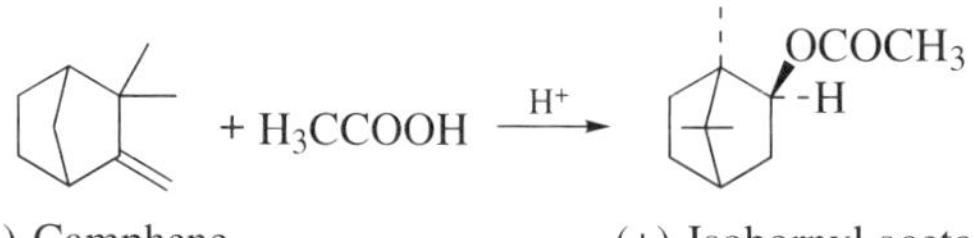

(−)-Camphene   + H₃CCOOH   →(H⁺)→   (+)-Isobornyl acetate

Isobornyl acetate is used in large amounts for perfuming soap, bath products, and air fresheners. However, the major use of isobornyl acetate is as an intermediate in the production of camphor.
FCT 1975 (**13**) p. 552.

Guaiyl acetate
$C_{17}H_{28}O_2$, M_r 264.41, $bp_{0.3 kPa}$ 118–123 °C, d_{25}^{25} 0.965–0.990, n_D^{20} 1.489–1.495, is obtained by esterification of guaiac wood oil with acetic anhydride and consists of the acetates of the natural sesquiterpene alcohols guaiol and bulnesol (see Guaiac Wood Oil). Guaiyl acetate is a yellow to amber liquid with a weak but lasting, woody odor. It is used in perfumery for tea-rose and wood nuances.
FCT 1974 (**12**) p. 903.

Cedryl acetate [*77-54-3*]

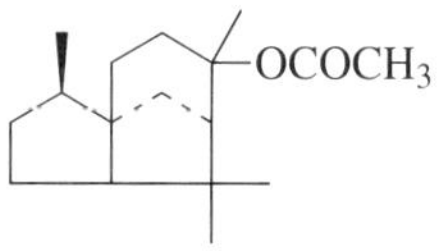

$C_{17}H_{28}O_2$, M_r 264.41, $bp_{0.4 kPa}$ 146–150 °C, d_{25}^{25} 0.966–1.012, n_D^{20} 1.495–1.506, occurs in cedarwood oils. The pure compound is crystalline (*mp* 80 °C). Commercial cedryl acetate is a colorless to amber liquid, with a cedarwood-like odor. It is prepared by esterification of the cedrol-rich fraction from cedarwood oil and is used in perfumery for wood and leather notes, and as a fixative.
FCT 1974 (**12**) p. 847.

Vetiveryl acetate [*62563-80-8*]
$bp_{0.3 kPa}$ 125–128 °C, d_{25}^{25} 0.979–1.015, n_D^{20} 1.5050–1.5180, is not a single compound; its main component is khusimyl acetate [*61474-33-7*].

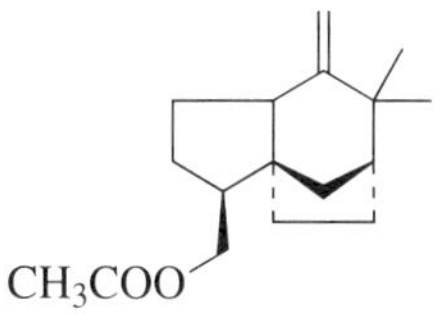

CH_3COO
Khusimyl acetate

Vetiveryl acetate is prepared by esterification of the sesquiterpene alcohols isolated from vetiver oils.
Vetiveryl acetate is a light yellow liquid with a dry, fresh-woody odor. It is a popular fragrance mixture that is frequently used in luxury perfumery; it is also used as a fixative in many fine fragrances.
FCT 1974 (**12**) p. 1011.

2.3.5 Miscellaneous Compounds

Of the few known terpene compounds that contain heteroatoms such as nitrogen or sulfur, the thiol 8-mercapto-*p*-menthan-3-one described below has qualitatively important applications as a fragrance and flavor substance. The second thiol, 1-*p*-menthene-8-thiol, is described because its odor threshold value is far lower than that of most other fragrance and flavor compounds.

8-Mercapto-*p*-menthan-3-one [*38462-22-5*]

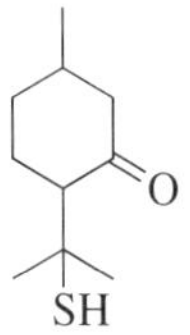

$C_{10}H_{18}OS$, M_r 186.31, $bp_{1\,Pa}$ 57 °C, d_{20}^{20} 1.002–1.007, n_D^{20} 1.493–1.497, is an essential odoriferous constituent of buchu leaf oil. The commercial product is synthesized from pulegone and is a liquid mixture of *cis–trans* isomers with a typical blackcurrant odor. 8-Mercapto-*p*-menthan-3-one is prepared by reacting pulegone with hydrogen sulfide in the presence of a base (e.g., triethylamine) [104]:

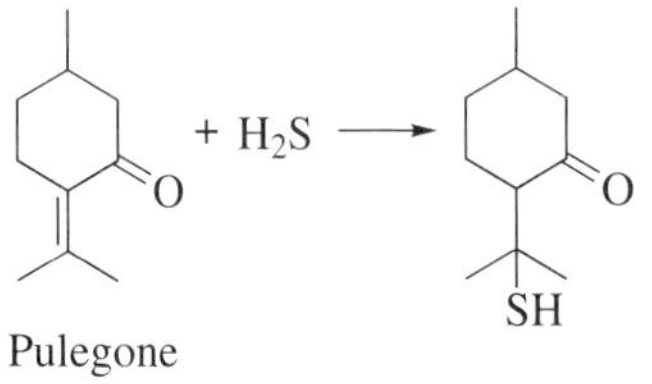

Pulegone

This powerful fragrance and flavor substance is used in perfume and aroma compositions.

Trade Name. Sulfox (Firmenich).

1-*p*-Menthene-8-thiol [*71159-90-5*]

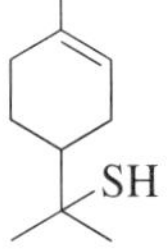

$C_{10}H_{18}S$ M_r 170.31, $bp_{0.13\,Pa}$ 40 °C, d_4^{20} 0.948, n_D^{20} 1.503, has been identified in grapefruit juice. It is a liquid with an extremely powerful, obnoxious odor; when diluted it has the typical aroma of fresh grapefruit juice. Its odor threshold value is extremely low: $2 \times 10^{-5}\,\mu g/kg$ for the $(+)$-R and $8 \times 10^{-5}\,\mu g/kg$ for the $(-)$-S isomer [105].

2.4 Other Cycloaliphatic Compounds

In addition to cyclic terpenoids, several other cycloaliphatic compounds have above-average importance as fragrance materials; some of them are structurally related to the terpenes.

Ketones are most widely represented and include cyclopentanone derivatives, such as the jasmin fragrance compounds, and cyclic ketones with 15–17-membered carbon rings, such as muscone and civetone, which are constituents of the extremely expensive animal products, musk and civet. Cyclopentadecanone, a natural musk fragrance, and the unsaturated 5-cyclohexadecen-1-one, which has not yet been found in nature, have odor characteristics similar to those of muscone and civetone and are more easily synthesized. They are, therefore, often used as substitutes.

Some alicyclic alcohols are important as synthetic sandalwood fragrances. A few alicyclic aldehydes are valuable perfume materials and are obtained by Diels–Alder reactions using terpenes and acrolein. Esters derived from hydrogenated aromatic compounds, such as *tert*-butylcyclohexyl and decahydro-β-naphthyl acetates, are also used in large amounts as fragrance materials.

2.4.1 Alcohols and Ethers

2-Methyl-4-(2,2,3-trimethyl-3-cyclopenten-1-yl)butanol [*72089-08-8*]

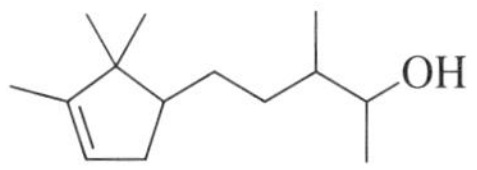

$C_{13}H_{24}O$, M_r 196.34, d_4^{20} 0.900–0.906, n_D^{20} 1.470–1.475, is a colorless liquid with a woody, tenacious sandalwood odor with a slight musk nuance. It is prepared by sequential aldol condensation of campholenaldehyde (2,2,3-trimethyl-3-cyclopenteneacetaldehyde, obtained by epoxidation of α-pinene and rearrangement of the epoxide) with propanal, hydrogenation and reduction [106].

It is used to perfume soaps and detergents.

Trade Name. Brahmanol (Dragoco).

2-Methyl-4-(2,2,3-trimethyl-3-cyclopenten-1-yl)-2-buten-1-ol [*28219-60-5*]

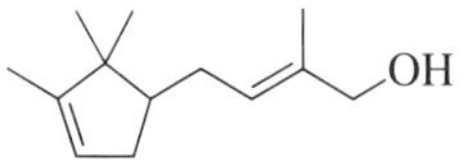

$C_{13}H_{22}O$, M_r 194.32, is not found in nature. It is a clear, colorless liquid, d_{20}^{20} 0.919–0.929, n_D^{20} 1.483–1.493, with a powerful sandalwood odor. It is prepared by condensation of campholenaldehyde with propionaldehyde and reduction of the formyl group.

It is used in perfume compositions for soaps and household products.

Trade Name. Sandalmysore Core (Kao).

5-(2,2,3-Trimethyl-3-cyclopenten-1-yl)-3-methylpentan-2-ol [*65113-99-7*] **(3)**

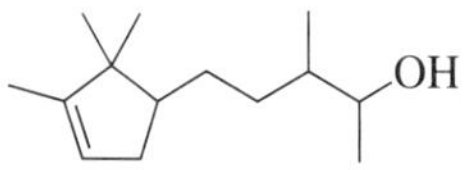

$C_{14}H_{26}O$, M_r 210.36, d_4^{20} 0.896–0.904, n_D^{20} 1.470–1.476, is a sandalwood-like fragrance ingredient that does not occur in nature. It is prepared by condensation of campholenaldehyde with methyl ethyl ketone, followed by selective hydrogenation of the resulting unsaturated ketone [107].

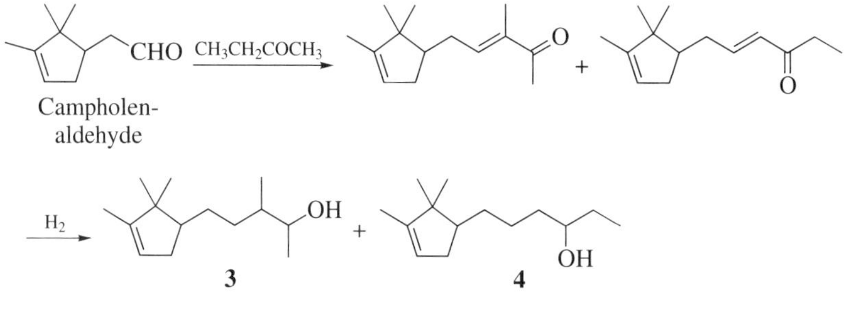

Campholen-
aldehyde

3

4

Compound (**3**) can be used either in a pure state or as a mixture with its byproduct (**4**) in perfume compositions and soap perfumes.

Trade Name. Sandalore (Giv.-Roure).

2-Ethyl-4-(2,2,3-trimethyl-3-cyclopenten-1-yl)-2-buten-1-ol [*28219-61-6*]

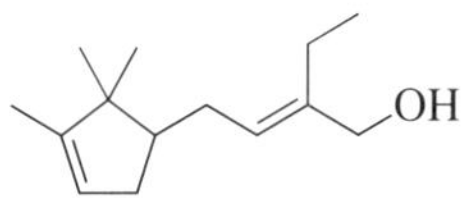

$C_{14}H_{24}O$, M_r 208.35, d_4^{20} 0.913–0.920, n_D^{20} 1.484–1.490, is a mixture of the *cis-* and *trans*-isomer. It does not occur in nature. It is a pale yellow liquid with a powerful sandalwood odor and a slight rose nuance. The mixture can be prepared starting from campholenaldehyde and butanal. The intermediate unsaturated aldehyde is partially hydrogenated to give the title alcohol.

It is used in perfume compositions for cosmetics, soaps and detergents.

Trade Names. Bacdanol (IFF), Bangalol (Quest), Sandolene (H&R), Sandranol (Dragoco).

3-Methyl-5-(2,2,3-trimethyl-3-cyclopenten-1-yl)-4-penten-2-ol [*67801-20-1*]

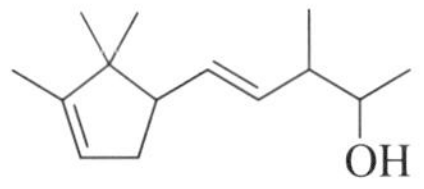

$C_{14}H_{24}O$, M_r 208.35, is a mixture of isomers and not reported as being found in nature. It is a pale yellow liquid with a powerful woody, sandalwood odor with musk aspect. The mixture is obtained by condensation of campholenaldehyde with 2-butanone and isomerization of the reaction mixture with potassium *tert*-butylate in dimethyl formamide. Subsequent reduction with $NaBH_4$ yields a mixture of chiefly four diastereomeric title alcohols [108].

It is used in fine fragrances as well as in functional products.

Trade Name. Ebanol (Giv.-Roure).

3,3-Dimethyl-5-(2,2,3-trimethyl-3-cyclopenten-1-yl)-4-penten-2-ol [*107898-54-4*]

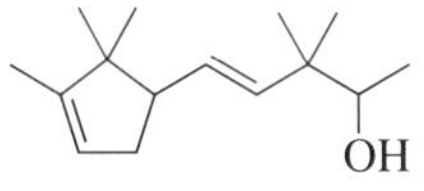

$C_{15}H_{26}O$, M_r 222.37, is not found in nature. It is a clear, colorless liquid, d_{25}^{25} 0.897–0.906, n_D^{20} 1.480–1.484, with a powerful, diffusive sandalwood odor and musk and cedarwood aspects. The compound is prepared starting from campholenaldehyde which is condensed with 2-butanone to give 3-methyl-5-(2,2,3-trimethyl-3-cyclopenten-1-yl)-3-penten-2-one. Methylation using phase transfer conditions gives the dimethylpentenone which is reduced with $NaBH_4$ to yield the title compound [109].

It is used in fine fragrances.

Trade Name. Polysantol (Firmenich).

3-*trans*-Isocamphylcyclohexanol [4105-12-8]

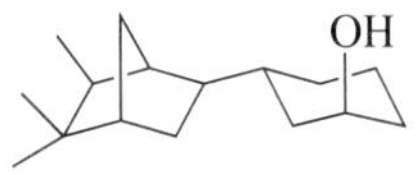

$C_{16}H_{28}O$, M_r 236.40, does not occur in nature. It is the component responsible for the sandalwood odor of a synthetic mixture of terpenylcyclohexanol isomers. A commercially available mixture containing 3-*trans*-isocamphylcyclohexanol is prepared by reacting camphene and guaiacol in the presence of an acidic catalyst (e.g., boron trifluoride), followed by catalytic hydrogenation of the resulting terpenylguaiacols. In the alkylation reaction, camphene rearranges to the isobornyl, isofenchyl, and isocamphyl skeletons. These substituents may be introduced in guaiacol at four positions. In the subsequent hydrogenation with simultaneous elimination of the methoxy group, additional possibilities for isomerism arise because the hydroxyl group may be either axial or equatorial to the terpenyl moiety. Therefore, the actual content of the desired isomer, 3-*trans*-isocamphylcyclohexanol, is low in most products. The other isomers are either weak in odor or odorless.

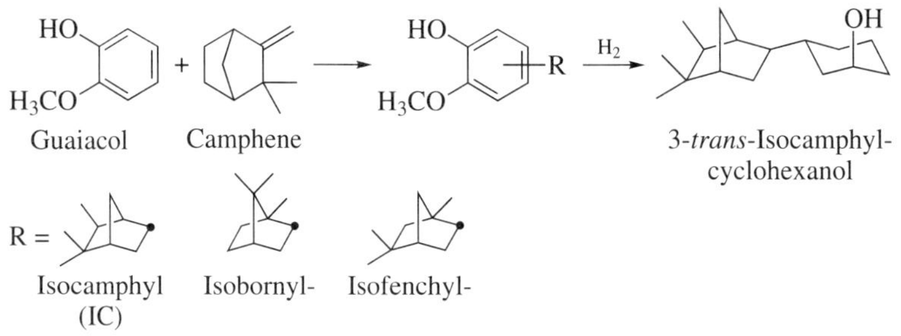

A process starting from catechol, instead of guaiacol, yields a mixture with a higher content of 3-*trans*-isocamphylcyclohexanol [110]. Moreover, the process starting from guaiacol has been improved by converting the main component formed in the first step, *p*-isocamphylguaiacol (**5**), into a mixed ether (**6**) by reaction with diethyl sulfate. Nucleophilic cleavage of the ether with alcoholates or Grignard reagents results in a high yield of *m*-isocamphylguethol (**7**), which is hydrogenated catalytically, with concomitant loss of the ethoxy group, to give a high yield of 3-*trans*-isocamphylcyclohexanol (**8**) [111].

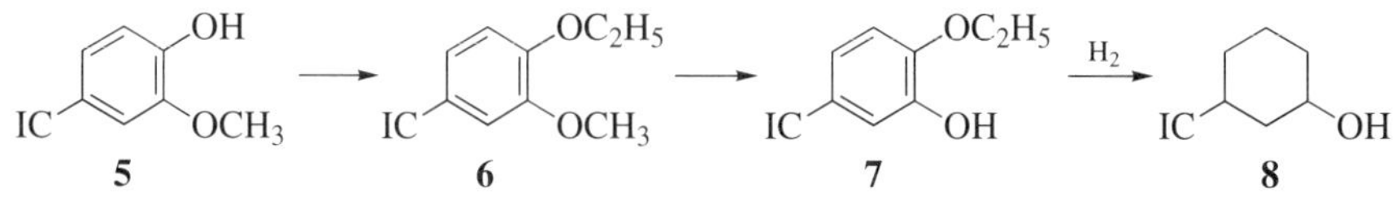

The mixture is used as such in large amounts as a replacement for sandalwood oil in soaps, cosmetics, and perfume compositions.
FCT 1976 (**14**) p. 801.

Trade Names. Sandela (Giv.-Roure), Sandel H&R (H&R), IBCH (Rhône-Poulenc), Santalex (Takasago).

1-(2,2,6-Trimethylcyclohexyl)hexan-3-ol [*70788-30-6*]

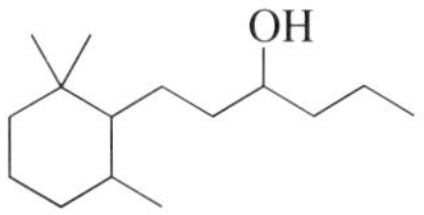

$C_{15}H_{30}O$, M_r 226.41, $bp_{0.0133\,kPa}$ 150 °C, d_4^{20} 0.896–0.902, n_D^{20} 1.470–1.476, is a colorless liquid with a highly diffusive, powdery-woody odor. It is a mixture of the *cis*-and *trans*-isomers. The *trans*-isomer has a more distinct animal and ambery character [112].

The title compound is prepared by condensation of citral (see p. 35) with 2-pentanone in the presence of bases to give 8,12-dimethyltrideca-5,7,11-trien-4-one, which is cyclized and hydrogenated [113].

A high *trans*-mixture can be prepared starting from β-cyclocitral which is hydrogenated to 2,2,6-trimethylcyclohexane carboxaldehyde. Condensation with 2-pentanone in the presence of sodium ethoxide yields the corresponding 3-hexenone. Hydrogenation with nickel–copper chromite as a catalyst gives a mixture with up to 95% of the *trans*-isomer [114].

It is used in perfume compositions for soaps, detergents and household products.

Trade Name. Timberol (Dragoco).

2,5,5-Trimethyl-1,2,3,4,4a,5,6,7-octahydronaphthalen-2-ol [*41199-19-3*]

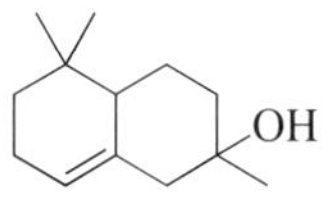

$C_{13}H_{22}O$, M_r 194.32, d_{20}^{20} 0.940–0.960, n_D^{20} 1.485–1.498, is a colorless to pale yellow liquid with an extremely powerful, amber, somewhat musty and animal odor. It is a constituent of ambergris (see p. 168). A synthesis starts with the thermolysis of β-ionone (see p. 61) which leads to dehydroambrinol. The title compound is obtained by hydrogenation over Raney nickel in methanolic solution [115].

It is used in small amounts in all perfume types for, for example, cosmetics, soaps, body care products, and detergents.

Trade Names. Ambrinol (Firmenich), Ambrinol S (H&R).

Cyclododecyl methyl ether [*2986-54-1*], **methoxycyclododecane**

$C_{13}H_{26}O$, M_r 198.35, does not occur in nature. It is a clear, colorless to pale yellowish liquid, d_{25}^{25} 0.910–0.915, n_D^{20} 1.472–1.475, with a woody, cedarlike odor. It can be prepared from dodecanol sodium by reaction with methyl halogenide. It is used as a stable wood fragrance in technical perfumery.

Trade Name. Palisandin (H&R).

(Ethoxymethoxy)cyclododecane [*58567-11-6*]

$C_{15}H_{30}O_2$, M_r 242.41, is not known in nature. It is a colorless liquid, $bp_{0.1\,Pa}$ 94 °C, d_{25}^{25} 0.931, n_D^{20} 1.463–1.467, with a noble woody odor with ambergris nuances. It is prepared by reaction of cyclododecanol with paraformaldehyde/hydrochloric gas to give cyclododecyl chloromethyl ether which is treated with sodium ethylate [116]. (Ethoxymethoxy)cyclododecane is stable in alkaline media and can be used in perfume compositions for soaps and detergents.
FCT 1988 (**26**) p. 325.

Trade Name. Boisambrene forte (Henkel).

2.4.2 Aldehydes

2,4-Dimethyl-3-cyclohexene carboxaldehyde [*68039-49-6*]

$C_9H_{14}O$, M_r 138.21, $bp_{4\,kPa}$ 94–96 °C, n_D^{25} 1.4696, is prepared as a mixture of its *cis* and *trans* isomers by a Diels–Alder reaction of 2-methyl-1,3-pentadiene and acrolein. It is a liquid with a strongly green, slightly herbaceous, citrus note. It is used for perfuming cosmetic preparations as well as household products.

Trade Names. Cyclal (Giv.-Roure), Ligustral (Quest), Triplal (IFF), Vertocitral (H&R).

4-(4-Methyl-3-penten-1-yl)-3-cyclohexene carboxaldehyde [*37677-14-8*]

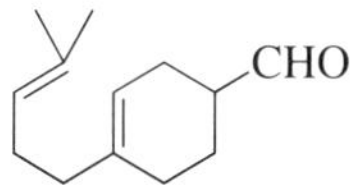

$C_{13}H_{20}O$, M_r 192.30, is prepared, together with its 3-isomer, by a Diels–Alder reaction of myrcene and acrolein. The mixture, d_4^{20} 0.927–0.935, n_D^{20} 1.488–1.492, has a fresh-fruity, slightly citrus-like odor and is used to perfume household products.
FCT 1976 (**14**) p. 803.

   Trade Names. Empetal (Quest), Myrac aldehyde (IFF), Myraldene (Giv.-Roure), Vertomugal (H&R).

4-(4-Hydroxy-4-methylpentyl)-3-cyclohexene carboxaldehyde [*31906-04-4*]

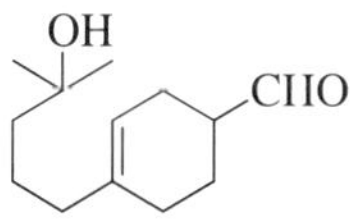

$C_{13}H_{22}O_2$ M_r 210.32, $bp_{0.13\,kPa}$ 120–122 °C, d_4^{20} 0.9941, n_D^{20} 1.4915, is a fragrance substance that does not occur in nature. It is a colorless, viscous liquid with a sweet odor reminiscent of lily of the valley. The aldehyde can be prepared by a Diels–Alder reaction of myrcenol and acrolein in the presence of a Lewis catalyst (e.g., zinc chloride) [117]:

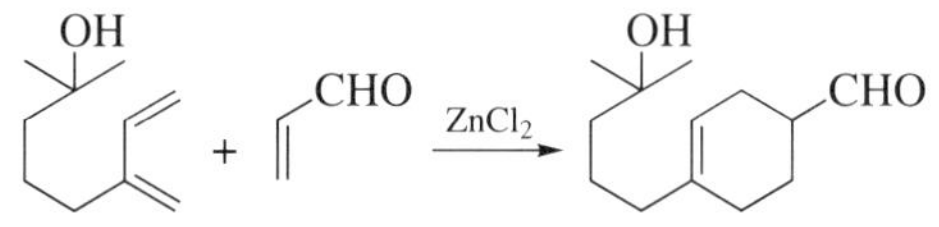

Myrcenol   Acrolein

   Reaction of myrcenol with acrolein at elevated temperatures, without a catalyst, yields a 70 : 30 mixture of the 4- and 3-substituted cyclohexene carboxaldehydes [118]. This mixture is a commercial product. The title compound has excellent fixative properties and is used especially in soap and cosmetics perfumery.
FCT 1992 (**30**) p. 49 S.

   Trade Names. Kovanol (Takasago), Lyral (IFF).

2-Methyl-4-(2,6,6-trimethyl-1-cyclohexen-1-yl)-2-butenal [*3155-71-3*]

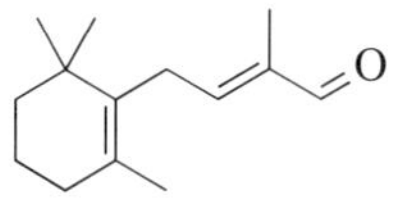

$C_{14}H_{22}O$, M_r 206.33, $bp_{0.01\,kPa}$ 100–103 °C, d_{25}^{25} 0.939–0.949, n_D^{20} 1.507–1.517, a clear yellowish liquid, does not occur in nature. Its floral odor is reminiscent of boronia absolute with violet accents. It can be prepared by reaction of β-ionone with ethyl chloroacetate and hydrolysis/decarboxylation of the intermediate glycidic ester. It is used in fine fragrances for ambra nuances in combination with sweet odor elements.

Trade Name. Boronal (H&R).

2.4.3 Ketones

2-Pentylcyclopentanone [*4819-67-4*]

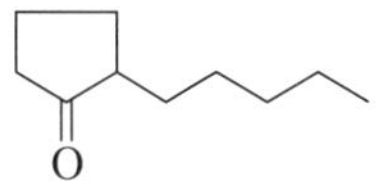

$C_{10}H_{18}O$, M_r 154.25, d_{20}^{20} 0.887–0.893, n_D^{20} 1.446–1.450, is a colorless liquid with a complex floral, aromatic and fruity odor, and a lactonic undertone. It is not found in nature.

2-Pentylcyclopentanone and its higher homologue 2-heptylcyclopentanone (see below) are prepared by condensation of cyclopentanone with the corresponding aliphatic aldehydes to give 2-alkylidenecyclopentanone and subsequent hydrogenation of the double bond. It is used in jasmin, herbal and lavender compositions.

Trade Name. Delphone (Firmenich).

2-Heptylcyclopentanone [*137-03-1*]

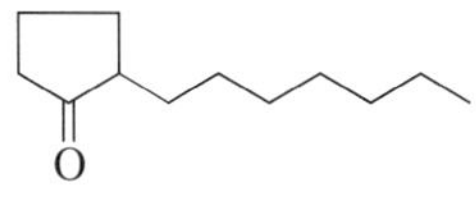

$C_{12}H_{22}O$, M_r 182.31, $bp_{1.3\,kPa}$ 130 °C, d^{20} 0.890, n_D^{20} 1.4530, is a colorless, viscous liquid with a fruity, slightly herbaceous, jasmin odor; it has not yet been found in nature.

2-Heptylcyclopentanone is used in, for example, jasmin, honeysuckle, and lavender compositions.
FCT 1975 (**13**) p. 452.

Trade Names. Alismone (Giv.-Roure), Fleuramone (IFF), Frutalone (PFW), Projasmon P (H&R).

2,2,5-Trimethyl-5-pentylcyclopentanone [*65443-14-3*]

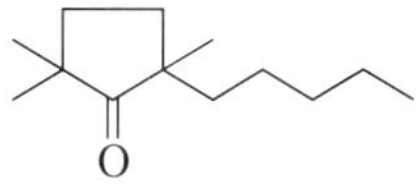

$C_{13}H_{24}O$, M_r 196.34, d_{20}^{20} 0.865–0.872, n_D^{20} 1.442–1.446, is a colorless to pale yellow liquid with a jasmin, lactonic and fruity odor. It is not found in nature. It is prepared by methylation of 2-pentylcyclopentanone with methyl iodide and sodium hydride in tetrahydrofuran [119].

Because of its stability combined with a soft floral note it has a wide use in jasmin and honeysuckle creations for, for example, body care products, soaps, and detergents.

Trade Name. Veloutone (Firmenich).

Dihydrojasmone [*1128-08-1*], **3-methyl-2-pentyl-2-cyclopenten-1-one**

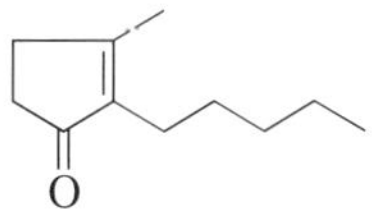

$C_{11}H_{18}O$, M_r 166.26, $bp_{2.7\,kPa}$ 87–88 °C, d^{25} 0.9157, n_D^{25} 1.4771, is a colorless, slightly viscous liquid with a typical jasmin odor, resembling that of the naturally occurring *cis*-jasmone. Dihydrojasmone can be synthesized by various routes. A preferred method is intramolecular aldol condensation of 2,5-undecanedione, which can be prepared from heptanal and 3-buten-2-one in the presence of a thiazolium salt, such as 5-(2-hydroxyethyl)-4-methyl-3-benzylthiazolium chloride [120]:

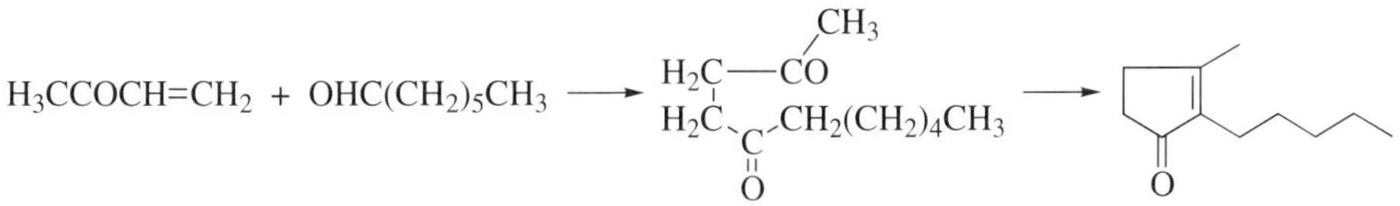

Dihydrojasmone is used in perfumery in jasmin bases and, more generally, in blossomy and fruity fragrances.
FCT 1974 (**12**) p. 523.

cis-**Jasmone** [*488-10-8*], **3-methyl-2-(2-*cis*-penten-1-yl)-2-cyclopenten-1-one**

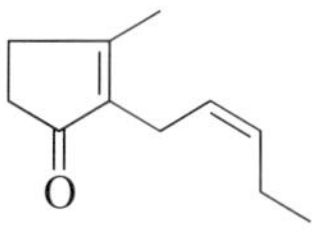

$C_{11}H_{16}O$, M_r 164.25, $bp_{1.6\,kPa}$ 78–79 °C, d^{20} 0.9423, n_D^{20} 1.4989, occurs in jasmin absolute and contributes to its typical jasmin odor. It is a pale yellow, viscous liquid with a strong jasmin odor. Various stereospecific syntheses for *cis*-jasmone have been reported. A patented method involves alkylation of 3-methyl-2-cyclopenten-1-one with *cis*-2-pentenyl chloride in an alkaline medium in the presence of a phase-transfer catalyst (e.g., tricaprylmethylammonium chloride) [121]:

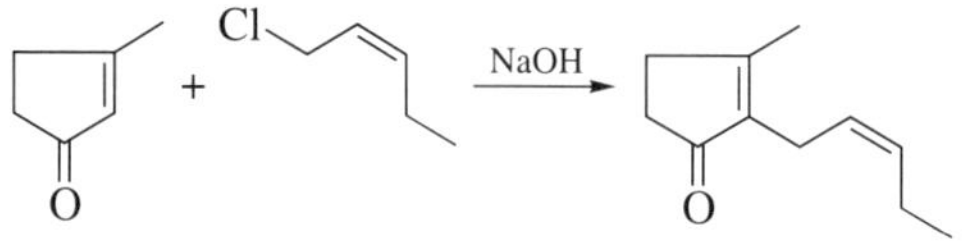

cis-Jasmone is used in perfumery in fine jasmin bases and floral compositions. FCT 1979 (**17**) p. 845.

4-*tert*-Pentylcyclohexanone [*16587-71-6*]

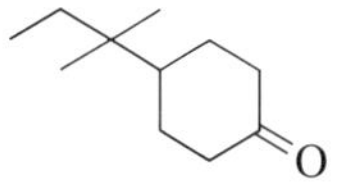

$C_{11}H_{20}O$, M_r 168.28, does not occur in nature. It is a colorless liquid, d_4^{20} 0.919–0.927, n_D^{20} 1.466–1.471, with a powerful orris type aroma. Hydrogenation of *p-tert*-amylphenol over palladium in the presence of borax is described for its synthesis [122].

It is useful in perfume compositions for, for example, laundry detergents. FCT 1974 (**12**) p. 819.

Trade Name. Orivone (IFF).

6,7-Dihydro-1,1,2,3,3-pentamethyl-4(5H)-indanone [*33704-61-9*]

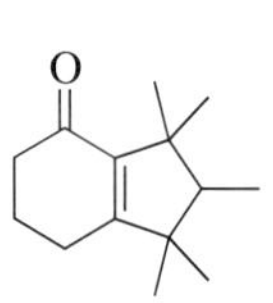

$C_{14}H_{22}O$, M_r 206.33, is not found in nature. It is a pale yellow liquid, d_4^{20} 0.954–0.962, n_D^{20} 1.497–1.502, with a long lasting diffusive, conifer-like musk odor. A process for its production starts with the corresponding pentamethyltetra-hydroindane which is treated with oxygen in the presence of cobalt naphthenate [123]. It is used in fine fragrances together with noble wood notes.

Trade Name. Cashmeran (IFF).

Cyclopentadecanone [*502-72-7*]

$C_{15}H_{28}O$, M_r 224.39, *mp* 65–67 °C, *bp*$_{7 Pa}$ 85 °C, is a musk fragrance found in the scent gland of the male civet cat.

A number of syntheses have been developed for its manufacture. In one method, 1,12-dodecanedial is reacted with 1,3-bis(dimethylphosphono)propan-2-one in the presence of a base, preferably in two steps via the intermediates 15-(dimethylpho-sphono)pentadec-12-en-14-on-1-al and 2,14-cyclopentadecadien-1-one. Subsequent hydrogenation yields cyclopentadecanone [124].

$$
\begin{array}{c}
CH_2-CHO \\
| \\
(CH_2)_8 \qquad + \ (CH_3O)_2\overset{O}{\overset{||}{P}}CH_2CO CH_2\overset{O}{\overset{||}{P}}(OCH_3)_2 \ \xrightarrow{\text{Base}} \\
| \\
CH_2-CHO
\end{array}
$$

$$
\begin{array}{c}
CH_2-CH=CHCOCH_2\overset{O}{\overset{||}{P}}(OCH_3)_2 \\
| \\
(CH_2)_8 \qquad\qquad \xrightarrow{\text{Base}} \\
| \\
CH_2CHO
\end{array}
\qquad
\begin{array}{c}
CH_2-CH=CH \\
| \qquad\qquad \diagdown \\
(CH_2)_8 \qquad\qquad CO \\
| \qquad\qquad \diagup \\
CH_2CH=CH
\end{array}
$$

$$
\xrightarrow{\text{H}_2/\text{Cat.}} \quad (CH_2)_{14} \quad C=O
$$

Cyclopentadecanone is used in fine fragrances.
FCT 1976 (**14**) p. 735.

Trade Name. Exaltone (Firmenich).

3-Methylcyclopentadecanone [*541-91-3*], **muscone**

$C_{16}H_{30}O$, M_r 238.42, d_{20}^{20} 0.918–0.925, n_D^{20} 1.477–1.482, is an odoriferous constituent of natural musk. It is a colorless liquid with very soft, sweet, musky odor and a perfumery, animal tonality. Numerous syntheses have been developed for its preparation [125].

Because of its excellent stability it can be used in a wide range of products to give elegant, warm, animal notes. It is important for the reconstitution of natural musk. FCT 1982 (**20**) p. 749.

Trade Name. Muscone (Firmenich).

5-Cyclohexadecen-1-one [*37609-25-9*]

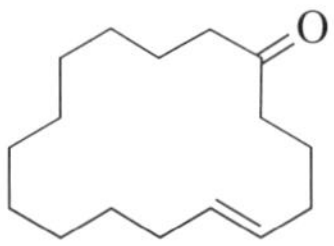

$C_{16}H_{28}O$, M_r 236.40, $bp_{0.01\,kPa}$ 121 °C, n_D^{25} 1.4865, is commercially available as a 40 : 60 mixture of its *cis* and *trans* isomers. It is a colorless liquid with an intense musk odor.

A three-step synthesis starts from cyclododecanone. Reaction with chlorine gives 2-chlorocyclododecanone which is reacted with 2 mol of vinylmagnesium chloride to give 1,2-divinylcyclododecan-1-ol. This is finally converted into 5-cyclohexadecen-1-one by an oxy-Cope rearrangement [126]:

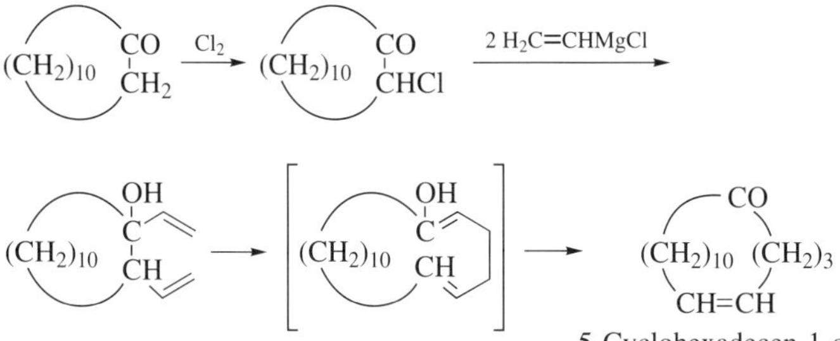

5-Cyclohexadecen-1-one can be added to perfume compositions as a substitute for the natural macrocyclic ketone musks.

Trade Name. Musk TM (Soda Aromatic).

9-Cycloheptadecen-1-one [*542-46-1*], **civetone**

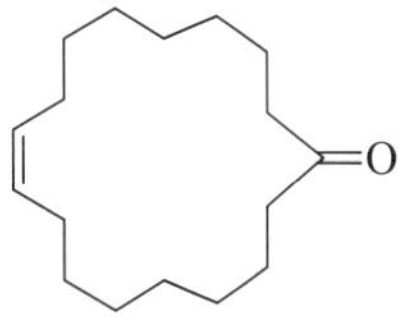

$C_{17}H_{30}O$, M_r 250.43, $bp_{0.007\,kPa}$ 103 °C, d_{20}^{20} 0.923–0.940, n_D^{20} 1.485–1.492, is a colorless liquid with warm sensual animal and musky odor and extreme tenacity. It is the main odoriferous constituent of civet. A multistep synthesis, starting from cyclohexanone enol acetate has been patented. Two acyclic C_6-fragments are linked to acetylene to give tetradeca-7-yne-1,14-dial. This is reacted with 1,3-bis(dimethylphosphono)propan-2-one. The resulting cyclohepta-2,16-dien-9-yn-1-one is hydrogenated over Pd/BaSO₄ to give civetone [127].

It is used in fine fragrance compositions for, for example, toiletries and body care products.
FCT 1976 (**14**) p. 727.

Trade Name. Civettone (Firmenich).

1-(5,5-Dimethyl-1-cyclohexen-1-yl)-4-penten-1-one [*56973-85-4*]

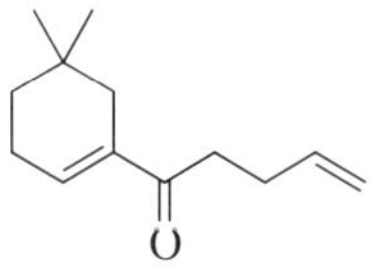

$C_{13}H_{20}O$, M_r 192.30, d_{20}^{20} 1.008–1.016, n_D^{20} 1.444–1.448, found not to occur in nature, is a colorless to pale yellow liquid with green, fruity, floral odor reminiscent of galbanum. One of several cited syntheses involves as a key step a double vinyl magnesium chloride addition to methyl 3,3-dimethyl-6-cyclohexene-1-carboxylate to give the title compound [128].

It can be used in all types of perfumes to give unique green, flora, and fruity effects. Its excellent stability permits the application in nearly all types of toiletries.

Trade Name. Dynascone 10 (Firmenich).

2,3,8,8-Tetramethyl-1,2,3,4,5,6,7,8-octahydro-2-naphthalenyl methyl ketone
[*54464-57-2*]

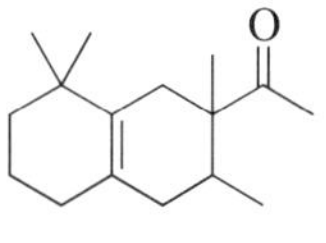

$C_{16}H_{26}O$, M_r 234.38, $bp_{0.37\,kPa}$ 134–135 °C, n_D^{20} 1.498–1.500, is a synthetic amber fragrance. It is prepared by a Diels–Alder reaction of myrcene and 3-methyl-3-penten-2-one in the presence of aluminum chloride and cyclization of the substituted

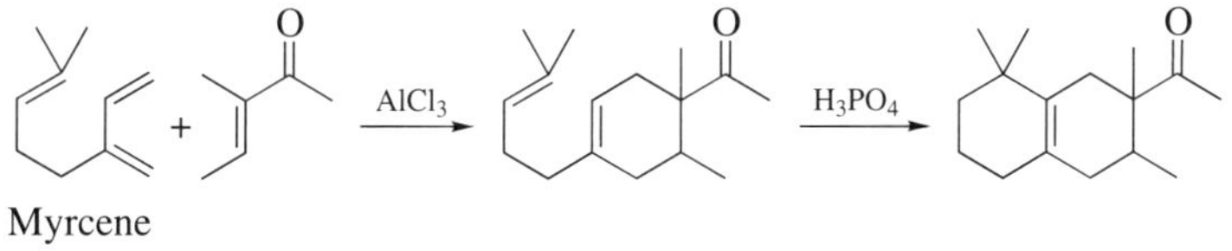

Myrcene

cyclohexenyl methyl ketone intermediate with phosphoric acid. Some of the corresponding 3-naphthalenyl methyl ketone is also formed [129].

The ketone is used in perfume bases for soaps, eau de cologne, and detergent compositions.

Trade Name. Iso E Super (IFF).

Methyl 2,6,10-trimethyl-2,5,9-cyclododecatrien-1-yl ketone [28371-99-5]

$C_{17}H_{26}O$, M_r 246.40, does not occur in nature. It is a pale yellow liquid, d_4^{20} 0.979–0.989, n_D^{20} 1.514–1.520, with a powerful, diffusive amber woody note with vetivert and smoky tobacco nuances. It is prepared from the corresponding trimethyl-cyclododecatriene by acetylation with acetic anhydride in the presence of boron trifluoride diethyl ether [130].

The product can be used to replace traditional materials such as sandalwood, vetivert and patchouli in perfume compositions.

Trade Name. Trimofix 'O' (IFF).

3-Methyl-2-cyclopenten-2-ol-1-one [80-71-7]

$C_6H_8O_2$, M_r 112.13, *mp* (monohydrate) 106 °C, occurs in beechwood tar and has a caramel-like odor. It has been identified as a flavor component in food. Crystals of the compound usually contain 1 mol of water. Synthetic routes of production are of limited importance in comparison with isolation from beechwood tar.

The compound is frequently used in flavor compositions for its caramel note, e.g., in beverages and in confectionery. It is rarely used in perfumery, and then mainly as an intensifier.
FCT 1976 (**14**) p. 809.

4-(1-Ethoxyvinyl)-3,3,5,5-tetramethylcyclohexanone [36306-87-3]

$C_{14}H_{24}O_2$, M_r 224.35, is not reported as being found in nature. It is a pale yellow liquid, $bp_{0.3\,kPa}$ 80–82 °C, n_D^{15} 1.4758, with a rich, warm, and long lasting amber/woody note. The title compound is the main component of a mixture which is obtained by cyclodimerization of mesityl oxide in the presence of boron trifluoride etherate and reaction with ethyl orthoformate [131]. It is used in large proportions in almost every type of perfume composition.

FCT 1982 (**20**) p. 835.

Trade Name. Kephalis (Giv.-Roure).

2.4.4 Esters

2-*tert*-Butylcyclohexyl acetate [*88-41-5*]

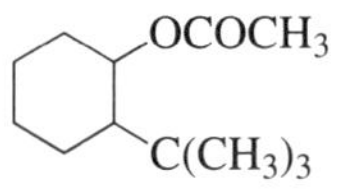

$C_{12}H_{22}O_2$, M_r 189.30, *mp* (pure *cis* isomer) 34.5–35.4 °C; commercial product: d_{25}^{25} 0.938–0.944, n_D^{20} 1.4500–1.4560; this compound does not occur in nature and exists in *cis* and *trans* forms. Pure 2-*cis*-*tert*-butylcyclohexyl acetate is a crystalline solid, with a fruity, agrumen-like odor. The commercial product is a colorless liquid and consists of a mixture of *cis* and *trans* isomers, which contains 60–95% of the *cis* ester. With an increasing percentage of the *trans* isomer, the odor becomes more woody-camphory.

The acetate is prepared by esterification of 2-*tert*-butylcyclohexanol, which is obtained from 2-*tert*-butylphenol. It is highly stable and is used for perfuming soap as well as bath and household products.

FCT 1992 (**30**) p. 13 S.

Trade Names. Agrumex (H&R), Ortholate (Quest), Verdox (IFF).

4-*tert*-Butylcyclohexyl acetate [*32210-23-4*]

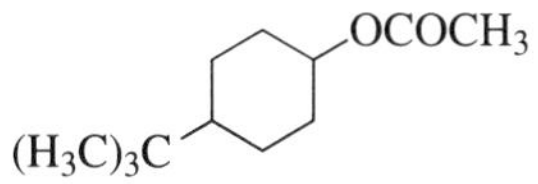

$C_{12}H_{22}O_2$, M_r 198.30, $bp_{1\,kPa}$ 89–95 °C, d_{25}^{25} 0.933–0.939, n_D^{20} 1.450–1.454, does not occur in nature and exists in *cis* and *trans* forms. The *trans* isomer has a rich, woody odor, while the odor of the *cis*–isomer is more intense and more flowery. Considerable variations in *cis*–*trans* ratios in commercial mixtures have little effect on the physical constants. Therefore, the composition of mixtures should be determined by gas chromatography.

The ester is prepared by catalytic hydrogenation of 4-*tert*-butylphenol followed by acetylation of the resulting 4-*tert*-butylcyclohexanol [132]. If Raney nickel is used as the catalyst, a high percentage of the *trans* isomer is obtained. A rhodium–carbon catalyst yields a high percentage of the *cis* isomer. The *trans* alcohol can be isomerized by alkaline catalysts; the lower-boiling *cis* alcohol is then removed continuously from the mixture by distillation [133].

4-*tert*-Butylcyclohexyl acetate is used particularly in soap perfumes.
FCT 1978 (**16**) p. 657.

Trade Names. Lorysia (Firmenich), Oryclon (H&R), Vertenex (IFF).

Decahydro-β-naphthyl acetate [*10519-11-6*]

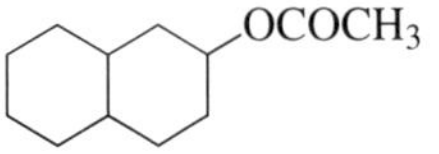

$C_{12}H_{20}O_2$, M_r 196.29, $bp_{0.65 kPa}$ 120 °C, d^{25} 1.005–1.015, n_D^{25} 1.475–1.482, is a fragrance substance that does not occur in nature. It consists of a mixture of several stereoisomers and is a colorless liquid with a sweet-fruity-herbaceous odor and a slight jasmin note.

Decahydro-β-naphthyl acetate is prepared by esterifieation of technical-grade decahydro-β-naphthol (e.g., with acetic anhydride). The acetate is used for perfuming household products.
FCT 1979 (**17**) p. 755.

4,7-Methano-3a,4,5,6,7,7a-hexahydro-5 (or 6)-indenyl acetate [*2500-83-6*] (or [*5413-60-5*])

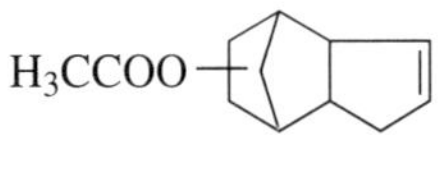

$C_{12}H_{16}O_2$, M_r 192.26, $bp_{1 kPa}$ 119–121 °C, d_4^{25} 1.0714, n_D^{25} 1.4935, is a colorless liquid with a herbal, fresh-woody odor. It consists of a mixture of isomers that is obtained by addition of acetic acid to dicyclopentadiene in the presence of an acid catalyst. It is used for perfuming soaps, detergents, and air fresheners.
FCT 1976 (**14**) p. 889.

Trade Names. Cyclacet (IFF), Greenylacetate (Dragoco), Herbaflorat (H&R), Verdylacetate (Giv.-Roure).

Allyl 3-cyclohexylpropionate [*2705-87-5*]

$C_{12}H_{20}O_2$, M_r 196.29, $bp_{0.13\,kPa}$ 91 °C, has not yet been found in nature. It is a colorless liquid with a sweet-fruity odor, reminiscent of pineapples. The ester is prepared by esterification of 3-cyclohexylpropionic acid (obtained by hydrogenation of cinnamic acid) with allyl alcohol. It is used in perfumery to obtain fruity top notes as well as pineapple and chamomile nuances.
FCT 1973 (11) p. 491.

Ethyl 2-ethyl-6,6-dimethyl-2-cyclohexenecarboxylate [57934-97-1], **and Ethyl 2,3,6,6-tetramethyl-2-cyclohexenecarboxylate** [77851-07-1]

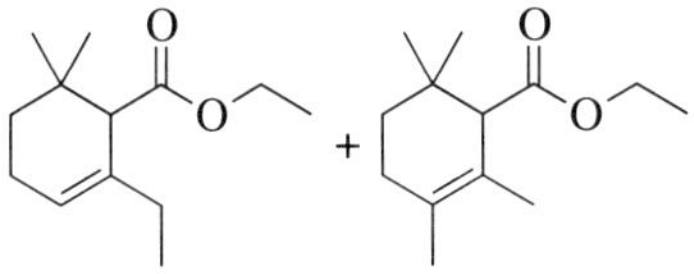

$C_{13}H_{22}O_2$, M_r 210.32 is a mixture of isomers, $bp_{0.8\,kPa}$ 102 °C, n_D^{20} 1.4626, a colorless to pale yellow liquid with rosy, spicy, fruity, and woody odor. For its preparation 3,6-dimethyl-6-hepten-2-one and 7-methyl-6-octen-3-one are treated with ethyl diethylphosphoryl acetate to give a mixture of octadienoic acid esters. Cyclisation with sulfuric/formic acid yields the title compounds as a mixture with isomers [134]. With its complex odor picture it is used in fine fragrances for shading.

Trade Name. Givescone (Giv.-Roure).

Allyl cyclohexyloxyacetate [68901-15-5], **cyclohexyloxyacetic acid allyl ester**

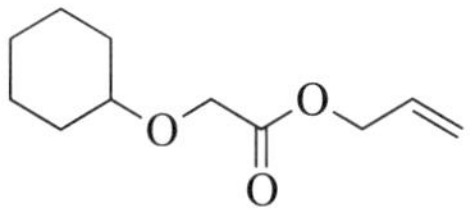

$C_{11}H_{18}O_3$, M_r 198.26, d_4^{20} 1.012–1.020, n_D^{20} 1.460–1.464, is a colorless to pale yellowish liquid with a strong, fruity, herbal-green odor reminiscent of galbanum. It is prepared by esterification of cyclohexyloxyacetic acid (from phenoxyacetic acid) with allyl alcohol and is used in fragrance compositions for toiletries and household products.

Trade Names. Cyclogalbanat (Dragoco), Isoananat (H&R).

cis-**Methyl jasmonate** [1211-29-6], **3-oxo-2-(2-*cis*-pentenyl)cyclopentaneacetic acid methyl ester**

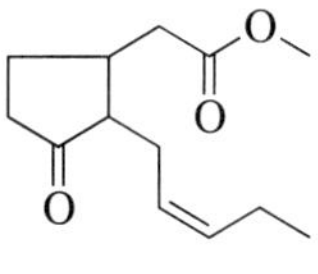

$C_{13}H_{20}O_3$, M_r 224.30, is a volatile component of jasmin flower absolute. It is a colorless to pale yellow liquid, $bp_{0.25\,kPa}$ 116–118 °C, d_{20}^{20} 1.022–1.028, n_D^{20} 1.473–1.477, possessing an odor reminiscent of the floral heart of jasmin. Synthesis of the title compound is accomplished by reaction of *cis*-buten-1-yl bromide with Li in the presence of CuI, the product is treated with a mixture of 2-methylene- and 4-methylene-3-oxocyclopentylacetic acid methyl ester. The aforementioned ester mixture is obtained by reaction of methyl 3-oxocyclopentylacetate with formaldehyde [135].

Methyl jasmonate is used in fine fragrances where it provides rich soft effects in jasmin and muguet compositions.

Trade Name. Methyl jasmonate (Firmenich).

Methyl dihydrojasmonate [*24851-98-7*]**, methyl (3-oxo-2-pentylcyclopentyl)acetate**

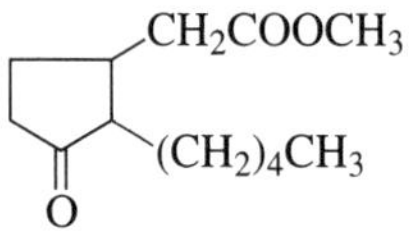

$C_{13}H_{22}O_3$, M_r 226.32, $bp_{1\,Pa}$ 85–90 °C, $d_4^{21.6}$ 1.003, $n_D^{20.2}$ 1.4589, is a jasmin fragrance that is closely related to methyl jasmonate, which occurs in jasmin oil. Methyl dihydrojasmonate has been identified in tea. It is a liquid with a typical fruity, jasmin-like blossom odor.

Of the four possible optical isomers, the (+)-(1*R*)-*cis*-isomer possesses the most characteristic jasmin odor. Methyl dihydrojasmonate is prepared by Michael addition of malonic acid esters to 2-pentyl-2-cyclopenten-1-one, followed by hydrolysis and decarboxylation of the resulting (2-pentyl-3-oxocyclopentyl) malonate, and esterification of the (2-pentyl-3-oxocyclopentyl)acetic acid [136].

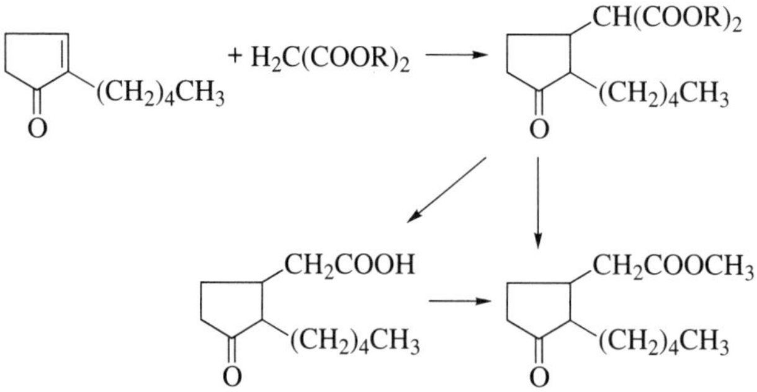

Dealkoxycarbonylation of the malonate can also be accomplished directly with water at elevated temperature [137].

A recently developed industrially feasible process for the synthesis of a mixture with high (+)-(1*R*)-*cis*-isomer content comprises the catalytical hydrogenation of the corresponding cyclopenteneacetic acid in the presence of a ruthenium(II) complex with chiral ligands and subsequent esterification [138].

Methyl dihydrojasmonate is used in perfumery for blossom fragrances, particularly in jasmin types.

FCT 1992 (**30**) p. 85 S.

Trade Name. Cepionate (NZ), Hedion (Firmenich), MDJ Super (Quest).

Methyl 2-hexyl-3-oxocyclopentanecarboxylate [*37172-53-5*]

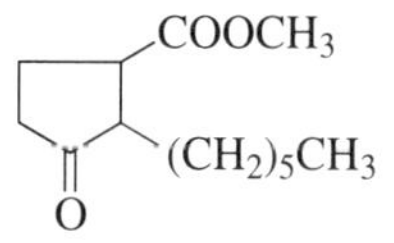

$C_{13}H_{22}O_3$, M_r 226.32, $bp_{8\,Pa}$ 85 °C, n_D^{20} 1.4562, is a colorless liquid with a long-lasting, floral, jasmin-like odor, that has only little of the fatty aspect characteristic of many jasmin fragrances. The product has not yet been found in nature.

The title compound can be prepared by condensing an alkyl α-bromocaprylate with a trialkyl propane-1,1,3-tricarboxylate to give a substituted cyclopentanone. Hydrolysis, decarboxylation, and esterification of the resulting monocarboxylic acid with methanol yields the desired ester [139]. Trialkyl propane-1,1,3-tricarboxylates can be prepared by Michael addition of dialkyl malonates to alkyl acrylates.

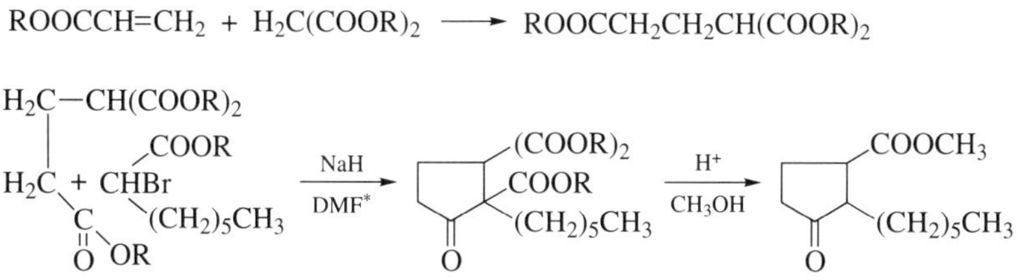

$$ROOCCH=CH_2 + H_2C(COOR)_2 \longrightarrow ROOCCH_2CH_2CH(COOR)_2$$

*DMF = dimethylformamide

The compound is used in perfumery in floral compositions.

Trade Name. Jasmopol (PFW).

2.5 Aromatic Compounds

2.5.1 Hydrocarbons

A few alkyl- and aralkyl-substituted aromatic hydrocarbons find limited use in perfumery. Examples include *p*-cymene [*99-87-6*], which is a component of many essential oils and when pure has a weak, citrus odor, as well as diphenylmethane [*101-81-5*] which has an odor like geranium:

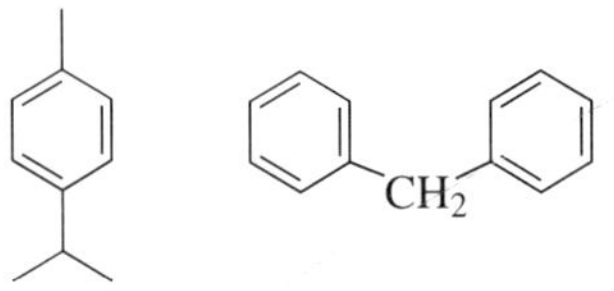

p-Cymene   Diphenylmethane

Slightly more important as a fragrance substance is the halogen-substituted hydrocarbon ω-bromostyrene.

ω-Bromostyrene [*103-64-0*]

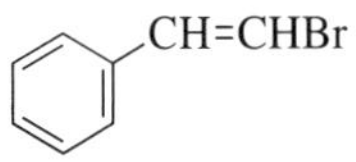

C_8H_7Br, M_r 183.05; trans isomer: *mp* 7 °C, $bp_{2.7\,kPa}$ 108 °C, d^{16} 1.4269, n_D^{20} 1.6093; the *trans* isomer has a strong hyacinth odor. It is prepared from cinnamic acid by addition of bromine and treatment of the resulting 3-phenyl-2,3-dibromopropionic acid with sodium carbonate. It is purified by fractional distillation. ω-Bromostyrene is used in soap perfumes.
FCT 1973 (**11**) p. 1043.

2.5.2 Alcohols and Ethers

Phenethyl alcohol is qualitatively and quantitatively one of the most important fragrance substances that belongs to the class of araliphatic alcohols. Its lower homologue (benzyl alcohol) and higher homologue (dihydrocinnamic alcohol) also have characteristic odor properties, but are more frequently used in the form of their esters. Cinnamic alcohol, the most important unsaturated araliphatic alcohol, is valuable for both fragrances and flavors.

The araliphatic alcohols mentioned above occur in many natural fragrances and flavors, but are generally not the main components. These alcohols are nearly always prepared synthetically for use in compositions. The branched-chain homologues of phenethyl and dihydrocinnamic alcohols (dimethyl benzyl carbinol and dimethyl phenethyl carbinol, respectively) are used in fairly large amounts as fragrance materials, but have not been found in nature.

Benzyl alcohol [*100-51-6*]

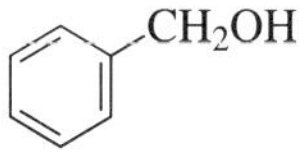

C_7H_8O, M_r 108.14, $bp_{101.3\,kPa}$ 205.4 °C, d_4^{20} 1.0419, n_D^{20} 1.5396, occurs in many essential oils and foods. It is a colorless liquid with a weak, slightly sweet odor. Benzyl alcohol can be oxidized to benzaldehyde, for example with nitric acid. Dehydrogenation over a copper–magnesium oxide–pumice catalyst also leads to the aldehyde. Esterification of benzyl alcohol results in a number of important fragrance and flavor compounds. Diphenylmethane is prepared by a Friedel–Crafts reaction of benzyl alcohol and benzene with aluminum chloride or concentrated sulfuric acid. By heating benzyl alcohol in the presence of strong acids or strong bases dibenzyl ether is formed.

Synthesis.   Benzyl alcohol is manufactured mainly by two processes.

1.   Benzyl chloride is hydrolyzed by boiling with aqueous solutions of alkali or alkaline earth hydroxides or carbonates. By-product in this process is dibenzyl ether (up to 10%).

2.   Toluene is, in low conversion, oxidized, with air, in the liquid phase to benzyl hydroperoxide, which yields mainly benzyl alcohol and some benzaldehyde upon hydrolysis, for example, in the presence of a cobalt salt. Benzyl alcohol thus obtained requires a more thorough purification for use in perfumes and flavors.

Because of its relatively weak odor, benzyl alcohol is used in fragrance and flavor compositions mainly as a solvent and for dilution. It is the starting material for a large number of benzyl esters, which are important fragrance and flavor substances.
FCT 1973 (**11**) p. 1011.

Phenethyl alcohol [*60-12-8*], **2-phenylethyl alcohol**

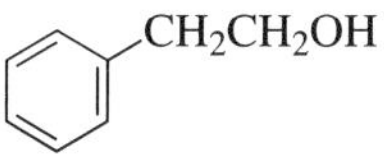

$C_8H_{10}O$ M_r 122.17, $bp_{101.3\,kPa}$ 219.8 °C, d_4^{20} 1.0202, n_D^{20} 1.5325, is the main component of rose oils obtained from rose blossoms. It occurs in smaller quantities in neroli oil, ylang-ylang oil, carnation oil, and geranium oils. Since the alcohol is rather soluble in water, losses occur when essential oils are produced by steam distillation.

Properties.   Phenethyl alcohol is a colorless liquid with a mild rose odor. It can be dehydrogenated catalytically to phenylacetaldehyde and oxidized to phenylacetic acid (e.g., with chromic acid). Its lower molecular mass fatty acid esters as well as some alkyl ethers, are valuable fragrance and flavor substances.

Production.   Many synthetic methods are known for preparing phenethyl alcohol; the following are currently of industrial importance:

1. *Friedel–Crafts Reaction of Benzene and Ethylene Oxide.* In the presence of molar quantities of aluminum chloride, ethylene oxide reacts with benzene to give an addition product, which is hydrolyzed to phenethyl alcohol:

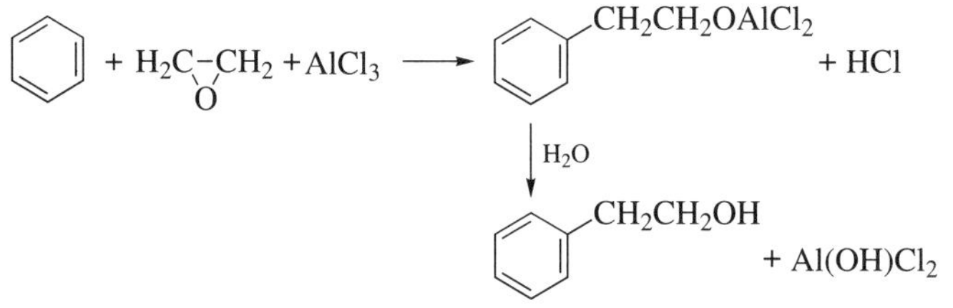

Formation of byproducts, such as 1,2-diphenylethane, is largely avoided by using an excess of benzene and low temperature. Special purification procedures are required to obtain a pure product that is free of chlorine and suitable for use in perfumery.

2. *Hydrogenation of Styrene Oxide.* Excellent yields of phenethyl alcohol are obtained when styrene oxide is hydrogenated at low temperature, using Raney nickel as a catalyst and a small amount of sodium hydroxide [140].

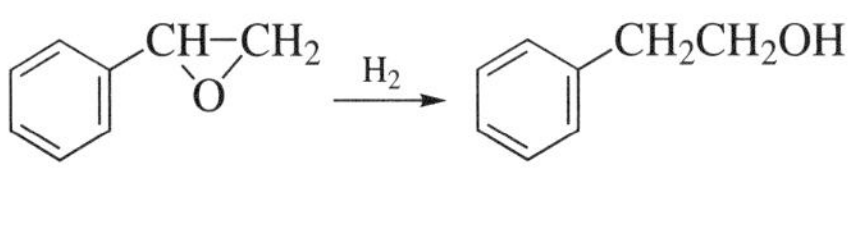

Uses.   Phenethyl alcohol is used frequently and in large amounts as a fragrance compound. It is a popular component in rose type compositions, but it is also used in other blossom notes. It is stable to alkali and, therefore, ideally suited for use in soap perfumes.
FCT 1975 (**13**) p. 903.

Phenethyl methyl ether [*3558-60-9*]

CH$_2$CH$_2$OCH$_3$

C$_9$H$_{12}$O, M_r 136.19, $bp_{94.6\,kPa}$ 185–186 °C, d^{27} 0.9417, n_D^{24} 1.4970, is a colorless liquid with a sharp, rosy-green odor. It is used in oriental type perfumes as well as in artificial keora oil.
FCT 1982 (**20**) p. 807.

Phenethyl isoamyl ether [*56011-02-0*]

CH$_2$CH$_2$OCH$_2$CH$_2$CHCH$_3$
                              CH$_3$

C$_{13}$H$_{20}$O, M_r 192.30, d_{25}^{25} 0.901–0.904, n_D^{20} 1.481–1.484, is a colorless liquid with a green, sweet-flowery odor of chamomile blossoms and a secondary, soapy note; it is used in perfumes.
FCT 1983 (**21**) p. 873.

   Trade Name. Anther (Quest).

Acetaldehyde ethyl phenethyl acetal [*2556-10-7*]

CH$_2$CH$_2$OCHOCH$_2$CH$_2$
                    CH$_3$

C$_{12}$H$_{18}$O$_2$, M_r 194.28, does not occur in nature. It is a colorless liquid, d_4^{20} 0.954–0.962, n_D^{20} 1.478–1.483, with powerful leafy-green, nasturtium, and hyacinth note. It can be synthesized by reaction of a 1:1 molar ratio of ethyl vinyl ether and phenethyl alcohol in the presence of cation exchange resin [141]. It imparts fresh, floral, green notes and is used in fine fragrances as well as in soap, cosmetics and detergents.
FCT 1992 (**30**) p. 1 S.

   Trade Names. Efetaal (Quest), Hyacinth Body (IFF).

1-Phenylethyl alcohol [*98-85-1*]**, styrallyl alcohol**

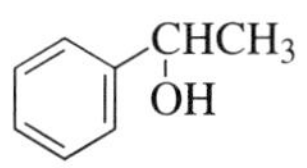

CHCH$_3$
OH

$C_8H_{10}O$, M_r 122.17, *mp* 20 °C, $bp_{101.3\,kPa}$ 203 °C, d_4^{20} 1.0135, n_D^{20} 1.5275, has been identified as a volatile component of food (e.g., in tea aroma and mushrooms). The alcohol is a colorless liquid with a dry, roselike odor, slightly reminiscent of hawthorn. It can be prepared by catalytic hydrogenation of acetophenone. 1-Phenylethyl alcohol is used in small quantities in perfumery and in larger amounts for the production of its esters, which are more important as fragrance compounds.

Dihydrocinnamic alcohol [*122-97-4*], **3-phenylpropanol**

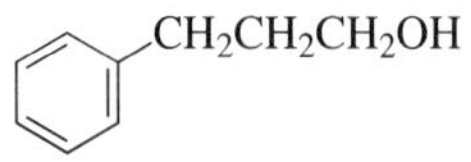

$C_9H_{12}O$, M_r 136.19, $bp_{100\,kPa}$ 237.5 °C, d_4^{20} 1.008, n_D^{20} 1.5278, occurs both in free and esterified form in resins and balsams (e.g., benzoe resin and Peru balsam). It has been identified in fruit and cinnamon.

Hydrocinnamic alcohol is a slightly viscous, colorless liquid with a blossomy-balsamic odor, slightly reminiscent of hyacinths. Esterification with aliphatic carboxylic acids is important because it leads to additional fragrance and flavor compounds.

Hydrocinnamic alcohol is prepared by hydrogenation of cinnamaldehyde. A mixture of hydrocinnamic alcohol and the isomeric 2-phenylpropanol can be obtained from styrene by a modified oxo synthesis. The two isomers can be separated by distillation [142].

Hydrocinnamic alcohol is used in blossom compositions for balsamic and oriental notes.
FCT 1979 (**17**) p. 893.

2,2-Dimethyl-3-(3-methylphenyl)propanol [*103694-68-4*]

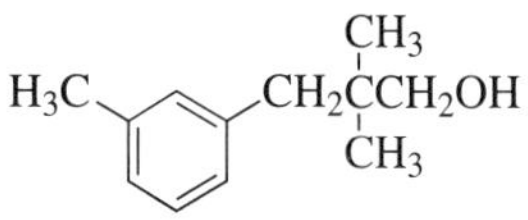

$C_{12}H_{18}O$, M_r 178.28, is not reported as being found in nature. It is a viscous liquid or crystalline mass, *mp* 22 °C, $bp_{0.013\,kPa}$ 74–76 °C, d_4^{25} 0.960, n_D^{20} 1.515–1.518, with a fresh floral odor, reminiscent of lily of the valley and linden blossoms. It is prepared by reaction of 3-methylbenzyl chloride with 2-methylpropanal in the presence of tetrabutylammonium iodide and reduction of the resulting aldehyde with $NaBH_4$ [143].

It can be used in perfume compositions for toiletries, soaps, and detergents.

Trade Name. Majantol (FR).

α,α-Dimethylphenethyl alcohol [*100-86-7*], **1-phenyl-2-methyl-2-propanol, α,α-dimethyl benzyl carbinol**

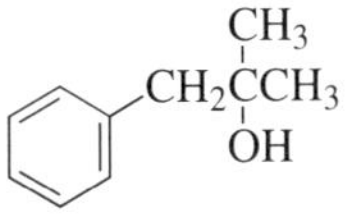

$$\text{C}_6\text{H}_5\text{—CH}_2\overset{\overset{\displaystyle CH_3}{|}}{\underset{\underset{\displaystyle OH}{|}}{C}}\text{CH}_3$$

$C_{10}H_{14}O$, M_r 150.22, *mp* 24 °C, $bp_{101.3\,kPa}$ 214–216 °C, d_4^{20} 0.9840, n_D^{20} 1.5170, has not yet been found in nature. The alcohol has a floral-herbaceous odor, reminiscent of lilac, and is prepared by a Grignard reaction of benzylmagnesium chloride and acetone. It is used in perfumery for various flower notes (e.g., lilac, hyacinth, mimosa). The alcohol is stable to alkali and, thus, is suited for soap perfumes. It is used to prepare a number of esters, which are also used as fragrance compounds. FCT 1974 (**12**) p. 531.

4-Phenyl-2-methyl-2-butanol [*103-05-9*], **α,α-dimethylphenethyl carbinol**

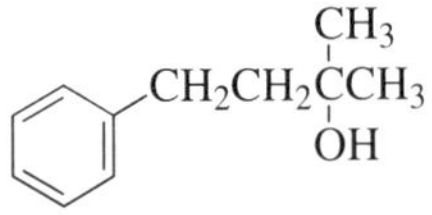

$$\text{C}_6\text{H}_5\text{—CH}_2\text{CH}_2\overset{\overset{\displaystyle CH_3}{|}}{\underset{\underset{\displaystyle OH}{|}}{C}}\text{CH}_3$$

$C_{11}H_{16}O$, M_r 164.25, $bp_{1.9\,kPa}$ 124–125 °C, $d_4^{20.7}$ 0.9626, $n_D^{20.7}$ 1.5077, is a colorless liquid with a dry-flowery, lily-like odor. It has been identified in cocoa aroma and is prepared by a Grignard reaction of benzylacetone and methylmagnesium chloride. It is used in blossom compositions. FCT 1974 (**12**) p. 537.

2-Methyl-5-phenylpentanol [*25634-93-9*]

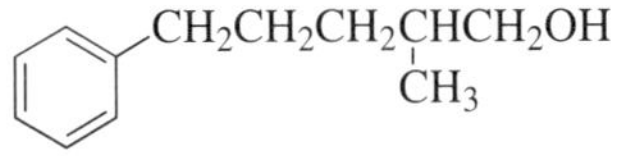

$$\text{C}_6\text{H}_5\text{—CH}_2\text{CH}_2\text{CH}_2\underset{\underset{\displaystyle CH_3}{|}}{C}\text{HCH}_2\text{OH}$$

$C_{12}H_{18}O$, M_r 178.28, is not reported to be found in nature. It is a colorless to pale yellow liquid, d_{25}^{25} 0.958–0.962, n_D^{20} 1.510–1.515, with a rose blossom, slightly waxy odor. It can be prepared by condensation of cinnamaldehyde with propanal and hydrogenation of the resulting unsaturated aldehyde.

Because of its excellent stability and relatively low price it is used in many perfumery fields.

Trade Name. Rosaphen (H&R).

3-Methyl-5-phenylpentanol [*55066-48-3*]

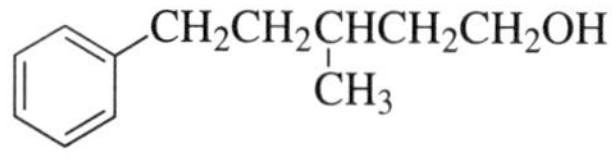

$$CH_2CH_2CHCH_2CH_2OH$$
$$CH_3$$

$C_{12}H_{18}O$, M_r 178.28, does not occur in nature. It is a colorless liquid, $bp_{0.013\,kPa}$ 86–91 °C, d_{20}^{20} 0.960–0.964, n_D^{20} 1.511–1.514, with a long lasting diffusive, fresh, floral, rose odor. The alcohol is prepared by hydrogenation of tetrahydro-4-methylene-5-phenylpyran which is obtained by cyclocondensation of benzaldehyde with 3-methyl-3-buten-1-ol in the presence of *p*-toluenesulfonic acid [144].

It can be used in all perfume types for a wide range of products, e.g., soaps, detergents and body care products.

Trade Names. Phenoxanol (IFF), Phenylhexanol (Firmenich).

1-Phenyl-3-methyl-3-pentanol [*10415-87-9*], **phenethyl methyl ethyl carbinol**

$$CH_3$$
$$CH_2CH_2CCH_2CH_3$$
$$OH$$

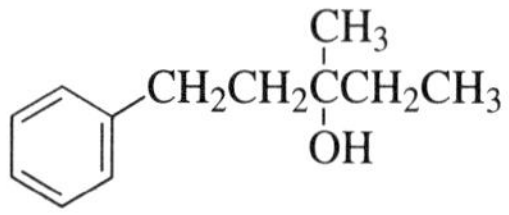

$C_{12}H_{18}O$, M_r 178.28, $bp_{1.7\,kPa}$ 129–130 °C, d_4^{25} 0.9582, n_D^{20} 1.509–1.513, has not yet been found in nature. It is a colorless liquid with a delicate peony, slightly fruity odor. Phenethyl methyl ethyl carbinol can be prepared from benzylacetone and ethylmagnesium chloride by a Grignard reaction. It is used to perfume soap, cosmetics, and detergents.
FCT 1979 (**17**) p. 891.

Cinnamic alcohol [*104-54-1*], **3-phenyl-2-propen-1-ol**

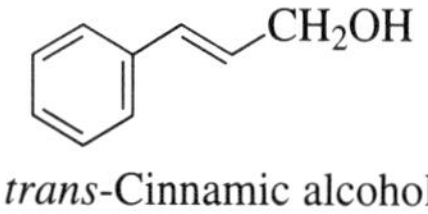

$$CH_2OH$$

trans-Cinnamic alcohol

$C_9H_{10}O$, M_r 134.18; *trans* isomer [*4407-36-7*]: *mp* 34 °C, $bp_{101.3\,kPa}$ 257.5 °C, d_4^{20} 1.0440, n_D^{20} 1.5819; this alcohol can exist in *cis* and *trans* forms. Although both isomers occur in nature, the *trans* isomer is far more abundant and is present, for example, in styrax oil. *trans*-Cinnamic alcohol is a colorless, crystalline solid with a hyacinth-like balsamic odor.

Cinnamic alcohol can be dehydrogenated to give cinnamaldehyde and oxidized to give cinnamic acid. Hydrogenation yields 3-phenylpropanol and/or 3-cyclohexylpropanol. Reaction with carboxylic acids or carboxylic acid derivatives results in the formation of cinnamyl esters, some of which are used as fragrance compounds.

Production.   Cinnamic alcohol is prepared on an industrial scale by reduction of cinnamaldehyde. Three methods are particularly useful:

1. In the *Meerwein–Ponndorf reduction*, cinnamaldehyde is reduced to cinnamic alcohol (yield ca. 85%) with isopropyl or benzyl alcohol in the presence of the corresponding aluminum alcoholate.

2. A 95% yield of cinnamic alcohol is obtained by selective hydrogenation of the carbonyl group in cinnamaldehyde with, for example, an osmium–carbon catalyst [145].

3. High yields of cinnamic alcohol can be obtained by reduction of cinnamaldehyde with alkali borohydrides. Formation of dihydrocinnamic alcohol is thus avoided [146].

Uses.   Cinnamic alcohol is valuable in perfumery for its odor and fixative properties. It is a component of many flower compositions (lilac, hyacinth, and lily of the valley) and is a starting material for cinnamyl esters, several of which are valuable fragrance compounds. In aromas, the alcohol is used for cinnamon notes and for rounding off fruit aromas. It is used as an intermediate in many syntheses (e.g., for pharmaceuticals such as the antibiotic chloromycetin).
FCT 1974 (**12**) p. 855.

2.5.3 Aldehydes and Acetals

Several araliphatic aldehydes are of special commercial importance as fragrance and flavor materials. These include cinnamaldehyde and its homologues in which the side-chain carries an alkyl substituent; α-amyl- and α-hexylcinnamaldehyde are particularly important. Other important members of this group are the substituted phenylpropanals, 4-isopropyl- and 4-*tert*-butyl-α-methyldihydrocinnamaldehyde. Arylacetaldehydes and arylpropionaldehydes are, in comparison, seldom used in compositions. The corresponding acetals are more stable and are used as well, although their odor is slightly different and significantly weaker.

The simplest araliphatic aldehyde, benzaldehyde and its 4-isopropyl homologue, cuminaldehyde, are used to a limited extent as fragrance and flavor compounds. However, both compounds are used in large quantity for the production of the corresponding cinnamic and dihydrocinnamic aldehydes.

Benzaldehyde [*100-52-7*]

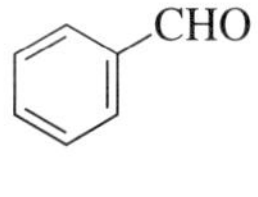

C_7H_6O, M_r 106.12, $bp_{101.3\,kPa}$ 178.1 °C, d_4^{15} 1.0415, n_D^{20} 1.5463, is the main, characteristic component of bitter almond oil. It occurs in many other essential oils and is a colorless liquid with a bitter almond odor. In the absence of inhibitors, benzaldehyde undergoes autoxidation to perbenzoic acid, which reacts with a second molecule of benzaldehyde to benzoic acid.

Hydrogenation of benzaldehyde yields benzyl alcohol, condensation with aliphatic aldehydes leads to additional fragrance substances or their unsaturated intermediates. Unsaturated araliphatic acids are obtained through the Perkin reaction, for example, the reaction with acetic anhydride to give cinnamic acid.

Benzaldehyde is prepared by hydrolysis of benzal chloride, for example in acidic media in the presence of a catalyst such as ferric chloride, or in alkaline media with aqueous sodium carbonate. Part of the commercially available benzaldehyde originates from a technical process for phenol. In this process, benzaldehyde is a by-product in the oxidation, with air, of toluene to benzoic acid.

Benzaldehyde is used in aroma compositions for its bitter almond odor. It is the starting material for a large number of araliphatic fragrance and flavor compounds.

FCT 1976 (**14**) p. 693.

Phenylacetaldehyde [*122-78-1*]

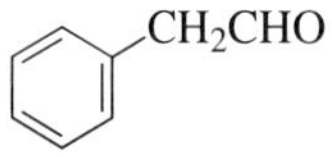

C_8H_8O, M_r 120.15, $bp_{101.3\,kPa}$ 195 °C, d_4^{20} 1.0272, n_D^{20} 1.5255, has been identified in many essential oils and as a volatile constituent of foods. It is a colorless liquid with a sweet-green odor, reminiscent of hyacinth. Since it readily undergoes oxidation and polymerizes, it must be stabilized by addition of antioxidants and by dilution with, for example, diethyl phthalate before use in compositions.

Phenylacetaldehyde can be obtained in high yield by vapor-phase isomerization of styrene oxide, for example, with alkali-treated silica-alumina [147]. Another process starts from phenylethane-1,2-diol, which can be converted into phenylacetaldehyde in high yield. The reaction is performed in the vapor phase in the presence of an acidic silica–alumina catalyst [148].

Phenylacetaldehyde is used in perfume compositions, in particular for hyacinth and rose notes.

FCT 1979 (**17**) p. 377.

Phenylacetaldehyde dimethyl acetal [*101-48-4*]

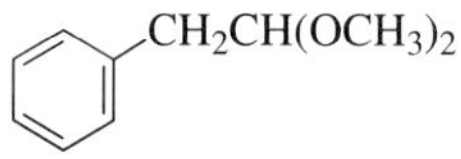

$C_{10}H_{14}O_2$, M_r 166.22, $bp_{100.2\,kPa}$ 219–221 °C, d^{18} 1.004, is a colorless liquid with a strong, rose-petal odor. The dimethyl acetal is more stable than phenylacetaldehyde itself. It imparts a herbal green note to many flower compositions.
FCT 1975 (13) p. 899.

Dihydrocinnamaldehyde [*104-53-0*], **3-phenylpropanal**

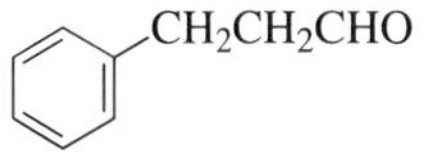

$C_9H_{10}O$, M_r 134.18, $bp_{2.7\,kPa}$ 112 °C, d_4^{20} 1.019, n_D^{20} 1.5266, occurs in Sri Lanka cinnamon oil, among others. The aldehyde is a colorless liquid with a strong, flowery, slightly balsamic, heavy hyacinth-like odor. It tends to undergo self-condensation.
   Dihydrocinnamaldehyde can be obtained with scarcely any byproducts by selective hydrogenation of cinnamaldehyde. It is used in perfumery for hyacinth and lilac compositions.
FCT 1974 (12) p. 967.

Hydratropaldehyde [*93-53-8*], **2-phenylpropanal**

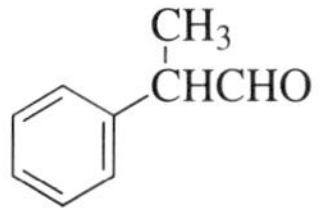

$C_9H_{10}O$, M_r 134.18, $bp_{1.5\,kPa}$ 92–94 °C, d_4^{20} 1.0089, n_D^{20} 1.5176, identified in dried mushroom, is a colorless liquid with a green hyacinth odor. Hydratropaldehyde can be hydrogenated to hydratropic alcohol, which is also used to a limited extent as a fragrance compound. Hydratropaldehyde is obtained from styrene by oxo synthesis; small quantities of the isomeric dihydrocinnamaldehyde are formed as a byproduct. Hydratropaldehyde is used in perfumery in blossom compositions.
FCT 1975 (13) p. 548.

Hydratropaldehyde dimethyl acetal [*90-87-9*]

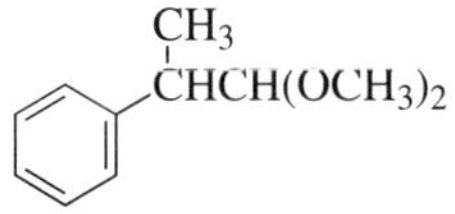

$C_{11}H_{16}O_2$, M_r 180.25, n_D^{20} 1.4938, is a liquid with a mushroom-like, earthy odor. It is used for green nuances in flower compositions.
FCT 1979 (17) p. 819.

4-Methylphenylacetaldehyde [*104-09-6*]

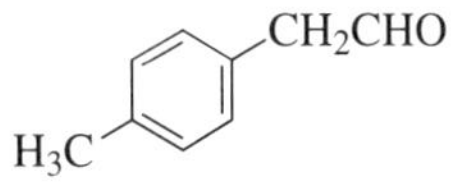

$C_9H_{10}O$, M_r 134.18, $bp_{101.3\,kPa}$ 221–222 °C, d_4^{20} 1.0052, n_D^{20} 1.5255, which has been identified in corn oil, is a colorless liquid with a strong green odor. It can be prepared by reaction of 4-methylbenzaldehyde with chloroacetates, followed by hydrolysis of the resulting glycidates and decarboxylation. The aldehyde is used in flower compositions for green notes.
FCT 1978 (**16**) p. 877.

3-(4-Ethylphenyl)-2,2-dimethylpropanal [*67634-15-5*]

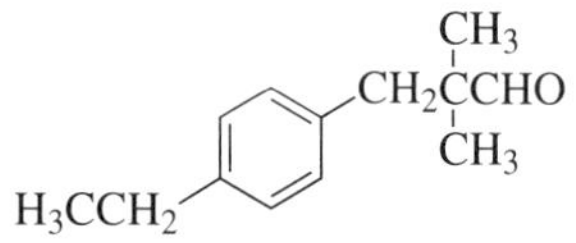

$C_{13}H_{18}O$, M_r 190.29, does not occur in nature. It is a colorless liquid, d_4^{20} 0.951–0.959, n_D^{20} 1.504–1.509, with a powerful, clean, fresh air tone, reminiscent of ocean breeze. It can be prepared by reaction of *p*-ethylbenzyl chloride and isobutyric aldehyde in the presence of catalysts. The product is used in perfumes for, e.g., household products.
FCT 1988 (**26**) p. 307.

   Trade Names. Floralozone (IFF), Florazon (Dragoco).

Cyclamenaldehyde [*103-95-7*], **2-methyl-3-(4-isopropylphenyl)propanal**

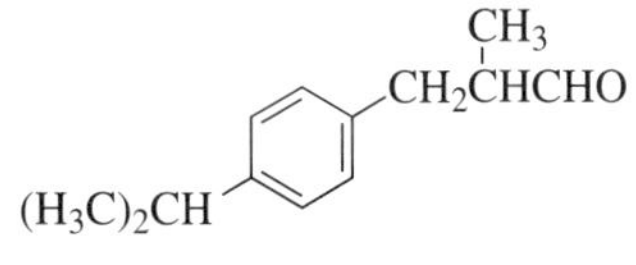

$C_{13}H_{18}O$, M_r 190.28, $bp_{0.3\,kPa}$ 108–108.5 °C, d_4^{20} 0.9502, n_D^{20} 1.5068, has been reported in nutmeg [149]. The commercially available racemate is a colorless to yellowish liquid with an intense flowery odor reminiscent of *Cyclamen europaeum* (cyclamen, sowbread).

Production. Two main processes are used for the industrial synthesis of cyclamenaldehyde:

1. Alkaline condensation of 4-isopropylbenzaldehyde and propanal results, via the aldol, in the formation of 2-methyl-3-(4-isopropylphenyl)-2-propenal. The

unsaturated aldehyde is hydrogenated selectively to the saturated aldehyde in the presence of potassium acetate and a suitable catalyst, such as palladium–alumina [150]:

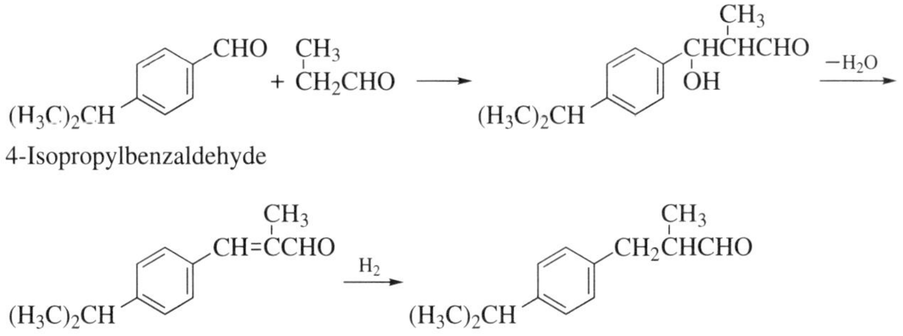

4-Isopropylbenzaldehyde

2. Friedel–Crafts reaction of isopropylbenzene and 2-methylpropenal diacetate (methacrolein diacetate) in the presence of titanium tetrachloride/boron trifluoride etherate gives cyclamenaldehyde enolacetate, which is hydrolyzed to the aldehyde [151]:

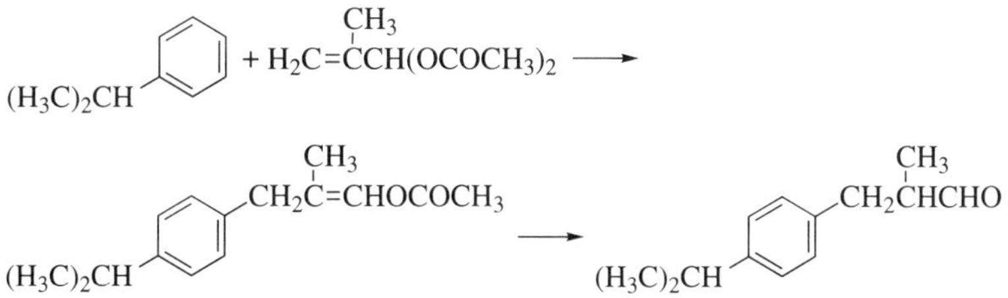

Uses. Cyclamenaldehyde is an important component for obtaining special blossom notes in perfume compositions, particularly the cyclamen type. Because of its fresh-flowery aspect, it is also used as the top note in many other blossom fragrances.
FCT 1974 (**12**) p. 397.

4-*tert*-Butyl-α-methyldihydrocinnamaldehyde [*80-54-6*],
2-methyl-3-(4-*tert*-butyl-phenyl)propanal

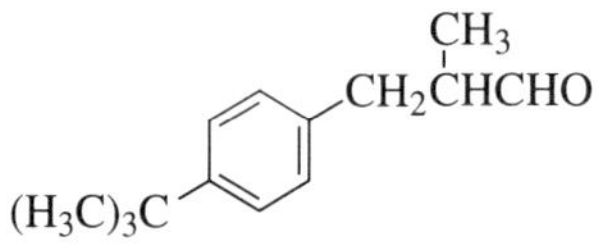

$C_{14}H_{20}O$, M_r 204.31, $bp_{0.8\,kPa}$ 126–127 °C, d_4^{20} 0.9390, n_D^{20} 1.5050, is a homologue of cyclamenaldehyde, but is not found in nature. The racemic compound is a colorless to slightly yellow liquid with a mild-flowery odor, reminiscent of cyclamen and lily of the valley.

The aldehyde is prepared by the same routes as cyclamenaldehyde.

More recent patents describe the following preparation from α-methylcinnamaldehyde. α-Methylcinnamaldehyde (from benzaldehyde and propionaldehyde) is hydrogenated to α-methyldihydrocinnamic alcohol. The alcohol is alkylated with *tert*-butyl chloride or isobutene to 4-*tert*-butyl-α-methyldihydrocinnamic alcohol, which is subsequently dehydrogenated to the desired aldehyde [152], [153].

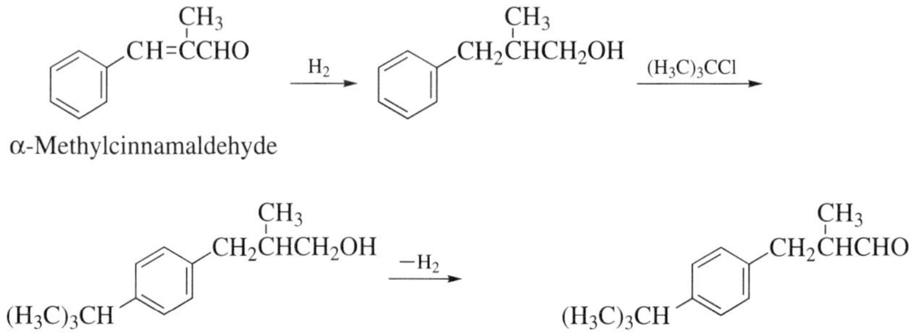

α-Methylcinnamaldehyde

The compound is more stable than cyclamenaldehyde and is a popular component of flower compositions, particularly lily of the valley and linden types, because of its mild, pleasant, blossom fragrance. Large quantities are used in soap and cosmetic perfumes.
FCT 1978 (**16**) p. 659.

Trade Names. Lilestralis (BBA), Lilial (Giv.-Roure), Lysmeral (BASF).

Cinnamaldehyde [*14371-10-9*], **3-phenyl-2-propenal**

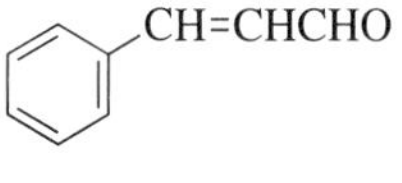

C_9H_8O, M_r 132.16, *trans* isomer: $bp_{101.3\,kPa}$ 253 °C, d_4^{20} 1.0497, n_D^{20} 1.6195, is the main component of cassia oil (ca. 90%) and Sri Lanka cinnamon bark oil (ca. 75%). Smaller quantities are found in many other essential oils. In nature, the *trans* isomer is predominant.

trans-Cinnamaldehyde is a yellowish liquid with a characteristic spicy odor, strongly reminiscent of cinnamon. Being an α,β-unsaturated aldehyde, it undergoes many reactions, of which hydrogenation to cinnamic alcohol, dihydrocinnamaldehyde, and dihydrocinnamic alcohol are important. Cinnamic acid is formed by autoxidation.

On an industrial scale, cinnamaldehyde is prepared almost exclusively by alkaline condensation of benzaldehyde and acetaldehyde. Self-condensation of acetaldehyde can be avoided by using an excess of benzaldehyde and by slowly adding acetaldehyde [154].

Cinnamaldehyde is used in many compositions for creating spicy and oriental notes (e.g., soap perfumes). It is the main component of artificial cinnamon oil. In addition, it is an important intermediate in the synthesis of cinnamic alcohol and dihydrocinnamic alcohol.

FCT 1979 (**17**) p. 253.

α-Amylcinnamaldehyde [*122-40-7*], **2-pentyl-3-phenyl-2-propenal**

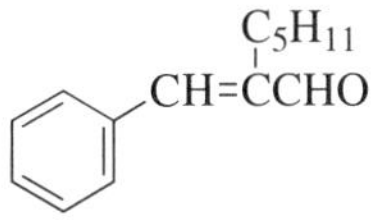

$$\text{CH=CCHO, with } C_5H_{11}$$

$C_{14}H_{18}O$, M_r 202.30, $bp_{0.7\,kPa}$ 140 °C, d_4^{20} 0.9710, n_D^{20} 1.5381, has been identified as an aroma volatile of black tea. It is a light yellow liquid with a flowery, slightly fatty odor, which becomes reminiscent of jasmin when diluted. The aldehyde is relatively unstable and must be stabilized by antioxidants.

It is prepared from benzaldehyde and heptanal in the same way as cinnamaldehyde.

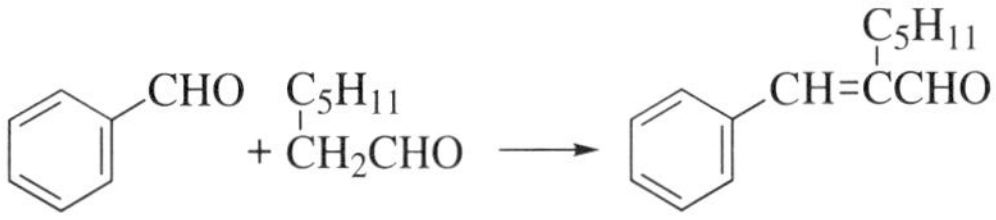

$$\text{CHO} + \text{CH}_2\text{CHO} \ (C_5H_{11}) \longrightarrow \text{CH=CCHO} \ (C_5H_{11})$$

α-Amylcinnamaldehyde is a very popular fragrance substance for creating jasmin notes. It is stable to alkali and long-lasting; large quantities are used, particularly in soap perfumes.

FCT 1973 (**11**) p. 855.

α-Hexylcinnamaldehyde [*101-86-0*], **2-hexyl-3-phenyl-2-propenal**

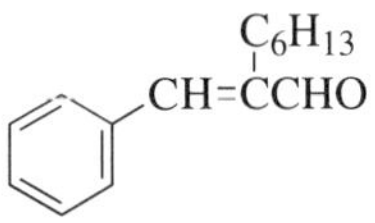

$$\text{CH=CCHO, with } C_6H_{13}$$

$C_{15}H_{20}O$, M_r 216.32, $bp_{2\,kPa}$ 174–176 °C, d^{24} 0.9500, n_D^{25} 1.5268, has been identified in rice. It is a yellow liquid with a mild, slightly fatty, flowery, somewhat herbal odor, and a distinct jasmin note. Like the α-amyl homologue, α-hexylcinnamaldehyde must be protected against oxidation by the addition of stabilizers. It is

prepared in a manner similar to α-amylcinnamaldehyde by alkaline condensation of excess benzaldehyde with octanal (instead of heptanal). α-Hexylcinnamalde-hyde is widely used in flower compositions (e.g., jasmin and gardenia) and, because of its stability to alkali, in soap perfumes.
FCT 1974 (**12**) p. 915

2.5.4 Ketones

The aromatic ketones that occur or are used as fragrance and flavor materials are predominantly aryl methyl ketones, which include acetophenones and β-naphthyl methyl ketone. Several acetylpolymethylindanes and -tetralins are commercially important as musk fragrances.

Acetophenone [*98-86-2*], **methyl phenyl ketone**

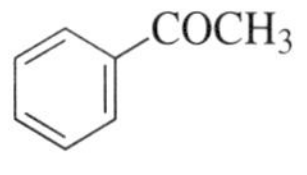

C_8H_8O, M_r 120.15, *mp* 20.5 °C, $bp_{101.3\,kPa}$ 202.0 °C, d_4^{20} 1.0281, n_D^{20} 1.5372, is a naturally occurring component of a large number of foods and essential oils. It is a colorless liquid with a penetrating sweet odor, reminiscent of orange blossom.

Acetophenone can be hydrogenated catalytically to l-phenylethanol. It is obtained as a by-product in the Hock phenol synthesis and is purified from the high-boiling residue by distillation. The quantitites obtained from this source satisfy the present demand.

Acetophenone is used for perfuming detergents and industrial products and is an intermediate in the synthesis of other fragrance compounds.
FCT 1973 (**11**) p. 99.

4-Methylacetophenone [*122-00-9*], **p-tolyl methyl ketone**

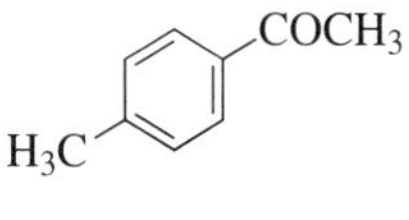

$C_9H_{10}O$, M_r 134.18, *mp* 28 °C, $bp_{101.3\,kPa}$ 226 °C, d_4^{20} 1.0051, n_D^{20} 1.5335, has been identified in Brazilian rosewood oil and in pepper. It occurs as colorless crystals with a flowery-sweet odor that is milder than that of acetophenone. 4-Methyl-acetophenone is prepared from toluene and acetic anhydride or acetyl chloride by a Friedel–Crafts reaction. It is used for blossom notes in mimosa and hawthorn type perfumes, especially in soap perfumes.
FCT 1974 (**12**) p. 933.

4-*tert*-Butyl-2,6-dimethylacetophenone [*2040-10-0*]

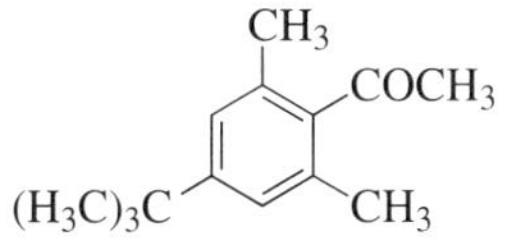

$C_{14}H_{20}O$, M_r 204.31, *mp* 47 °C, $bp_{0.3\,kPa}$ 150 °C, forms an off-white solid mass. It has a strong dry woody odor with musky background. It is prepared by Friedel–Crafts acetylation of 1,3-dimethyl-5-*tert*-butylbenzene.

It has an excellent stability and can be used in perfuming, e.g., all kinds of household cleaners.

Trade Name. Orinox (PFW).

Benzylacetone [*2550-26-7*], **4-phenyl-2-butanone**

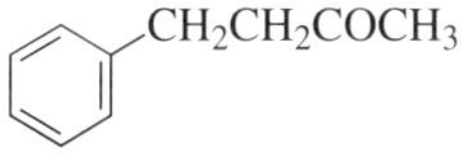

$C_{10}H_{12}O$, M_r 148.20, $bp_{101.3\,kPa}$ 233–234 °C, d_4^{22} 0.9849, n_D^{22} 1.5110, has been identified as a volatile component of cocoa. Benzylacetone is a sweet-flowery smelling liquid, which can be prepared by selective hydrogenation of benzylidene acetone (from benzaldehyde and acetone). It is used in soap perfumes.
FCT 1983 (**21**) p. 647.

Methyl β-naphthyl ketone [*93-08-3*]

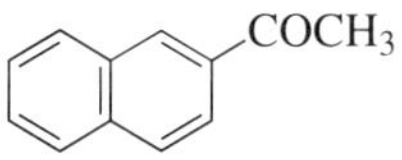

$C_{12}H_{10}O$, M_r 170.21, $bp_{1.7\,kPa}$ 171–173 °C, d_4^{20} 1.171, n_D^{20} 1.6752, has been identified in some essential oils. It smells like orange blossom and is a colorless crystalline solid (*mp* 56 °C). It is usually prepared by Friedel–Crafts acetylation of naphthalene (with acetylchloride, acetic anhydride, etc.) in the presence of aluminum chloride. In polar solvents (e.g., nitrobenzene), the percentage of the simultaneously formed α-isomer is lower. Methyl β-naphthyl ketone is used in eau de cologne, soap perfumes, and detergents. It is a good fixative.
FCT 1975 (**13**) p. 876.

Benzophenone [*119-61-9*], **diphenyl ketone**

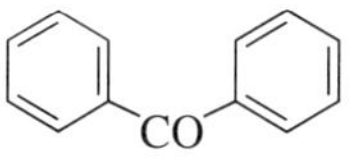

$C_{13}H_{10}O$, M_r 182.22, $bp_{6.3\,kPa}$ 200.5 °C, d^{50} 1.976, has been identified as a flavor component of grapes. It is a colorless crystalline solid (*mp* 48.1 °C) with a rosy, slightly geranium-like odor. It can be prepared in several ways, for example, by Friedel–Crafts reaction of benzene and benzoyl chloride with aluminum chloride, or of benzene and carbon tetrachloride, and subsequent hydrolysis of the resulting α,α-dichlorodiphenylmethane. Benzophenone can also be prepared by oxidation of diphenylmethane. It is used in flower compositions and as a fixative.
FCT 1973 (**11**) p. 873.

5-Acetyl-1,1,2,3,3,6-hexamethylindane [*15323-35-0*]

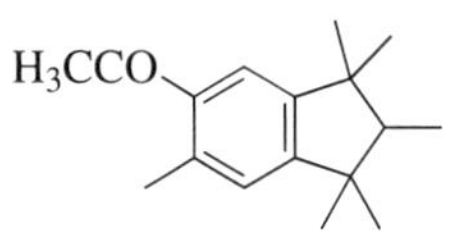

$C_{17}H_{24}O$, M_r 244.38, *mp* 35 °C, is a synthetic musk fragrance. It is prepared by Friedel–Crafts acetylation of 1,1,2,3,3,5-hexamethylindane, which can be obtained as a 70:30 mixture with l,1,3,5-tetramethyl-3-ethylindane by reacting α,p-dimethyl-styrene with amylenes or 2-methyl-2-butanol in a mixture of acetic acid and concentrated sulfuric acid [155]:

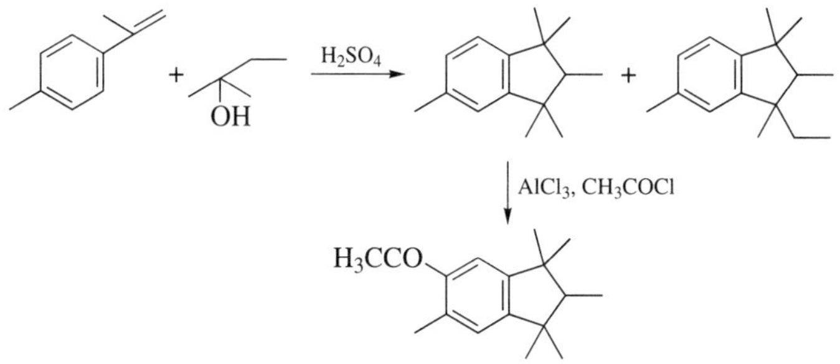

The indane ketone is a musk fragrance that is stable to light and in soap. It is used in perfumes and cosmetics for its fixative properties as well as its fragrance.
FCT 1975 (**13**) p. 693.

Trade Name. Phantolide (PFW).

4-Acetyl-1,1-dimethyl-6-*tert*-butylindane [*13171-00-1*]

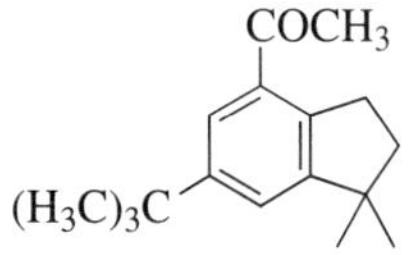

$C_{17}H_{24}O$, M_r 244.38, *mp* 76.7–77.2 °C, is a musk fragrance that does not occur in nature. It is prepared by reacting *tert*-butylbenzene with isoprene in the presence of sulfuric acid, followed by acetylation of the resulting 1,1-dimethyl-6-*tert*-butylindane [156]:

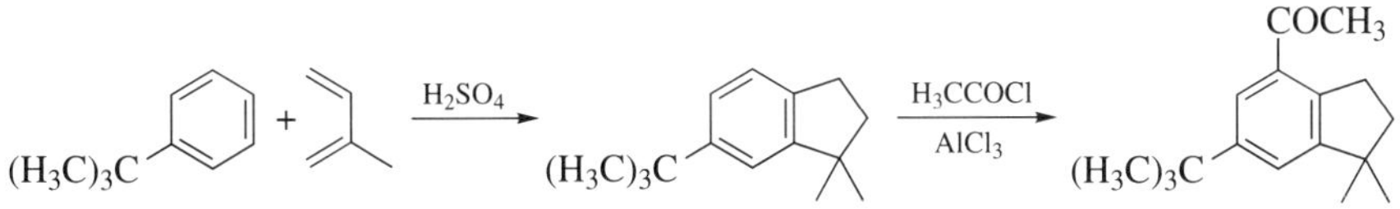

The indane is light-stable and is mainly used for perfuming soap and cosmetics. FCT 1976 (**14**) p. 699.

Trade Names. Celestolide (IFF), Crysolide (Giv.-Roure), Musk DTI (Firmenich).

5-Acetyl-1,1,2,6-tetramethyl-3-isopropylindane [68140-48-7]

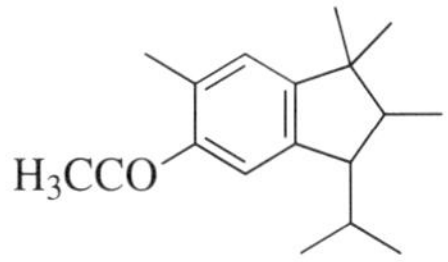

$C_{18}H_{26}O$, M_r 258.40, $bp_{0.13\,kPa}$ 144–146 °C, n_D^{20} 1.5301, is also a musk fragrance that does not occur in nature. It is prepared from toluene and isobutyryl chloride by a Friedel–Crafts reaction that yield *p*-tolyl isopropyl ketone; the ketone is reduced to the corresponding alcohol. Chlorination and treatment with 2-methyl-2-butene yield 1-isopropyl-2,3,3,5-tetramethylindane, which by a Friedel–Crafts reaction with acetyl chloride gives the title compound [157]:

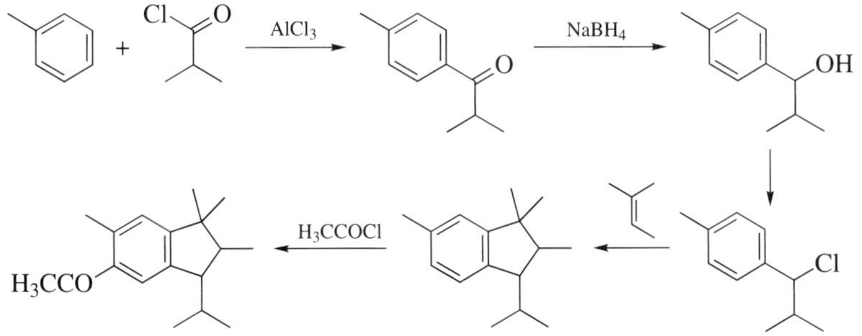

It is used in perfume compositions for soap and detergents. FCT 1983 (**21**) p. 645.

Trade Name. Traseolide (Quest).

6-Acetyl-1,1,2,4,4,7-hexamethyltetralin [*1506-02-1*]

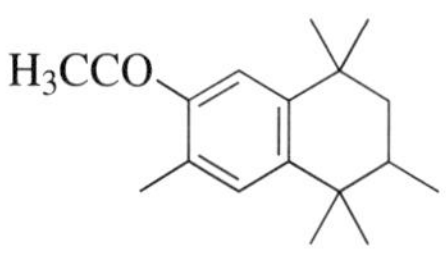

$C_{18}H_{26}O$, M_r 258.40, *mp* 55.5 °C, $bp_{0.25\,kPa}$ 119 °C, is a synthetic musk fragrance. It is prepared from 1,1,2,4,4,7-hexamethyltetralin, which is obtained by one of the following routes:

1. Reaction of α,p-dimethylstyrene with tetramethylethene [158] or 2,3-dimethyl-butan-2-ol [155] in an acetic acid–sulfuric acid mixture yields the desired hexamethyltetralin, in addition to polymerized starting materials:

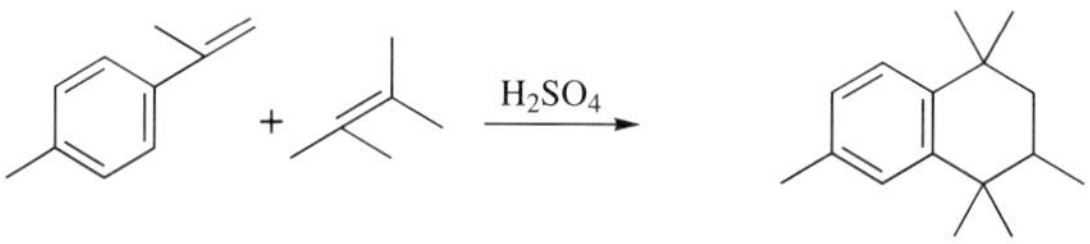

1,1,2,4,4,7-Hexamethyltetralin

2. Reaction of *p*-cymene with 3,3-dimethyl-1-butene (neohexene) and a tertiary alkylhalide (as a hydrogen scavenger) in the presence of catalytic amounts of anhydrous aluminum halide in inert solvents produces a high yield of the hexamethyltetralin [159].

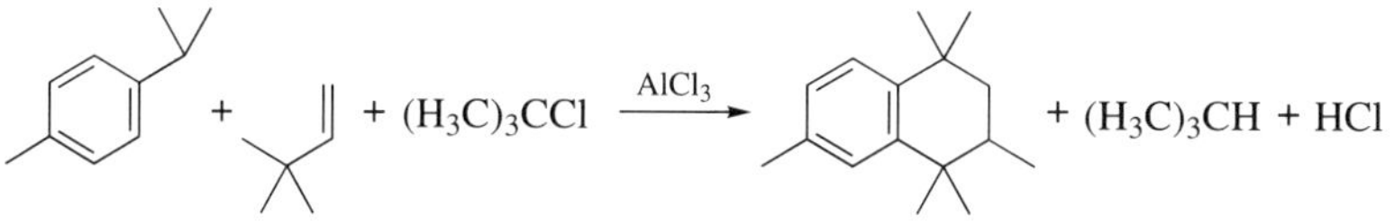

1,1,2,4,4,7-Hexamethyltetralin is subsequently acetylated to 6-acetyl-1,1,2,4,4,7-hexamethyltetralin, e.g., with acetyl chloride and aluminum chloride.

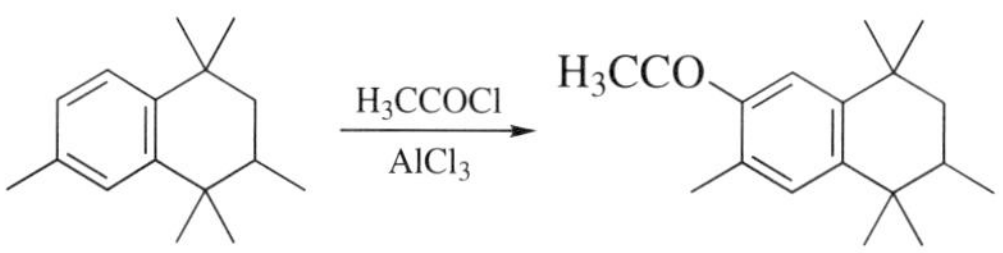

The product is a light-stable, versatile musk fragrance that is used in soap and cosmetics.

Trade Names. Fixolide (Giv.-Roure), Ganolide (Agan), Tetralide (BBA), Tonalide (PFW).

2.5.5 Esters of Araliphatic Alcohols and Aliphatic Acids

Esters of araliphatic alcohols and aliphatic acids are interesting as flavors and fragrances because of their characteristic odor properties. Acetates are the most popular esters. Benzyl acetate is particularly important commercially and occupies a prominent position in the fragrance and flavor industry.

Esters of other fatty acids are used to a lesser extent. In addition to benzyl esters and phenethyl esters, isomeric homologues with substituted side-chains are used in fairly large amounts in perfume compositions because of their special blossom odors. Not all have yet been found in nature. The esters are prepared from the corresponding alcohols via the customary routes.

Benzyl esters of lower molecular mass fatty acids occur widely in nature. The following are important fragrance and flavor materials:

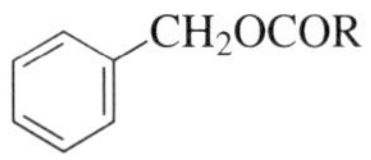

Benzyl acetate [*140-11-4*]
$R = CH_3$, $C_9H_{10}O_2$, M_r 150.18, $bp_{101.3\,kPa}$ 215 °C, d_4^{20} 1.0550, n_D^{20} 1.5232, is the main component of jasmin absolute and gardenia oils. It occurs as a minor component in a large number of other essential oils and extracts. It is a colorless liquid with a strong, fruity, jasmin odor. Benzyl acetate is prepared by esterification of benzyl alcohol with acetic anhydride (e.g., with sodium acetate as a catalyst) or by reaction of benzyl chloride with sodium acetate. In terms of volume, benzyl acetate is one of the most important fragrance and flavor chemicals.
FCT 1973 (**11**) p. 875.

Benzyl propionate [*122-63-4*]
$R = CH_2CH_3$, $C_{10}H_{12}O_2$, M_r 164.20, $bp_{101.3\,kPa}$ 219–220 °C, d_4^{20} 1.0335, n_D^{20} 1.4996, is a liquid with a sweet-fruity odor, which is used in perfumery for floral-fruity notes and in fruit flavor compositions.
FCT 1975 (**13**) p. 723.

Benzyl isovalerate [*103-38-8*]
$R = CH_2CH(CH_3)_2$, $C_{12}H_{16}O_2$, M_r 192.26, $bp_{101.3\,kPa}$ 245 °C, d_4^{20} 0.9900, n_D^{20} 1.4878, is a liquid with a heavy, flowery odor, which is used in perfumery for oriental and heavy blossom odors.
FCT 1974 (**12**) p. 829.

The most commonly used *phenethyl esters* are the following:

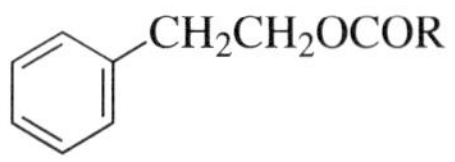

Phenethyl acetate [*103-45-7*]

$R = CH_3$, $C_{10}H_{12}O_2$, M_r 164.20, $bp_{101.3 \, kPa}$ 232.6 °C, d_4^{20} 1.0883, n_D^{20} 1.5171, occurs in a number of essential oils and is a volatile aroma component of many fruits and alcoholic beverages. Phenethyl acetate is a colorless liquid with a fine rose scent and a secondary, sweet, honey note. It is used in perfumery as a modifier of phenethyl alcohol, e.g., in rose and lilac compositions. In addition, it is used in a large number of aromas, in keeping with its natural occurrence.
FCT 1974 (**12**) p. 957.

Phenethyl isobutyrate [*103-48-0*]

$R = CH(CH_3)_2$, $C_{12}H_{16}O_2$, M_r 192.26, $bp_{2 \, kPa}$ 122–124 °C, d^{15} 0.9950, n_D^{20} 1.4871, occurs in peppermint oils. It has a heavy, fruity, blossom odor and is used accordingly in perfume and flavor compositions.
FCT 1978 (**16**) p. 847.

Phenethyl isovalerate [*140-26-1*]

$R = CH_2CH(CH_3)_2$, $C_{13}H_{18}O_2$, M_r 206.28, $bp_{7.2 \, kPa}$ 141–145 °C, d^{15} 1.9845, n_D^{20} 1.4855, has been identified as a volatile aroma component of peppermint oils. The fruitiness of its odor is even more pronounced than that of the isobutyrate. It is used in small quantities for the same purposes as phenethyl isobutyrate.
FCT 1974 (**12**) p. 961.

1-Phenylethyl acetate [*50373-55-2*], **styrallyl acetate**

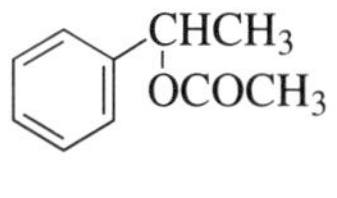

$C_{10}H_{12}O_2$, M_r 164.20, $bp_{1.3 \, kPa}$ 92.5 °C, d_4^{20} 1.0277, n_D^{20} 1.4954, has not been reported as occurring in nature. It is a liquid with a dry, fruity-green, blossom odor, reminiscent of gardenia. It can occur in the form of optically active enantiomers, but only the racemate is used in perfumery. Styrallyl acetate is a key ingredient in gardenia fragrances and is added to many other blossom compositions, particularly for dry top notes.
FCT 1976 (**14**) p. 611.

α-Trichloromethylbenzyl acetate [*90-17-5*]

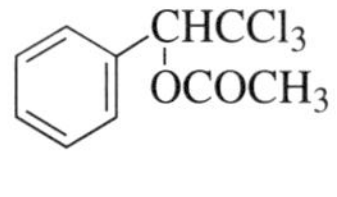

$C_{10}H_9Cl_3O_2$, M_r 267.54, is a fragrance substance that does not occur in nature. It forms white crystals (*mp* 88 °C) and has a weak, very natural, lasting rose odor.

The ester is prepared from α-trichloromethylbenzyl alcohol, for example, by reaction with acetic anhydride. The alcohol can be prepared by one of the following methods:

1. Addition of trichloroacetaldehyde (chloral) to benzene in the presence of aluminum chloride [160].

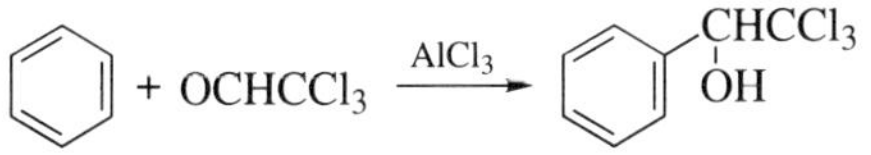

2. Reaction of benzaldehyde with chloroform in the presence of potassium hydroxide [161].

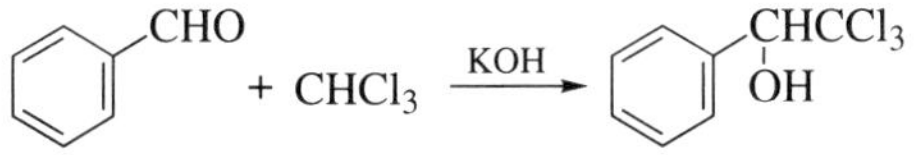

α-Trichloromethylbenzyl acetate is a stable rose fragrance with excellent fixative properties. It is preferentially used in soap, powders, and bath salts.

Trade Names. Rosacetat (H&R), Rosacetol (Giv.-Roure), Rosone (BBA).

The following α,α-*dimethylphenethyl esters* are commercially important:

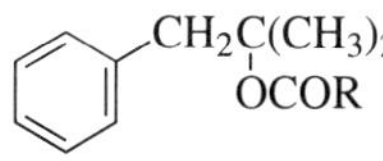

α,α-**Dimethylphenethyl acetate** [*151-05-3*]
$R=CH_3$, $C_{12}H_{16}O_2$, M_r 192.26, *mp* ca. 30 °C, $bp_{0.4\,kPa}$ 90 °C, d_{25}^{25} 0.998–1.000, n_D^{20} 1.4923, is a colorless liquid with a flowery-woody odor. The ester is used in blossom compositions, e.g., lily of the valley, rose, and geranium.
FCT 1974 (**12**) p. 533.

α,α-**Dimethylphenethyl butyrate** [*10094-34-5*]
$R=(CH_2)_2CH_3$, $C_{14}H_{20}O_2$, M_r 220.31, $bp_{0.4\,kPa}$ 96 °C, d_{25}^{25} 0.971–0.974, n_D^{20} 1.4860–1.4900, is a colorless liquid with a slightly herbal, strongly fruity odor, reminiscent of prune and apricot. It is used in perfumery as a modifier of the alcohol and for oriental notes.
FCT 1980 (**18**) p. 667.

Cinnamyl acetate [*21040-45-9*]

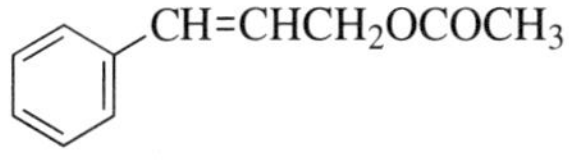

$C_{11}H_{12}O_2$, M_r 176.21, $bp_{1.3\,kPa}$ 139–140 °C, d^{22} 1.0520, n_D^{20} 1.5420, is the only ester of cinnamic alcohol of any importance. *trans*-Cinnamyl acetate occurs in cassia oil and is a colorless liquid with a sweet-flowery-fruity, slightly balsamic odor. It is a good fixative and is used in blossom compositions (e.g., lilac and jasmin) and for oriental notes. In aroma compositions, it is used for cinnamon-fruity effects. FCT 1973 (**11**) p. 1063.

2.5.6 Aromatic Acids

Aromatic acids (e.g., benzoic acid) and araliphatic acids (e.g., phenylacetic, cinnamic, and dihydrocinnamic acids) occur in numerous essential oils and have also been identified in the aromas of many foods. However, phenylacetic acid is the only acid that is used in significant quantities as a fragrance and flavor substance.

Phenylacetic acid [*103-82-2*]

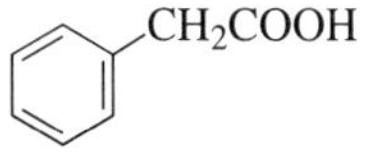

$C_8H_8O_2$, M_r 136.15, $bp_{101.3\,kPa}$ 265.5 °C, $d_4^{79.8}$ 1.0809, occurs in Japanese peppermint oil, in neroli oil, and in traces in rose oils. It is a volatile aroma constituent of many foods (e.g., honey). It forms colorless crystals (*mp* 78 °C) that have a honey odor.

The common route to phenylacetic acid is conversion of benzyl chloride into benzyl cyanide by reaction with sodium cyanide, followed by hydrolysis.

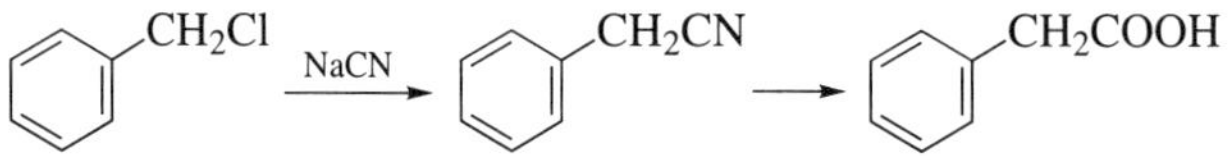

Because of its intense odor, phenylacetic acid is added to perfumes in small quantities for rounding off blossom odors. Addition to fruit aromas imparts a sweet honey note. FCT 1975 (**13**) p. 901.

2.5.7 Esters Derived from Aromatic and Araliphatic Acids

The acid moiety generally determines the odor of esters derived from aromatic or araliphatic acids. Unless stated otherwise, the esters are prepared from the corresponding acids or acid derivatives and alcohols by the customary methods.

2.5.7.1 Benzoates

The following benzoates are used in fairly large quantities as perfumery materials.

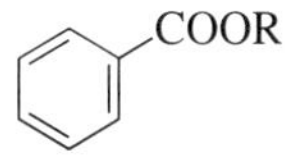

Methyl benzoate [*93-58-3*]
R $= CH_3$, $C_8H_8O_2$, M_r 136.15, $bp_{101.3\,kPa}$ 199.6 °C, d_4^{20} 1.0888, n_D^{20} 1.5164, has been found in essential oils (e.g., ylang-ylang oil). It is a colorless liquid with a strong, dry-fruity, slightly phenolic odor. Methyl benzoate can be converted simply into other benzoates by transesterification. Since methyl benzoate is a fairly large by-product in the manufacture of terylene, earlier synthetic routes such as those starting from benzoic acid or benzoyl chloride have largely been abandoned.

Methyl benzoate is used in perfume bases, such as ylang-ylang and tuberose types.
FCT 1974 (**12**) p. 937.

Hexyl benzoate [*6789-88-4*]
R $= (CH_2)_5CH_3$, $C_{13}H_{18}O_2$, M_r 206.28, $bp_{102.6\,kPa}$ 272 °C, is a liquid with a balsamic-green, melon-like odor. It is used in perfumery.
FCT 1979 (**17**) p. 813.

Benzyl benzoate [*120-51-4*]
R $= CH_2C_6H_5$, $C_{14}H_{12}O_2$, M_r 212.25, $bp_{2.0\,kPa}$ 170–171 °C, d_4^{25} 1.1121, n_D^{20} 1.5680, is the main component of Peru balsam oil. It occurs in fairly large amounts in a number of blossom concretes and absolutes (e.g., tuberose and hyacinth). It forms either a viscous liquid or solid flakes (*mp* 21–22 °C) and has a weak, sweet-balsamic odor. It is prepared either by transesterification of technical methyl benzoate with benzyl alcohol, or from benzyl chloride and sodium benzoate. A third process starts with benzaldehyde which is converted in high yield into benzyl benzoate in the presence of sodium or aluminum benzylate (Tishchenko reaction).

Benzyl benzoate is used in perfumery as a fixative and as a modifier in heavy blossom fragrances.
FCT 1973 (**11**) p. 1015.

2.5.7.2 Phenylacetates

Of the phenylacetates, the following are particularly important fragrance and flavor substances:

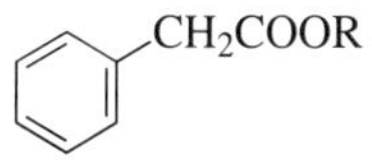

Ethyl phenylacetate [*101-97-3*]
R = CH_2CH_3, $C_{10}H_{12}O_2$, M_r 164.20, $bp_{101.3\,kPa}$ 227 °C, d_4^{20} 1.0333, n_D^{20} 1.4980, is a volatile aroma component of fruit and honey. It is a colorless liquid with a strong, sweet odor reminiscent of honey. Small amounts are used in flower perfumes and in fruit flavors.
FCT 1975 (**13**) p. 99.

Geranyl phenylacetate [*102-22-7*]
R = $CH_2CH = C(CH_3)CH_2CH_2CH = C(CH_3)_2$, $C_{18}H_{24}O_2$, M_r 272.39, has not been found in nature. It is a yellow liquid with a mild rose odor and a secondary honey note. It is used as a fixative in rose compositions and heavy perfumes.
FCT 1974 (**12**) p. 895.

Phenethyl phenylacetate [*102-20-5*]
R = $CH_2CH_2C_6H_5$, $C_{16}H_{16}O_2$, M_r 240.30, $bp_{0.6\,kPa}$ 177–178 °C, d_{25}^{25} 0.880, n_D^{20} 1.5496–1.5504, has been identified in e.g., the flower concrete of *Michelia champaca* L. It is a colorless liquid or crystals (*mp* 26.5 °C), which have a heavy, sweet, rose or hyacinth odor and a distinct honey note. The ester is used particularly in flowery fragrance compositions and as a fixative.
FCT 1975 (**13**) p. 907.

2.5.7.3 Cinnamates

Three cinnamates are of some importance in perfurnery:

CH=CHCOOR

Methyl cinnamate [*1754-62-7*]
R = CH_3, $C_{10}H_{10}O_2$, M_r 162.19, $bp_{101.3\,kPa}$ 261.9 °C, d_4^{20} 1.0911, n_D^{21} 1.5766, occurs in essential oils, mostly as the *trans* isomer. It is the main component of oils isolated from *Alpinia* species (content up to 80%) and *Ocimum canum* varieties (>50 %). It has also been identified as a volatile aroma component of cinnamon and strawberries.

Methyl cinnamate is a colorless crystalline solid (*mp* 36.5 °C) with a fruity, sweet-balsamic odor. In addition to the common esterification methods, it can be prepared by Claisen condensation of benzaldehyde and methyl acetate in the presence of sodium. Methyl cinnamate is used in soap perfumes, as well as in blossom and oriental perfumes, and is sometimes added to aromas.
FCT 1975 (**13**) p. 849.

Benzyl cinnamate [*103-41-3*]
R = $CH_2C_6H_5$, $C_{16}H_{14}O_2$, M_r 238.29, occurs in balsams and balsam oils. It forms white, sweet-balsamic-smelling crystals (*mp* 35–36 °C). Benzyl cinnamate is used as a fixative in perfumes and as a component of heavy, oriental perfumes.
FCT 1973 (**11**) p. 1017.

Phenethyl cinnamate [*103-53-7*]
$R = CH_2CH_2C_6H_5$, $C_{17}H_{16}O_2$, M_r 252.31, occurs in extracts from *Populus balsamifera* buds. It is a crystalline solid (*mp* 65–68°) with a heavy, rosy-balsamic odor. It is used as a fixative in blossom fragrances.
FCT 1978 (**16**) p. 845.

2.5.8 Miscellaneous Compounds

Of the few aromatic, nitrogen-containing fragrance substances, the nitro musks, musk xylol and musk ketone, are still considerably important commercially as inexpensive and adhesive musk fragrances. Musk ambrette, another artificial nitro musk, has lost its significance on account of its toxical properties. Methyl anthranilate and its *N*-methyl derivative are also aromatic, nitrogen-containing compounds that are used as fragrances and flavors in fairly large amounts. Schiff's bases of methyl anthranilate are of increasing importance as perfumery ingredients. A number of aromatic nitriles have been introduced; they are stable to alkali and, therefore, used in soap perfumes.

Musk xylol [*81-15-2*], **2,4,6-trinitro-1,3-dimethyl-5-*tert*-butylbenzene**

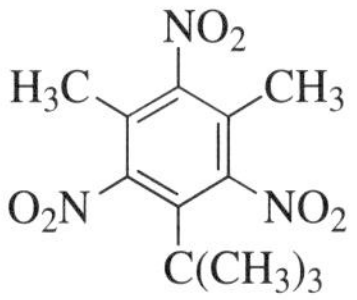

$C_{12}H_{15}N_3O_6$, M_r 297.27, *mp* 114 °C, does not occur in nature. It forms yellow crystals with a persistent musk odor. Musk xylol is prepared by alkylation of *m*-xylene with isobutene and subsequent nitration with a sulfuric acid–nitric acid mixture.

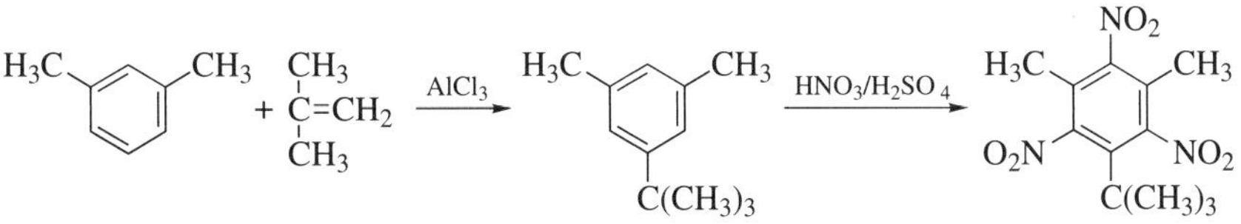

Musk xylol is used in large quantities in inexpensive perfumes for soap and household products.
FCT 1975 (**13**) p. 881.

Musk ketone [*81-14-1*], **3,5-dinitro-2,6-dimethyl-4-*tert*-butylacetophenone**

H3C $\quad$ COCH3 $\quad$ CH3

O2N $\quad$ NO2

C(CH3)3

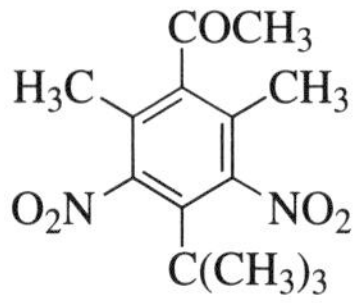

$C_{14}H_{18}N_2O_5$, M_r 294.31, *mp* 137 °C, does not occur in nature. It forms yellowish crystals with a sweet, very persistent, slightly animal musk odor. Musk ketone is prepared by Friedel–Crafts acetylation of 1,3-dimethyl-5-*tert*-butylbenzene, and nitration of the resulting 2,6-dimethyl-4-*tert*-butylacetophenone with nitric acid.

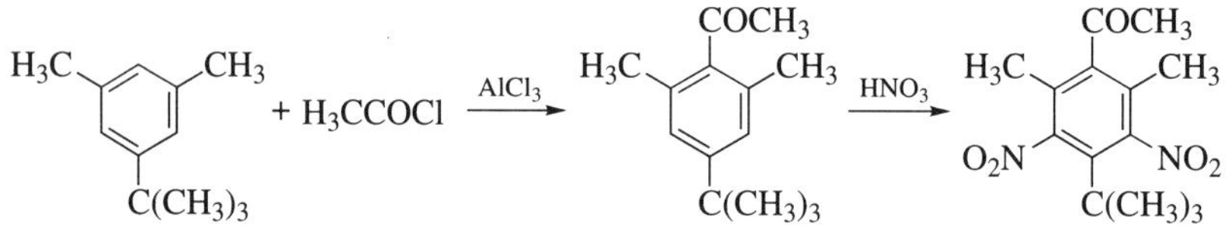

H3C $\quad$ CH3

+ H3CCOCl $\xrightarrow{AlCl_3}$ $\quad$ H3C $\quad$ COCH3 $\quad$ CH3 $\xrightarrow{HNO_3}$ $\quad$ H3C $\quad$ COCH3 $\quad$ CH3

C(CH3)3 $\qquad\qquad$ C(CH3)3 $\qquad\qquad$ O2N $\quad$ NO2 $\quad$ C(CH3)3

Musk ketone is widely used as a fixative in blossom and phantasy compositions. FCT 1975 (**13**) p. 877.

Cinnamonitrile [*4360-47-8*]

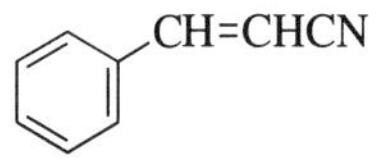

CH=CHCN

C_9H_7N, M_r 129.16, $bp_{1.7\,kPa}$ 135–135.5 °C, d^{25} 1.0244, n_D^{25} 1.6001; *trans*-cinnamonitrile is a colorless crystalline solid (*mp* 23.5–24 °C) or a colorless viscous liquid with a spicy, slightly flowery odor.

Cinnamonitrile can be prepared by one of the common routes to nitriles, e.g., by dehydration of cinnamaldoxime. It is stable to alkali and is used for perfuming soap and detergents.
FCT 1976 (**14**) p. 721.

Trade Name. Cinnamalva (IFF).

5-Phenyl-3-methyl-2-pentenonitrile [*93893-89-1*]

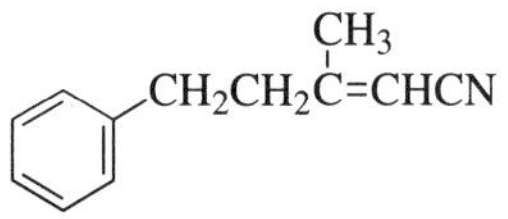

CH3

CH2CH2C=CHCN

$C_{12}H_{13}N$, M_r 171.24, $bp_{0.02 kPa}$ 82–88 °C, d^{25} 0.979, n_D^{20} 1.5340, is a colorless liquid with a citrus-fruity, slightly balsamic odor. The commercial product is a 2:3 mixture of the *cis* and *trans* isomers.

The nitrile is prepared by condensation of benzylacetone with cyanoacetic acid in the presence of pyridine and by elimination of carbon dioxide. The mixture is used in soap and detergent perfumes [162].

Trade Name. Citronitril (H&R).

Methyl anthranilate [134-20-3]

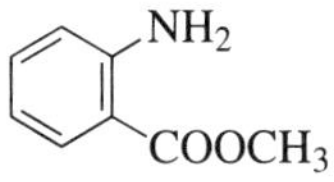

$C_8H_9NO_2$, M_r 151.16, $bp_{2 kPa}$ 135.5 °C, d_4^{19} 1.1682, n_D^{20} 1.5815, occurs in a large number of blossom essential oils (e.g., neroli, ylang-ylang, and jasmin oils), grapes, and citrus oils. It occurs as white crystals (*mp* 24–25 °C), or a yellowish liquid, that show blue fluorescence and have an orange blossom odor. Methyl anthranilate is prepared by esterification of anthranilic acid with methanol or by reaction of isatoic anhydride with methanol [163].

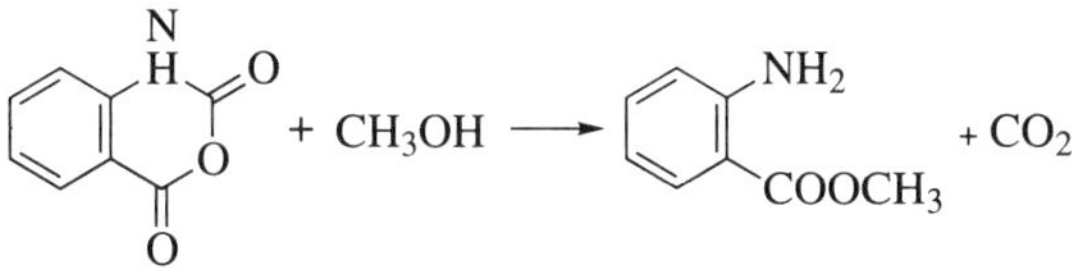

It is used in a large number of blossom fragrances. However, its use in perfumes for soap and cosmetics is limited because it causes discoloration. It is used in aroma compositions (e.g., in grape and citrus flavors).
FCT 1974 (**12**) p. 935

Methyl *N*-methylanthranilate [85-91-6]

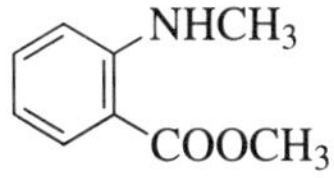

$C_9H_{11}NO_2$, M_r 165.19, $bp_{1.6 kPa}$ 130–131 °C, d_4^{20} 1.1295, n_D^{20} 1.5796, is the main component of petitgrain oils from mandarin leaves and is also found in mandarin oil. It is a pale yellow, fluorescent liquid with a delicate mandarin odor. The ester can be prepared by methylation of methyl anthranilate. It is used in soap and cosmetic perfumes as well as in aromas, particularly for mandarin flavors.
FCT 1975 (**13**) p. 791.

Some Schiff's bases of methyl anthranilate are interesting fragrance materials:

(a) with hydroxycitronellal [*89-43-0*]

> **Trade Names.** Auralva (IFF), Aurantesin (H&R), Aurantha (Takasago), Aurantiol Pure (Giv.-Roure), Aurantion (Quest),

(b) with 2,4-dimethyl-3-cyclohexene carbaldehyde [*68845-02-3*]

> **Trade Names.** Ligantraal (Quest), Vertosine (H&R),

(c) with 4-*tert*-butyl-α-methyldihydrocinnamaldehyde [*91-51-0*]

> **Trade Name.** Verdantiol (Giv.-Roure).

All of them have a heavy blossom odor and a high tenacity.

2.6 Phenols and Phenol Derivatives

2.6.1 Phenols, Phenyl Esters, and Phenyl Ethers

Of the phenols and phenyl ethers used as fragrance and flavor compounds, 4-allyl- (**9**, R = H) and 4-propenylphenols (**10** and **11**, R = H) and their methyl ethers (**9–11**, R = CH$_3$) occur particularly frequently in essential oils.

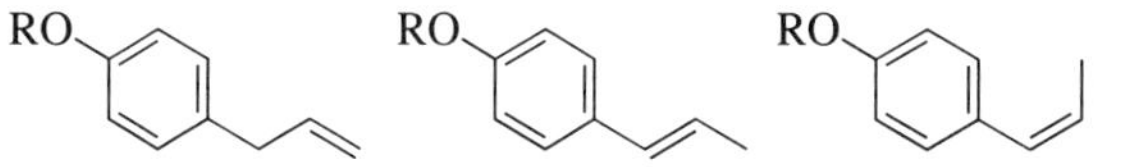

| 9 | 10 | 11 |

A second hydroxyl or methoxyl substituent is often present; 2-methoxy-4-allylphenol and 2-methoxy-4-propenylphenol are the most important compounds belonging to this category.

Although 1,2-methylenedioxy-4-allylbenzene and 1,2-methylenedioxy-4-propenylbenzene are really heterocyclic compounds, they are discussed here because of their close biogenetic relationship to the 2-methoxy-4-alkenylphenols.

Diphenyl ether [*101-84-8*]

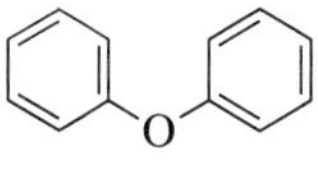

$C_{12}H_{10}O$, M_r 170.21, $bp_{1.34\,kPa}$ 121 °C, d^{20} 1.0748, has not been observed in nature. It is a colorless liquid or a crystalline solid (*mp* 26.8 °C) with an odor reminiscent of geranium leaves. Diphenyl ether is obtained as a byproduct in the production of phenol by high-pressure hydrolysis of chlorobenzene. Because of its stability and low price, diphenyl ether is used in large quantities in soap perfumes. However, its main application is as a heat-transfer medium (eutectic mixture with diphenyl). FCT 1974 (**12**) p. 707.

Phenoxyacetic acid 2-propenyl ester [*7493-74-5*], **phenoxyacetic acid allyl ester**

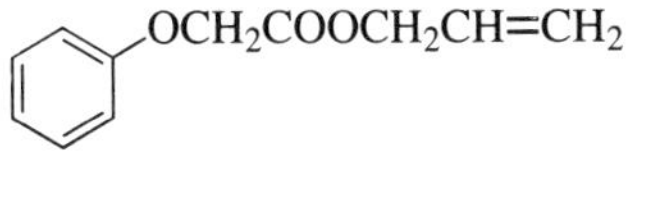

$C_{11}H_{12}O_3$ M_r 192.22, d^{25}_{25} 1.100–1.105, n^{20}_{D} 1.514–1.517, is a clear, colorless to yellowish liquid with a green, sweet, herbal-fruity odor with nuances of galbanum and pineapple.

It is prepared by reaction of phenoxyacetic acid alkali salts with allyl halogenide and used in technical perfumery.

Thymol [*89-83-8*], **2-isopropyl-5-methylphenol**

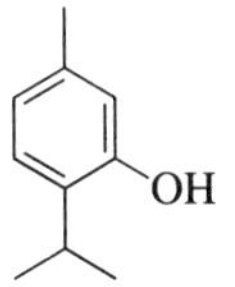

$C_{10}H_{14}O$, M_r 150.22, $bp_{101.3\,kPa}$ 232.5 °C, d^{20}_{4} 0.9756, n^{20}_{D} 1.5227, is the main constituent of thyme and some origanum oils; it also occurs in many other essential oils. It forms colorless crystals (*mp* 51.5 °C) with a spicy-herbal, slightly

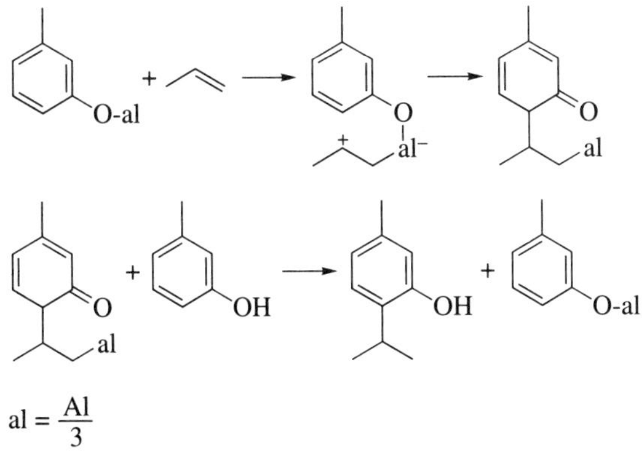

al = $\dfrac{Al}{3}$

medicinal odor reminiscent of thyme. Thymol is prepared on a technical scale in a continuous high-temperature, high-pressure, liquid-phase, ortho-alkylation process, from *m*-cresol and propylene, in the presence of activated aluminum oxide hydrate.

The crude thymol mixture, consisting of approximately 60% thymol, unreacted *m*-cresol (ca. 25%), and other (iso)propyl-substituted products, is separated by fractional distillation. Most of the by-products are recycled.

Thymol is used as a dry top note in lavender compositions, in men's fragrances, and as a disinfectant in oral hygiene products. It is also important as a starting material for the production of racemic menthol.

Anethole [*104-46-1*], **1-methoxy-4-(1-propenyl)benzene**

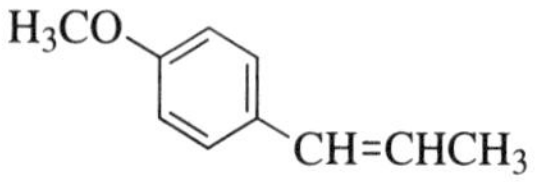

$C_{10}H_{12}O$, M_r 148.20, *trans* isomer: $bp_{101.7\,kPa}$ 234 °C, $bp_{1.6kPa}$ 115 °C, d_4^{20} 0.9883, n_D^{20} 1.5615, occurs both as its *cis* and *trans* isomers in nature; however, *trans*-anethole is always the main isomer. Anethole occurs in anise oil (80–90%), star anise oil (>90%), and fennel oil (80%).

trans-Anethole [*4180-23-8*] forms colorless crystals (*mp* 21.5 °C) with an anise-like odor and a sweet taste. Anethole is oxidized to anisaldehyde (e.g., with chromic acid); when hydrogenated it is converted into 1-methoxy-4-propylbenzene.

Production.   Anethole is isolated from anethole-rich essential oils as well as from sulfate turpentine oils.

1. Anethole can be crystallized from oils in which it occurs as a major component (star anise and sweet fennel oils), and estragole containing oils (e.g., basilicum oil).

2. A fraction of American sulfate turpentine oil (0.5% of the total) consists mainly of an azeotropic mixture of anethole and caryophyllene. *trans*-Anethole can be isolated from this mixture by crystallization.

3. Another fraction of American sulfate turpentine oil (1% of the total) consists essentially of an azeotropic mixture of estragole (1-methoxy-4-allylbenzene $bp_{101.3\,kPa}$ 216 °C) and α-terpineol. Treatment with potassium hydroxide yields a mixture of anethole isomers and α-terpineol, which can be separated by fractional distillation.

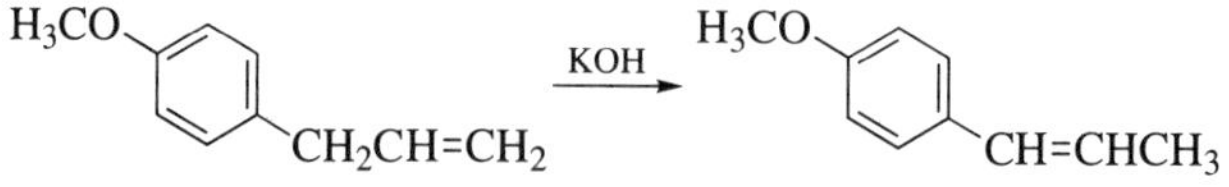

Uses. Anethole is used in large quantities in the alcoholic beverage industry (Pernod, Ouzo) and in oral hygiene products. Some crude anethole is converted into anisaldehyde.
FCT 1973 (**11**) p. 863.

2-Phenoxyethyl isobutyrate [*103-60-6*]

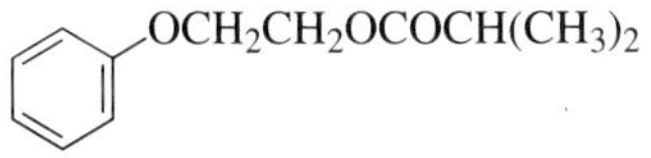

$C_{12}H_{16}O_3$, M_r 208.26, $bp_{0.53\,kPa}$ 125–127 °C, d_{25}^{25} 1.044–1.050, n_D^{20} 1.492–1.496, is a fragrance compound that does not occur in nature. It is a colorless liquid with a sweet, flowery-fruity odor.

The ester is prepared by esterification of 2-phenoxyethanol with isobutyric acid and is used as a fixative in perfumes (rose and lavender types) as well as for fruity notes.
FCT 1974 (**12**) p. 955.

Trade Name. Phenirat (H&R).

The *β-naphthyl alkyl ethers* described below are used in perfumery, especially in soap perfumes. The ethers are prepared by *O*-alkylation of *β*-naphthol. They have not been observed in nature.

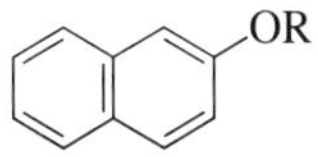

β-Naphthyl methyl ether [*93-04-9*]
R = CH_3, $C_{11}H_{10}O$ M_r 158.20, $bp_{1.3\,kPa}$ 138 °C, forms white crystals (*mp* 73–74 °C) with an intense orange blossom odor.
FCT 1975 (**13**) p. 885.

β-Naphthyl ethyl ether [*93-18-5*]
R = CH_2CH_3, $C_{12}H_{12}O$, M_r 172.23, $bp_{1.3\,kPa}$ 148 °C, forms white crystals (*mp* 37–38 °C), with a mild, long-lasting, orange blossom fragrance.
FCT 1975 (**13**) p. 883.

β-Naphthyl isobutyl ether [*2173-57-1*]
R = $CH_2CH(CH_3)_2$, $C_{14}H_{16}O$, M_r 200.28, forms white crystals (*mp* 33–33.5 °C) with a fruity, orange blossom odor.
FCT 1992 (**30**) p. 95 S.

Hydroquinone dimethyl ether [*150-78-7*], 1,4-dimethoxybenzene

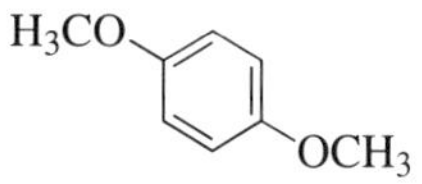

$C_8H_{10}O_2$, M_r 138.17, $bp_{2.7\,kPa}$ 109 °C, occurs in hyacinth oil and has also been identified in tea. It is a white crystalline solid (*mp* 57–58 °C) with an intensely sweet, somewhat herbal, nutlike odor.

Hydroquinone dimethyl ether is prepared by etherification of hydroquinone and is used in soap perfumes.

FCT 1978 (16) p. 715.

Isoeugenol [*97-54-1*], 2-methoxy-4-(1-propenyl)phenol

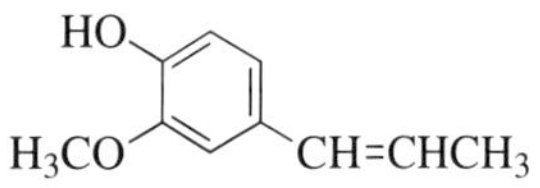

$C_{10}H_{12}O_2$, M_r 164.22; *cis* isomer [*5912-86-7*]: $bp_{1.7\,kPa}$ 134–135 °C, d_4^{20} 1.0837, n_D^{20} 1.5726; *trans* isomer [*5932-68-3*]: *mp* 33–34 °C, $bp_{1.7\,kPa}$ 141–142 °C. d_4^{20} 1.0852, n_D^{20} 1.5784. Isoeugenol occurs in many essential oils, mostly with eugenol, but not as the main component. Commercial isoeugenol is a mixture of *cis* and *trans* isomers, in which the *trans* isomer dominates because it is thermodynamically more stable. Isoeugenol is a yellowish, viscous liquid with a fine clove odor, that of the crystalline *trans* isomer being the more delicate.

Isoeugenol can be hydrogenated catalytically to form dihydroeugenol. Vanillin was formerly prepared by oxidation of isoeugenol. Additional fragrance compounds are prepared by esterification or etherification of the hydroxyl group.

Production.   Starting materials for the synthesis of isoeugenol are eugenol and guaiacol.

1. *Synthesis from Eugenol.* The sodium or potassium salt of eugenol is isomerized to isoeugenol by heating. Isomerization can also be carried out catalytically in the presence of ruthenium [164] or rhodium [165] compounds.

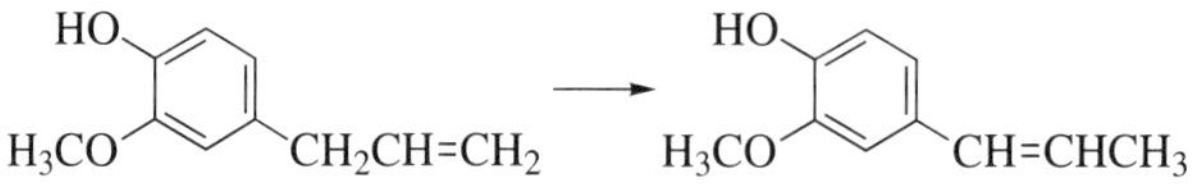

2. *Synthesis from Guaiacol.* In a process developed in the former Soviet Union, guaiacol is esterified with propionic acid, and the resulting guaiacyl propionate

rearranges in the presence of aluminum chloride to give 4-hydroxy-3-meth-oxypropiophenone. Reduction of the ketone to the corresponding secondary alcohol and dehydration finally yield isoeugenol [166]–[168].

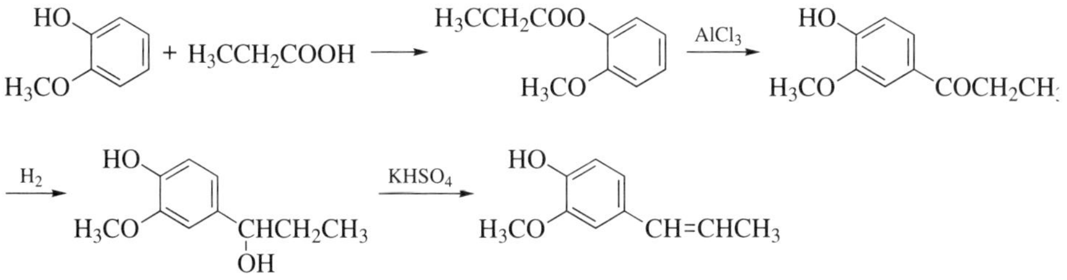

Uses. Isoeugenol is used in perfumery in a large number of blossom compositions, mostly for clove and carnation types, but also in oriental perfumes. Small amounts are employed in aromas and in reconstituted essential oils. FCT 1975 (**13**) p. 815.

Isoeugenol methyl ether [*93-16-3*]

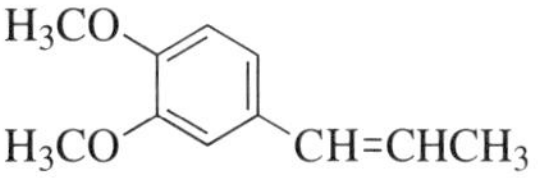

$C_{11}H_{14}O_2$, M_r 178.23, *cis* isomer [*6380-24-1*]: $bp_{0.9\,kPa}$ 137–137.5 °C, d_4^{20} 1.0530, n_D^{20} 1.5628; *trans* isomer [*6379-72-2*]: $bp_{0.7\,kPa}$ 126 °C, d_4^{20} 1.0556, n_D^{20} 1.5699, occurs in small quantities in several essential oils. It is a colorless to pale yellow liquid with a mild clove odor.

Isoeugenol methyl ether is used in perfumery in clove and carnation bases and as a fixative in spicy-floral compositions. FCT 1975 (**13**) p. 865.

Eugenol [*97-53-0*], **2-methoxy-4-allylphenol**

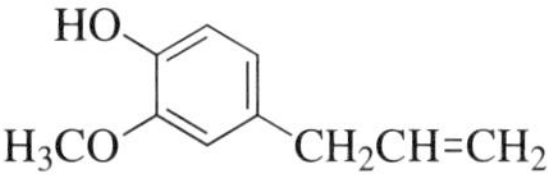

$C_{10}H_{12}O_2$, M_r 164.20, $bp_{1.3\,kPa}$ 121 °C, d_4^{20} 1.0652, n_D^{20} 1.5409, is the main component of several essential oils; clove leaf oil and cinnamon leaf oil may contain >90 %. Eugenol occurs in small amounts in many other essential oils. It is a colorless to slightly yellow liquid with a spicy, clove odor.

Catalytic hydrogenation (e.g., in the presence of noble-metal catalysts) yields dihydroeugenol. Isoeugenol is obtained from eugenol by shifting the double bond. Esterification and etherification of the hydroxyl group of eugenol yield valuable fragrance and flavor materials (e.g., eugenyl acetate and eugenyl methyl ether).

Production. Since sufficient eugenol can be isolated from cheap essential oils, synthesis is not industrially important. Eugenol is still preferentially isolated from clove leaf and cinnamon leaf oil (e.g., by extraction with sodium hydroxide solution). Nonphenolic materials are then removed by steam distillation. After the alkaline solution is acidified at low temperature, pure eugenol is obtained by distillation.

Uses. Eugenol is used in perfumery in clove and carnation compositions as well as for oriental and spicy notes. It is a common component of clove and other aroma compositions. In dentistry, it is used as an antiseptic.
FCT 1975 (**13**) p. 545.

Eugenol methyl ether [*93-15-2*]

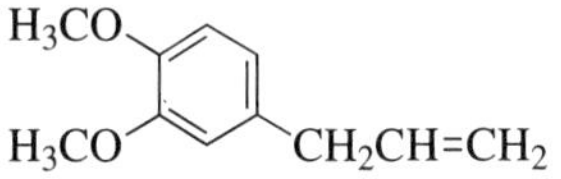

H$_3$CO
H$_3$CO   CH$_2$CH=CH$_2$

$C_{11}H_{14}O_2$, M_r 178.23, $bp_{1.5\,kPa}$ 127–129 °C, d_4^{20} 1.0396, n_D^{20} 1.5340, occurs in numerous essential oils, sometimes at a very high concentration; leaf and wood oil from *Dacrydium franklinii* Hook. (Huon pine oil) contain more than 90% . The ether is an almost colorless liquid with a mild-spicy, slightly herbal odor. It is prepared by methylation of eugenol and is used in perfumery (e.g., in carnation and lilac compositions) and in flavor compositions.

Eugenyl acetate [*93-28-7*]

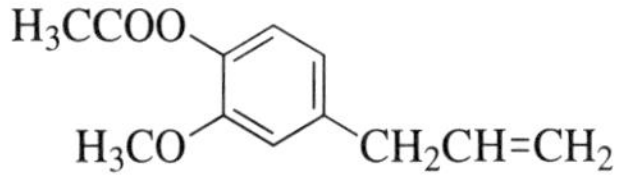

H$_3$CCOO
H$_3$CO   CH$_2$CH=CH$_2$

$C_{12}H_{14}O_3$, M_r 206.24, $bp_{0.4\,kPa}$ 120–121 °C, d_4^{20} 1.0806, n_D^{20} 1.5205, occurs in clove oil, together with eugenol. It is a crystalline solid (*mp* 29 °C) or yellowish liquid with a slightly fruity, clove odor. Eugenyl acetate is prepared by acetylation of eugenol with acetic anhydride and is used in clove compositions to accentuate flowery character.
FCT 1974 (**12**) p. 877.

Propenylguethol [*94-86-0*], 2-ethoxy-5-(1-propenyl)phenol

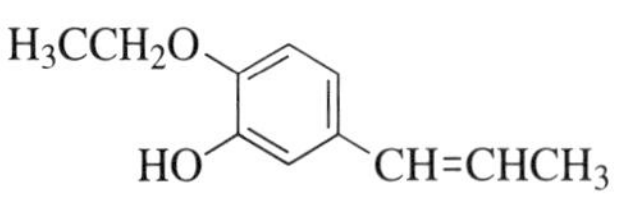

H$_3$CCH$_2$O
HO   CH=CHCH$_3$

$C_{11}H_{14}O_2$, M_r 178.23, exists in *cis* (*mp* 35–36 °C) and in *trans* (*mp* 86 °C) forms. The *trans* isomer has a sweet vanilla-like odor. Propenylguethol can be prepared from isosafrole by reaction with methylmagnesium chloride or by ethylation of isoeugenol followed by selective demethylation with alkali [169].

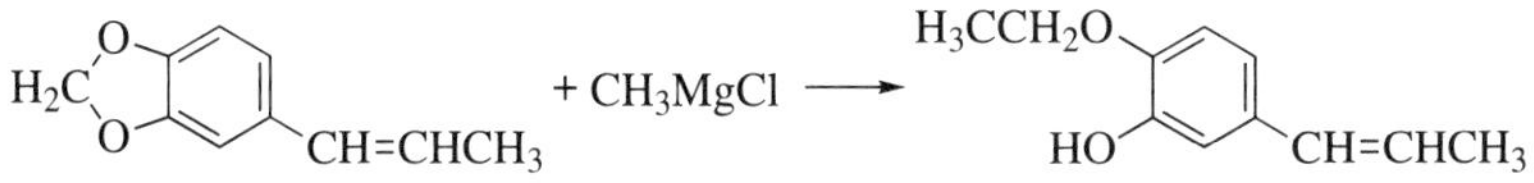

It is used in perfumery, for example, in soap and cosmetics, to create or enhance vanilla notes.

Isosafrole [*120-58-1*], **1,2-methylenedioxy-4-(1-propenyl)benzene**

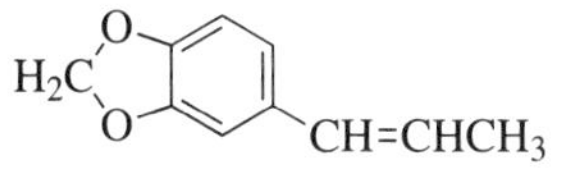

$C_{10}H_{10}O_2$, M_r 162.19, $bp_{0.45\,kPa}$ 85–86 °C, d_4^{20} 1.1206, n_D^{20} 1.5782 (*trans* isomer: *mp* 8.2 °C), has been identified in a few essential oils in the form of its *cis* [*17627-76-8*] and more stable *trans* [*4043-71-4*] isomers. It is a colorless, viscous liquid with a sweet, anise-like odor. Isosafrole is prepared by alkali-catalyzed isomerization of safrole (1,2-methylenedioxy-4-allylbenzene), which is the main component of Brazilian sassafras and brown camphor oils.

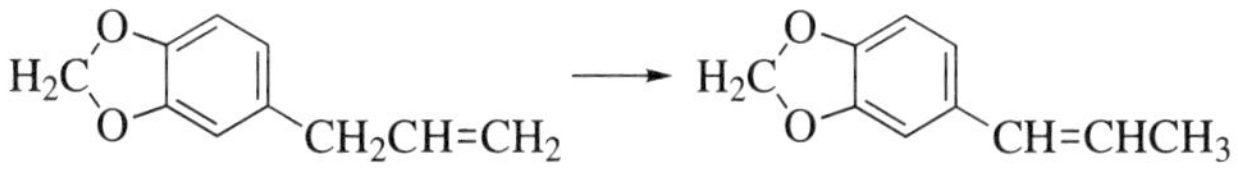

Isosafrole is used in toilet soap perfumes, but its main use is as a starting material for the synthesis of heliotropin.
FCT 1976 (**14**) p. 329.

***p*-Cresyl phenylacetate** [*101-94-0*]

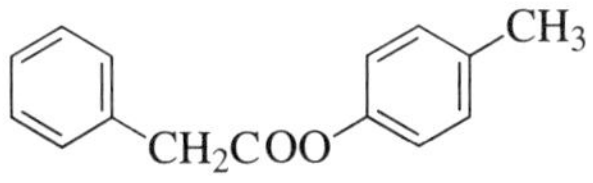

$C_{15}H_{14}O_2$, M_r 226.27, is prepared by esterification of *p*-cresol with phenylacetic acid. It forms crystals (*mp* 75–76 °C) with a narcissus odor and a honey note. It is used in blossom compositions with a slight animal note.
FCT 1975 (**13**) p. 775.

2.6.2 Phenol Alcohols and their Esters

In comparison with the araliphatic alcohols discussed in Section 2.5.2, very few phenol alcohols are used as fragrance and flavor materials. Neither the alcohols corresponding to vanillin, ethylvanillin, and heliotropin nor their esters have special organoleptic properties. Anise alcohol and its acetate are the only products that are used to some extent in perfume and aroma compositions.

Anise alcohol [*105-13-5*], **4-methoxybenzyl alcohol**

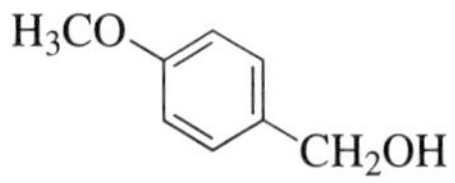

H$_3$CO — ⬡ — CH$_2$OH

$C_8H_{10}O_2$, M_r 138.17, $bp_{1.3\,kPa}$ 136 °C, d_4^{20} 1.1140, n_D^{25} 1.5420, occurs in vanilla pods and in anise seeds. It is a colorless liquid with a sweet-flowery, slightly balsamic odor.

Pure anise alcohol for perfumery and flavor purposes is prepared by hydrogenation of anisaldehyde. It is used in perfumery in blossom compositions (e.g., lilac and gardenia types) and in flavors for confectionery and beverages.
FCT 1974 (**12**) p. 825.

Anisyl acetate [*104-21-2*]
$C_{10}H_{12}O_3$, M_r 180.20, $bp_{1.5\,kPa}$ 133 °C, d_4^{20} 1.1084, has been found in several types of berries. It is a colorless liquid with a fruity, slightly balsamic blossom odor and is used occasionally in sweet-flowery compositions, but more frequently in flavor compositions for fruity notes.

2.6.3 Phenol Aldehydes

Phenol aldehydes are generally pleasant-smelling products. Some of them are particularly important as fragrance and flavor compounds. Anisaldehyde and certain derivatives of protocatechu aldehyde (3,4-dihydroxybenzaldehyde) are well-known representatives. The monomethyl ether of protocatechu aldehyde, vanillin, is perhaps the most widely used flavor compound. Other important derivatives of this aldehyde are veratraldehyde (dimethyl ether) and heliotropin (formaldehyde acetal derivative); they are not only used as fragrance and flavor substances, but also are intermediates in many industrial processes.

p-**Anisaldehyde** [*123-11-5*], **4-methoxybenzaldehyde**

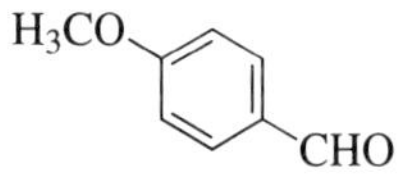

H$_3$CO — ⬡ — CHO

$C_8H_8O_2$, M_r 136.15, $bp_{1.85\,kPa}$ 132 °C, d_4^{25} 1.1192, n_D^{25} 1.5703, occurs in many essential oils, often together with anethole. It is a colorless to slightly yellowish liquid with a sweet, mimosa, hawthorn odor. *p*-Anisaldehyde can be hydrogenated to anise alcohol and readily oxidizes to anisic acid when exposed to air. Synthetic routes to anisaldehyde usually involve the oxidation of *p*-cresyl methyl ether. Manganese dioxide and sulfuric acid are usually used for oxidation. In a Russian process, *p*-cresyl methyl ether is oxidized with alkali peroxysulfates in the presence of silver salts [170].

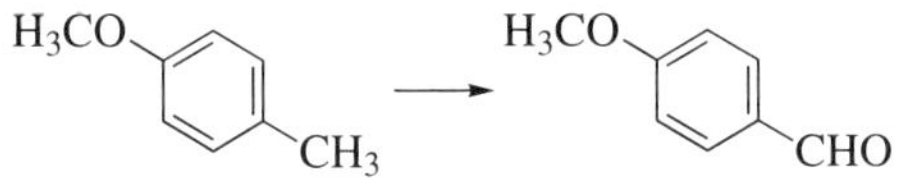

Other industrial processes are the liquid-phase oxidation in the presence of cobalt catalysts [171] and the electrochemical oxidation in the presence of lower aliphatic alcohols via the corresponding anisaldehyde dialkyl acetal [172].

p-Anisaldehyde is frequently used in sweet blossom compositions (e.g., in lilac and hawthorn types) as well as in flavor compositions for confectioneries and beverages. *p*-Anisaldehyde is an intermediate in many industrial processes. Its hydrogen sulfite derivative is used as a brightener for metals in galvanic baths. FCT 1974 (**12**) p. 823.

2-Methyl-3-(4-methoxyphenyl)propanal [*5462-06-6*]

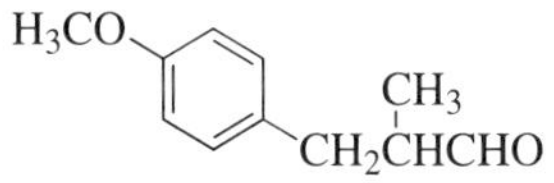

$C_{11}H_{14}O_2$, M_r 178.23, d_4^{20} 1.039–1.047, n_D^{20} 1.517–1.522, is a pale yellow liquid with licorice, anise note with slight fruity modification. It does not occur in nature. It can be prepared by condensation of anisaldehyde (see above) with propanal and selective hydrogenation of the resulting 2-methyl-3-(4-methoxyphenyl)-2-propenal. It fits well with flowery notes and is used in fine fragrances and cosmetics. FCT 1988 (**26**) p. 377.

Trade Name. Canthoxal (IFF).

Vanillin [*121-33-5*], **4-hydroxy-3-methoxybenzaldehyde**

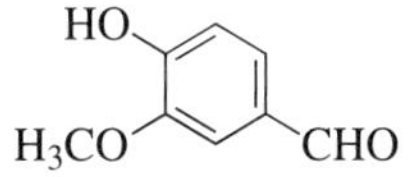

$C_8H_8O_3$, M_r 152.15, $bp_{1.3\,kPa}$ 155 °C, d_4^{20} 1.056, is found in many essential oils and foods, but is often not essential for their odor or aroma. However, it does determine the odor of essential oils and extracts from *Vanilla planifolia* and *V. tahitensis* pods, in which it is formed during ripening by enzymatic cleavage of glycosides.

Properties. Vanillin is a colorless crystalline solid (*mp* 82–83 °C) with a typical vanilla odor. Because it possesses aldehyde and hydroxyl substituents, it undergoes many reactions. Additional reactions are possible due to the reactivity of the aromatic nucleus. Vanillyl alcohol and 2-methoxy-4-methylphenol are obtained by catalytic hydrogenation; vanillic acid derivatives are formed after oxidation and protection of the phenolic hydroxyl group. Since vanillin is a phenol aldehyde, it is stable to autoxidation and does not undergo the Cannizzaro reaction. Numerous derivatives can be prepared by etherification or esterification of the hydroxyl group and by aldol condensation at the aldehyde group. Several of these derivatives are intermediates, for example, in the synthesis of pharmaceuticals.

Production.   Commercial vanillin is obtained by processing waste sulfite liquors or is synthesized from guaiacol. Preparation by oxidation of isoeugenol is of historical interest only.

1.   *Preparation from Waste Sulfite Liquors.* The starting material for vanillin production is the lignin present in sulfite wastes from the cellulose industry. The concentrated mother liquors are treated with alkali at elevated temperature and pressure in the presence of oxidants. The vanillin formed is separated from the by-products, particularly acetovanillone (4-hydroxy-3-methoxyacetophenone), by extraction, distillation, and crystallization.

    A large number of patents describe various procedures for the (mainly) continuous hydrolysis and oxidation processes, as well as for the purification steps required to obtain high-grade vanillin [173]. Lignin is degraded either with sodium hydroxide or with calcium hydroxide solution and simultaneously oxidized in air in the presence of catalysts. When the reaction is completed, the solid wastes are removed. Vanillin is extracted from the acidified solution with a solvent (e.g., butanol or benzene) and reextracted with sodium hydrogen sulfite solution. Reacidification with sulfuric acid followed by vacuum distillation yields technical-grade vanillin, which must be recrystallized several times to obtain food-grade vanillin. Water, to which some ethanol may be added, is used as the solvent in the last crystallization step.

2.   *Preparation from Guaiacol and Glyoxylic Acid.* Several methods can be used to introduce an aldehyde group into an aromatic ring. Condensation of guaiacol with glyoxylic acid followed by oxidation of the resulting mandelic acid to the corresponding phenylglyoxylic acid and, finally, decarboxylation continues to be a competitive industrial process for vanillin synthesis.

    Currently, guaiacol is synthesized from catechol, which is prepared by acid-catalyzed hydroxylation of phenol with hydrogen peroxide. Glyoxylic acid is obtained as a by-product in the synthesis of glyoxal from acetaldehyde and can also be produced by oxidation of glyoxal with nitric acid. Condensation of guaiacol with glyoxylic acid proceeds smoothly at room temperature and in weakly alkaline media. A slight excess of guaiacol is maintained to avoid formation of disubstituted products; excess guaiacol is recovered. The alkaline solution containing 4-hydroxy-3-methoxymandelic acid is then oxidized in air in the presence of a catalyst until the calculated amount of oxygen is consumed [174]. Crude vanillin is obtained by acidification and simultaneous decarboxylation of the (4-hydroxy-3-methoxyphenyl)glyoxylic acid solution. Commercial grades are obtained by vacuum distillation and subsequent recrystallization as described under method 1.

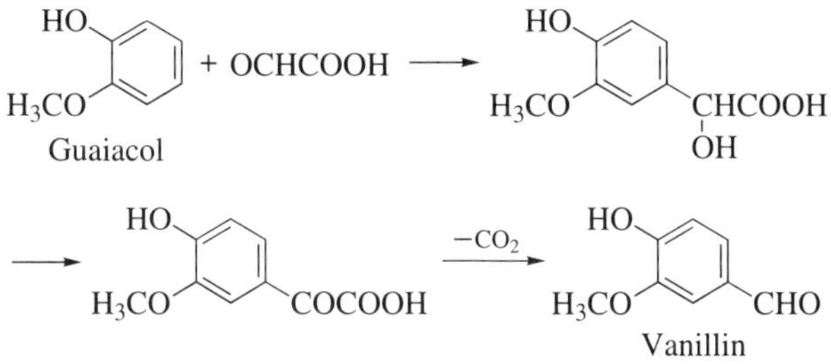

This process has the advantage that, under the reaction conditions, the glyoxyl radical enters the aromatic guaiacol ring almost exclusively *para* to the phenolic hydroxyl group. Tedious separation procedures are thus avoided.

Much research and development for the biotechnological manufacture of vanillin has been done [175].

Uses.   The main application of vanillin is the flavoring of foods (e.g., ice cream, chocolate, bakery products, and confectioneries). Small quantities are used in perfumery to round and fix sweet, balsamic fragrances. Vanillin is also used as a brightener in galvanotechnical processes and is an important intermediate in, for example, the production of pharmaceuticals such as L-3,4-dihydroxyphenylalanine (L-DOPA) and methyldopa.
FCT 1977 (**15**) p. 633.

Veratraldehyde [*120-14-9*], **3,4-dimethoxybenzaldehyde**

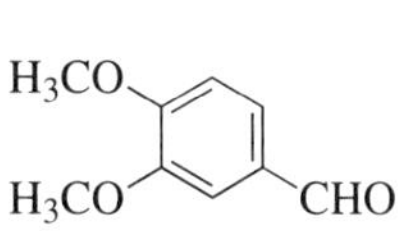

$C_9H_{10}O_3$, M_r 166.18, occurs in a few essential oils and is a crystalline solid (*mp* 44.5–45 °C) with a woody, vanilla-like odor.

Veratraldehyde can be prepared by methylation of vanillin. It is used in oriental and warm-woody fragrances, as well as in flavor compositions for vanilla notes. It is an intermediate in, for example, the synthesis of pharmaceuticals.
FCT 1975 (**13**) p. 923.

Ethylvanillin [*121-32-4*], **3-ethoxy-4-hydroxybenzaldehyde**

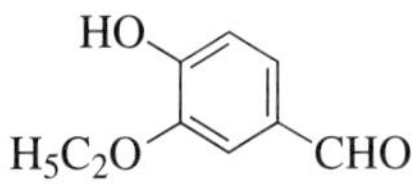

$C_9H_{10}O_3$, M_r 166.18, *mp* 77–78 °C, does not occur in nature. Its odor resembles that of vanillin but is approximately three times as strong. Ethylvanillin can be prepared by method 2 as described for vanillin, using guethol instead of guaiacol as the starting material.

Ethylvanillin is used in the chocolate and confectionery industry. It gives a sweet, balsamic note to flowery and fruity perfume compositions.
FCT 1975 (**13**) p. 103.

Heliotropin [*120-57-0*], **piperonal, 3,4-methylenedioxybenzaldehyde**

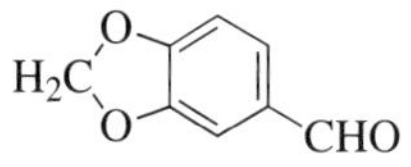

$C_8H_6O_3$, M_r 150.13, $bp_{1.6\,kPa}$ 139.4 °C, $d_4^{43.2}$ 1.2792, occurs in a number of essential oils, but never as the main component. It forms white crystals (*mp* 37 °C) with a sweet-flowery, slightly spicy, heliotrope-like odor.

Production.   Heliotropin is produced by two main routes:

1. *From Isosafrole.* For many years, oxidative cleavage of isosafrole was the only route applicable on an industrial scale. Examples of oxidants that give good yields of heliotropin are chromium(VI) salts, oxygen, and ozone.

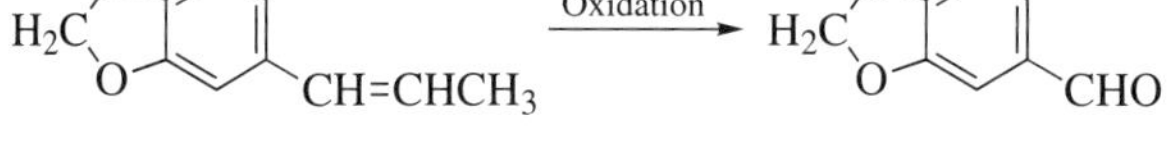

This method is still used currently because safrole (the starting material for isosafrole) can be isolated from essential oils relatively inexpensively and in sufficient quantity.

2. *From Catechol.* Several routes have recently been developed for the synthesis of heliotropin from catechol. In one such route, catechol is converted into 3,4-dihydroxymandelic acid with glyoxylic acid in an alkaline medium in the presence of aluminum oxide. 3,4-Dihydroxymandelic acid is oxidized to the corresponding keto acid (e.g. with copper(II) oxide), which is decarboxylated to 3,4-dihydroxybenzaldehyde [176]. The latter product is converted into heliotropin, for example, by reaction with methylene chloride in the presence of quaternary ammonium salts [177].

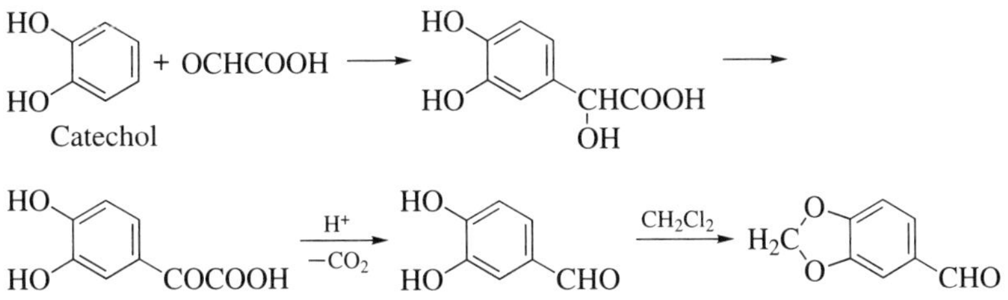

In another route, catechol is first reacted with methylene chloride and converted into 1,2-methylenedioxybenzene [177]. Reaction with glyoxylic acid in strongly acidic media yields 3,4-methylenedioxymandelic acid [178]. Subsequent oxidation and decarboxylation with nitric acid affords heliotropin.

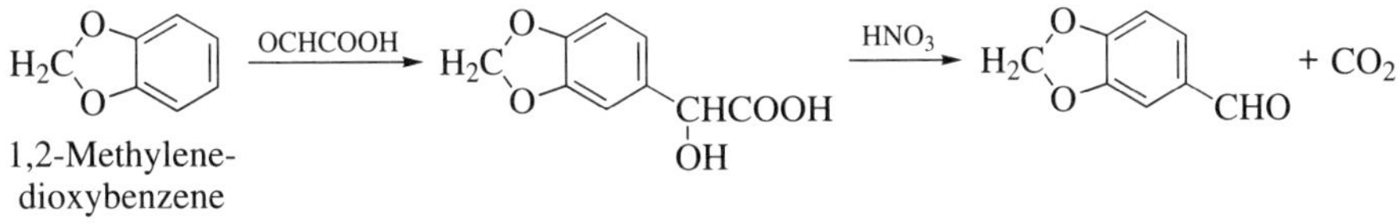

Uses. Heliotropin is used in many flowery-spicy fine fragrances and is also an important ingredient of flavor compositions.
FCT 1974 (**12**) p. 907.

2-Methyl-3-(3,4-methylenedioxyphenyl)propanal [*1205-17-0*]

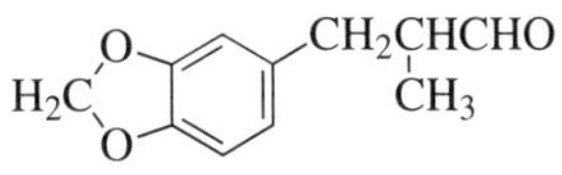

$C_{11}H_{12}O_3$, M_r 192.22, is not found in nature. It is a colorless to slightly yellow liquid, d_4^{20} 1.159–1.167, n_D^{20} 1.531–1.536, with green, floral odor with top-notes of ozone and new mown hay. It can be prepared by condensation of heliotropin (see previous page) with propanal and partial hydrogenation of the intermediately formed unsaturated aldehyde.

Title compound can be used in perfumes for toiletries, e.g., shave creams and detergents.

Trade Names. Heliofolal (H&R), Helional (IFF).

2.6.4 Phenol Ketones

Few of the phenol derivatives that have a keto substituent in their side-chain are of interest as fragrance or flavor substances. A number of phenols and phenyl ethers acetylated in the benzene ring have been identified as volatile components of foods. 4-Methoxyacetophenone is of some interest as a fragrance compound. 4-Hydroxybenzylacetone, a higher mass phenol ketone, has a characteristic raspberry aroma.

4-Methoxyacetophenone [*100-06-1*], **acetanisole**

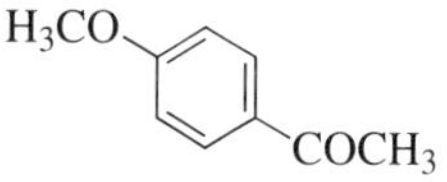

H$_3$CO⟷COCH$_3$

C$_9$H$_{10}$O$_2$, M_r 150.18, d^{41} 1.0818, n_D^{41} 1.5470, occurs in anise oil. It forms white crystals (*mp* 38 °C) with a sweet odor, reminiscent of hawthorn. 4-Methoxyacetophenone is prepared by Friedel–Crafts acetylation of anisole and is used in soap perfumes.
FCT 1974 (**12**) p. 927.

4-(4-Hydroxyphenyl)-2-butanone [*5471-51-2*], **raspberry ketone**

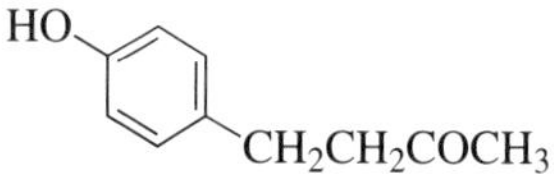

HO⟷CH$_2$CH$_2$COCH$_3$

C$_{10}$H$_{12}$O$_2$, M_r 164.20, is a highly characteristic component of raspberry aroma. It forms colorless crystals (*mp* 82–83 °C) with a sweet-fruity odor strongly reminiscent of raspberries.

Raspberry ketone is prepared by alkali-catalyzed condensation of the alkali salt of 4-hydroxybenzaldehyde and acetone, followed by selective hydrogenation of the double bond in the resulting 4-hydroxybenzalacetone. Other syntheses start from phenol which is converted into 4-(4-hydroxyphenyl)-2-butanone with methyl vinyl ketone (e.g., in the presence of phosphoric acid) [179] or with 4-hydroxy-2-butanone in the presence of concentrated sulfuric acid [180].

The ketone is used in fruit flavors, particularly in raspberry compositions.
FCT 1978 (**16**) p. 781.

Trade Names. Frambinon (Dragoco), Oxyphenylon (IFF).

2.6.5 Phenolcarboxylates

Alkyl and aralkyl salicylates, are sensorially important phenolcarboxylates that are used in flavors and fragrances. The following salicylates are used in perfume and flavor compositions and can be prepared by esterification of salicylic acid.

Methyl salicylate [*119-36-8*]
$R = CH_3$, $C_8H_8O_3$, M_r 152.15, $bp_{1.6 kPa}$ 98 °C, d_4^{25} 1.1782, n_D^{25} 1.5350, is the main component of wintergreen oil and occurs in small quantities in other essential oils and fruit. It is a colorless liquid with a sweet, phenolic odor. Methyl salicylate is used in perfumery as a modifier in blossom fragrances and as a mild antiseptic in oral hygiene products.
FCT 1978 (**16**) p. 821.

Isoamyl salicylate [*87-20-7*]
$R = CH_2CH_2CH(CH_3)_2$, $C_{12}H_{16}O_3$, M_r 208.26, $bp_{2 kPa}$ 151–152 °C, d_4^{20} 1.0535, n_D^{20} 1.5065, has been found in a number of fruit aromas. It is a colorless liquid with a sweet, clover-like odor and is used in perfumery for floral and herbal notes, particularly in soap perfumes.
FCT 1973 (**11**) p. 859.

Hexyl salicylate [*6259-76-3*]
$R = (CH_2)_5CH_3$, $C_{13}H_{18}O_3$, M_r 222.28, $bp_{1.6 kPa}$ 167–168 °C, d_{25}^{25} 1.035, n_D^{25} 1.5049, has been reported in carnation flower absolute [181]. It is a colorless liquid with a green, flowery-spicy odor, reminiscent of azaleas. It is used for blossom and herbal notes in perfumes, e.g., in soap, personal hygiene products, and detergents.
FCT 1975 (**13**) p. 807.

***cis*-3-Hexenyl salicylate** [*65405-77-8*]
$R = cis\text{-}(CH_2)_2CH = CHCH_2CH_3$, $C_{13}H_{16}O_3$, M_r 220.27, $bp_{0.15 kPa}$ 125 °C, d_{25}^{25} 1.0589, n_D^{20} 1.5210, has been identified in carnation flower absolute. It is a colorless liquid with a long-lasting, sweet, green balsamic odor. It is used in fine fragrances and for scenting soaps, cosmetics, and detergents.
FCT 1979 (**17**) p. 373.

Cyclohexyl salicylate [*25485-88-5*]
$R = cyclo\text{-}C_6H_{11}$, $C_{13}H_{16}O_3$, M_r 220.27 is not found in nature. It is a colorless liquid, $bp_{4 Pa}$ 115 °C, d_{25}^{25} 1.112, n_D^{20} 1.532–1.536, with aromatic, floral balsamic odor. It is used in perfumery instead of benzyl salicylate.

Benzyl salicylate [*118-58-1*]
$R = CH_2C_6H_5$, $C_{14}H_{12}O_3$, M_r 228.25, $bp_{1.3 kPa}$ 186–188 °C, d_4^{20} 1.1799, n_D^{25} 1.5805, which occurs in several essential oils, is a colorless, viscous liquid with a weak, sweet, slightly balsamic odor. Benzyl salicylate is used as a fixative in flowery-spicy perfume compositions and in flavors.
FCT 1973 (**11**) p. 1029.

Phenethyl salicylate [*87-22-9*]
$R = CH_2CH_2C_6H_5$, $C_{15}H_{14}O_3$, M_r 242.27, which has been reported to occur in some essential flower oils, is a crystalline solid (*mp* 44 °C) with a weak, long-lasting, balsamic, blossom odor, reminiscent of rose and hyacinth. It is used in perfumery for spicy and balsamic blossom compositions.
FCT 1978 (**16**) p. 849.

Two *alkyl-substituted resorcylic acid esters* are important as oakmoss fragrance substances.

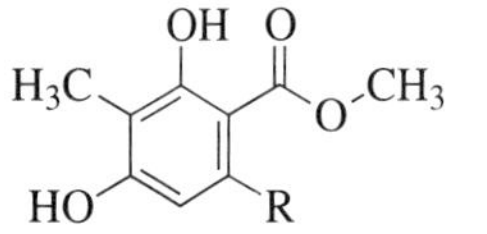

They can be prepared from acyclic compounds. In an industrial process, dimethyl malonate is condensed with 4-alken-3-ones (or a mixture of the respective ketones with 5-chloroalkan-3-ones) to give a substituted 3-hydroxy-2-cyclohexenone. Aromatization, in good yield is achieved by reaction of the hydroxycyclohexenones with a suitable *N*-haloamide. The intermediate 3-hydroxy-2-cyclohexenones can also be obtained by condensation of methyl 3-oxoalkanoate with methyl crotonate [182].

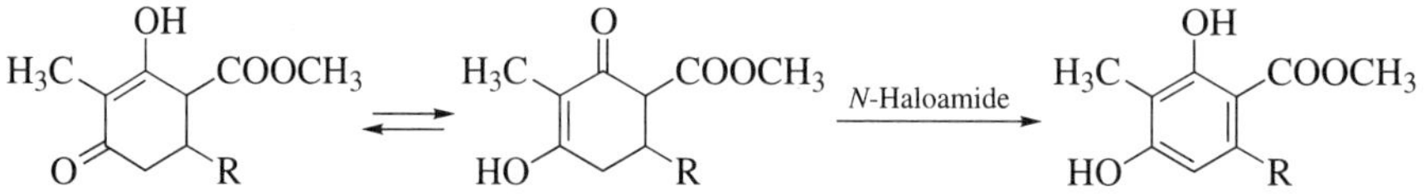

Another route starts from the corresponding (di)methyl-1,3-dihydroxybenzene which is carboxylated, and the resulting dihydroxymethylbenzoic acid is esterified.

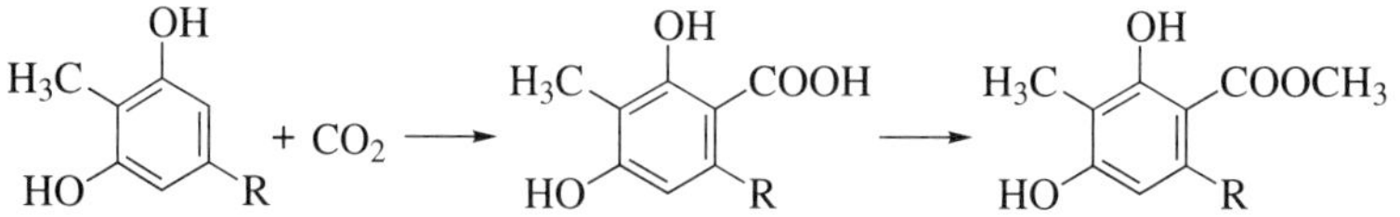

Methyl 3-methylresorcylate [*33662-58-7*], **methyl 2,4-dihydroxy-3-methylbenzoate**
$R = H$, $C_9H_{10}O_4$, M_r 182.18, *mp* 130.5–131.6 °C, forms an off-white powder with

long-lasting moss odor, slightly reminiscent of oakmoss and strong seaside in character. It is used in small amounts to provide marine effects to perfumes.

Trade Name. Seamoss (PFW).

Methyl 3,6-dimethylresorcylate [*4707-47-5*],
methyl 2,4-dihydroxy-3,6-dimethylbenzoate
R = CH_3, $C_{10}H_{12}O_4$, M_r 196.20, is an odor determining constituent of oakmoss absolute extract and forms colorless crystals (*mp* 145 °C) with a mossy-earthy odor. It is used as a substitute for oakmoss extract in fine fragrances, soap, and cosmetics.

Trade Names. Atralone (Agan), Evernyl (Giv.-Roure), Veramoss (IFF).

2.7 O- and O,S-Heterocycles

2.7.1 Cyclic Ethers

Cyclic ethers used as fragrances include a number of terpenoid compounds. Some of them, such as 1,4-cineole [*470-67-7*] and 1,8-cineole, occur in essential oils in significant quantities. Others are only minor components; examples are rose oxide, nerol oxide [*1786-08-9*], and rose furan [*15186-51-3*], which contribute to the specific fragrance of rose oil. Caryophyllene oxide [*1139-30-6*], which has a woody, slightly ambergris-like odor can be prepared by treatment of β-caryophyllene with organic peracids. α-Cedrene oxide [*11000-57-0*] is another wood-fragrance compound, that can be easily prepared by epoxidation of cedarwood oil hydrocarbons.

Except for some of the above-mentioned compounds, only a few other cyclic ethers are important, for instance, 4,6,6,7,8,8-hexamethyl-1,3,4,6,7,8-hexahydropenta[*g*]benzopyran, a musk fragrance that is used in large amounts.

Numerous furan and pyran derivatives, many of which originate from heat treatment of carbohydrates, largely determine the odor of processed foods. Of this group, 2,5-dimethyl-4-hydroxy-2*H*-furan-3-one and maltols are used in fairly large quantities in flavors. The following compounds are used in relatively small amounts in flavor compositions:

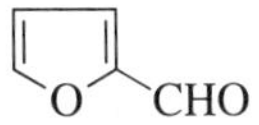

2-furaldehyde [*98-01-1*]: freshly baked bread odor

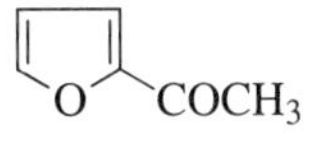

2-acetylfuran [*1192-62-7*]: sweet balsamic odor

methyl 2-furoate [*611-13-2*]: fruity, mushroom-like odor

2-methylfuran-3-thiol [*28588-74-1*]: roast beef aroma

2,5-dimethyl-3(2*H*)-furanone [*14400-67-0*]: roast, coffee odor

4-hydroxy-5-methyl-3(2*H*)-furanone [*19322-27-1*]: roast, meat odor

2,5-diethyltetrahydrofuran [*41239-48-9*]: fruity, herbal-minty note

2-methyl-4-propyl-1,3-oxathiane [*67715-80-4*]: typical sulfury note of tropical fruits

1,8-Cineole [*470-82-6*], 1,8-epoxy-*p*-menthane, eucalyptol

$C_{10}H_{18}O$, M_r 154.25, $bp_{101.8\,kPa}$ 176–177 °C, *fp* 1 °C, d_4^{20} 0.9267, n_D^{20} 1.4586, occurs in many terpene-containing essential oils, sometimes as the main component. For example, eucalyptus oils contain up to 85% 1,8-cineole and laurel leaf oil contains up to 70% . It is a colorless liquid with a characteristic odor, slightly reminiscent of camphor.

1,8-Cineole is one of the few fragrance materials that is obtained exclusively by isolation from essential oils, especially eucalyptus oils. Technical-grade 1,8-cineole with a purity of 99.6–99.8% is produced in large quantities by fractional distillation of *Eucalyptus globulus* oil. A product essentially free from other products can be obtained by crystallization of cineole-rich eucalyptus oil fractions.

1,8-Cineole has a fresh odor and is used in large quantities in fragrances as well as in flavors (e.g., in oral hygiene products).
FCT 1975 (**13**) p. 105.

Rose oxide [*16409-43-1*], **4-methyl-2-(2-methyl-1-propenyl)tetrahydropyran**

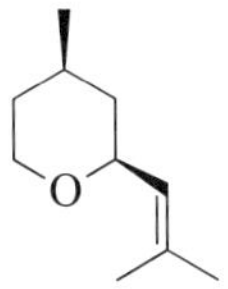

$C_{10}H_{18}O$, M_r 154.25, $bp_{1.6\,kPa}$ 70 °C, d_4^{20} 0.875, n_D^{20} 1.4570, $[\alpha]_D$ for the optically pure (−)-*cis* form −58.1 °, occurs in small quantities, mainly the levorotatory *cis* form, in essential oils (e.g., Bulgarian rose oil and geranium oil). Commercial synthetic products are either optically active or inactive mixtures of the *cis* and *trans* isomers. Their physical constants, particularly the optical rotation, depend on the starting material and the method of synthesis. They are colorless liquids with a strong odor reminiscent of geranium oil and carrot leaves.

Rose oxide is usually prepared from citronellol which can be converted into a mixture of two allyl hydroperoxides (e.g., by photosensitized oxidation with oxygen). Reduction of the hydroperoxides with sodium sulfite yields the corresponding diols [183]. Treatment with dilute sulfuric acid results in allylic rearrangement and spontaneous cyclization of one of the isomers; a mixture of diastereoisomeric rose oxides is thus formed. The unreacted diol isomer is separated by distillation. (−)- Citronellol as the starting material yields approximately a 1:1 mixture of (−)-*cis*- and (−)-*trans*-rose oxide.

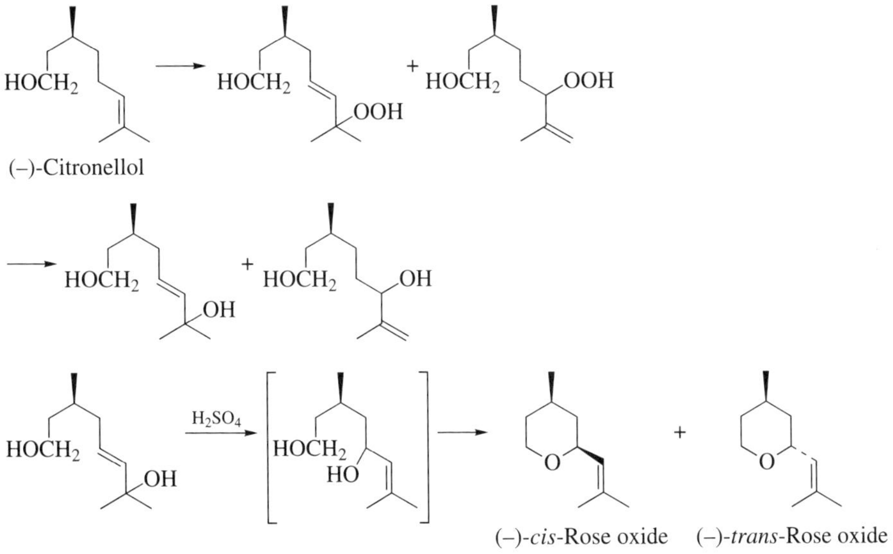

(−)-cis-Rose oxide   (−)-trans-Rose oxide

Rose oxide is used in rose and geranium perfumes.
FCT 1976 (**14**) p. 855.

Menthofuran [*494-90-6*], **3,6-dimethyl-4,5,6,7-tetrahydrobenzofuran**

$C_{10}H_{14}O$, M_r 150.22, $bp_{1.3\,kPa}$ 78–79 °C, d_4^{20} 0.9676, n_D^{20} 1.4855, $[\alpha]_D^{20}$ + 94.6 °, occurs mainly as the (+) isomer, in numerous essential oils (e.g., *Mentha* oils). It is a colorless liquid with a minty odor.

(+)-Menthofuran [*17957-94-7*] is isolated from *Mentha* oils or is prepared synthetically, for example, by treatment of (+)-pulegone with fuming sulfuric acid in acetic anhydride and pyrolysis of the resulting sultone.

(+)-Pulegone

Menthofuran is used mainly in peppermint oil reconstitutions.

Linalool oxide, 2-methyl-2-vinyl-5-(α-hydroxyisopropyl)tetrahydrofuran

$C_{10}H_{18}O_2$, M_r 170.25, $bp_{101.3\,kPa}$ 188 °C, d_4^{20} 0.939–0.944, n_D^{20} 1.451–1.455, has been identified in essential oils and in fruit aromas. Commercial linalool oxide is a mixture of the *cis* and *trans* forms, [*5989-33-3*] and [*34995-77-2*], respectively. It is a liquid with an earthy-flowery, slightly bergamot-like odor.

Linalool oxide is prepared by oxidation of linalool, e.g., with peracids. The isomeric compound 2,2,6-trimethyl-6-vinyltetrahydro-2*H*-pyran-3-ol [*14049-11-7*], which also occurs in nature, is formed as a by-product:

Linalool oxide is used in perfumery (e.g., for lavender notes) and for reconstitution of essential oils.

A dehydrated linalool oxide, 2-methyl-2-vinyl-5-isopropenyltetrahydrofuran [*13679-86-2*], occurs naturally; it has a minty eucalyptol odor and is used in perfumery.
FCT 1983 (**21**) p. 863.

1,5,9-Trimethyl-13-oxabicyclo[10.1.0]trideca-4,8-diene [*13786-79-3*]

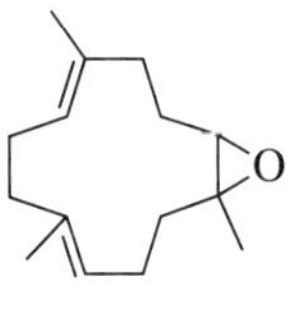

$C_{15}H_{24}O$, M_r 220.36, d_{20}^{20} 0.962–0.980, n_D^{20} 1.504–1.509, is a colorless to pale yellow liquid with powerful complex woody and amber odor. It does not occur in nature. It is prepared by monoepoxidation of 1,5,9-trimethyl-1,5,9-dodecatriene with, e.g., peracids.

It is used in perfumery for cosmetics and detergents.

Trade Name. Cedroxyde (Firmenich).

3a,6,6,9a-Tetramethyldodecahydronaphtho[2,1-*b*]furan [*6790-58-5*]

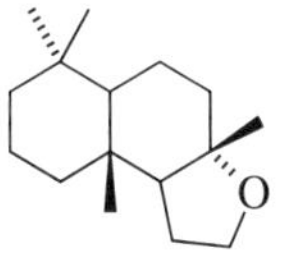

$C_{16}H_{28}O$, M_r 236.40, *mp* 75–76 °C, is a crystalline autoxidation product of ambrein (see Ambergris) with a typical ambergris odor. It is prepared from sclareol, a diterpene alcohol obtained from plant waste in the production of clary sage oil (see Sage Oils). Oxidative degradation to a lactone, hydrogenation of the latter to the corresponding diol and dehydration yield.

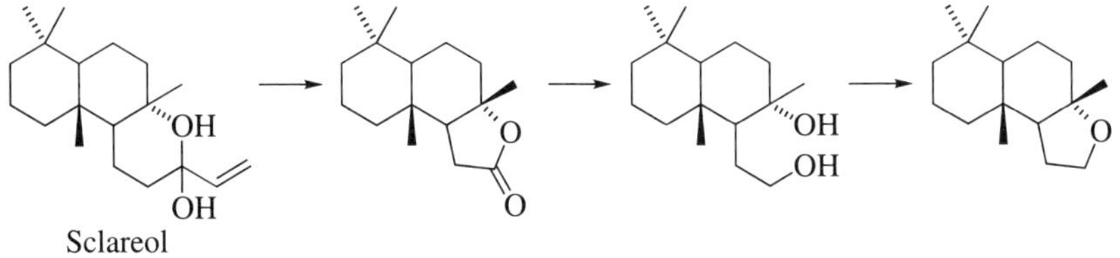

Sclareol

The lactone intermediate is prepared in another industrial process by cyclization of homofarnesic acid in the presence of $SnCl_4$ as a catalyst [184]. Pure diastereomers are obtained by acid cyclization of *trans*- and *cis*-4-methyl-6-(2,6,6-trimethylcyclohex-1(2)enyl)-3-hexen-1-ol, prepared from 2-methyl-4-(2,6,6-trimethylcyclohex-1(2)-enyl)-2-butenal [185].

The product is used in perfumery for creating ambergris notes.

Trade Names. Compound starting from natural sclareol: Ambermore (Aromor), Ambrox (Firmenich), Ambroxan (Henkel), Ambroxid (H&R); compound starting from homofarnesenic acid derivatives: Ambrox DL (Firmenich), Synambrane (FR); compound starting from 2-methyl-4-(2,6,6-trimethylcyclohex-1(2)enyl)-2-butenal: Cetalox (Firmenich).

2,5-Dimethyl-4-hydroxy-2*H*-furan-3-one [*3658-77-3*]

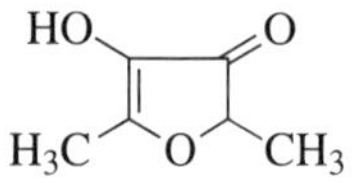

$C_6H_8O_3$, M_r 128.13, is a constituent of pineapple and strawberry aroma and is also found in other foods. It forms colorless crystals (*mp* 77–79 °C) with a relatively weak, nonspecific odor. Dilute solutions develop a pineapple, strawberry-like odor. It can be prepared by cyclization of hexane-2,5-diol-3,4-dione in the presence of an acidic catalyst [186]. The dione is the ozonization product of 2,5-hexynediol, which is obtained by ethynylation of acetaldehyde.

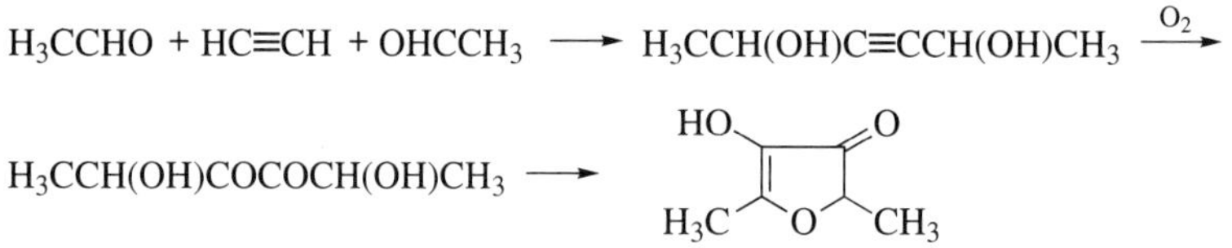

In another process, a dialkyl α-methyldiglycolate (formed from an alkyl lactate and an alkyl monochloroacetate) is reacted with dialkyl oxalate in the presence of a sodium alkoxide and dimethylformamide. The reaction product is cyclized, alkylated, hydrolyzed, and decarboxylated [187].

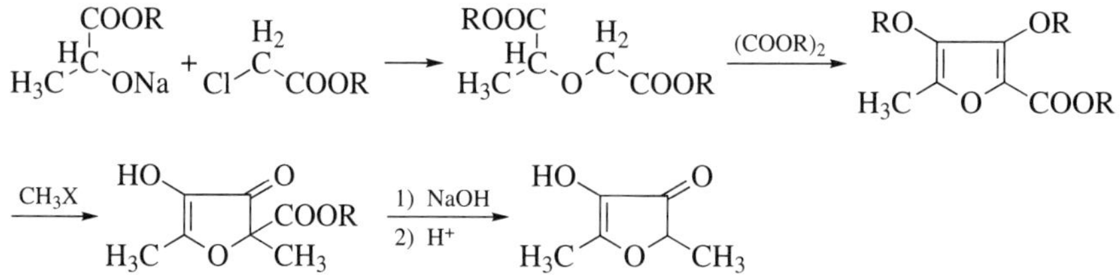

It is also manufactured in a multistep bioprocess from rhamnose [188].

The compound is used in the flavoring of foods.

Trade Name. Furaneol (Firmenich).

2-Ethyl-4-hydroxy-5-methyl-3(2H)-furanone [*27538-10-9*] **and 5-Ethyl-4-hydroxy-2-methyl-3(2H)-furanone** [*27538-09-6*]

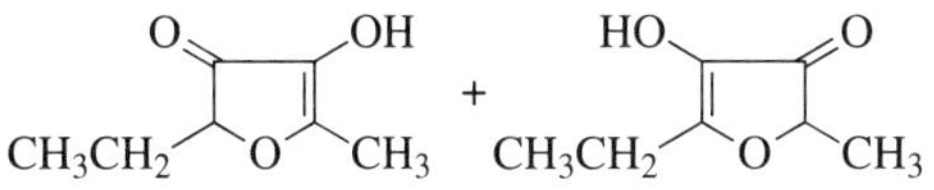

$C_7H_{10}O_3$, M_r 142.16, has been identified in, e.g., coffee and melon. The tautomer mixture is a clear slightly yellowish liquid, $bp_{0.02kPa}$ 82–83 °C, d_4^{20} 1.137, n_D^{20} 1.511 with sweet, caramel, fruity, bread-like odor. One commercially applied synthesis is the condensation of 2-pentene nitrile with ethyl lactate followed by oxidation of the intermediate 4-cyano-5-ethyl-2-methyldihydro-3(2H)-furanone with monoperoxysulfate [189].

It is used in fruit flavors as well as flavor compositions with caramel, coffee, meat, or bread character.

Trade Name. Homofuronol (Giv.-Roure).

Maltol [*118-71-8*], **3-hydroxy-2-methyl-4H-pyran-4-one**

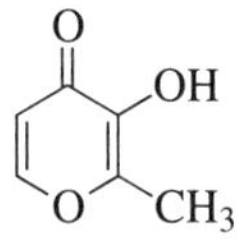

$C_6H_6O_3$, M_r 126.11, occurs in pine needles and the bark of young larch trees. It is produced when cellulose or starch are heated and is a constituent of wood tar oils. It forms crystals (*mp* 162–164 °C) with a caramel-like odor, reminiscent of freshly baked cakes.

Although many routes are known for its synthesis, maltol is still isolated mainly from beechwood tar. It is used in aroma compositions with a caramel note and as a taste intensifier, for example, in fruit flavors (particularly in strawberry flavor compositions).
FCT 1975 (**13**) p. 841.

Ethylmaltol [*4940-11-8*], **2-ethyl-3-hydroxy-4H-pyran-4-one**

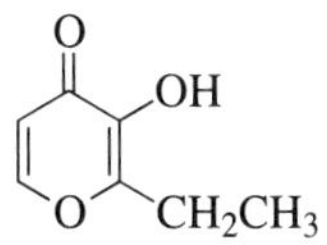

$C_7H_8O_3$, M_r 140.14, does not occur in nature. It forms white crystals (*mp* 90–91 °C) with very sweet caramel-like odor, four to six times more potent than maltol.

Several syntheses have been developed for its preparation. In a one pot process, e.g., α-ethylfurfuryl alcohol is treated with halogen to give 4-halo-6-hydroxy-2-ethyl-2*H*-pyran-3(6*H*)-one, which need not be isolated and can be converted to ethylmaltol by aqueous hydrolysis [190]

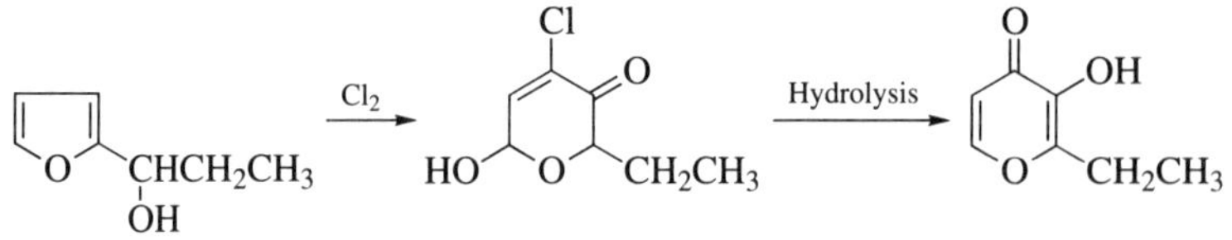

Ethylmaltol is used in aroma compositions and as a flavor enhancer in food, beverages and tobacco.
FCT 1975 (**13**) p. 805.

2-Isobutyl-4-methyltetrahydro-2*H*-pyran-4-ol [*63500-71-0*]

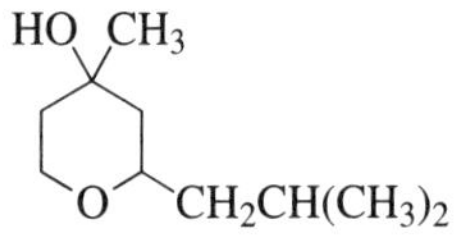

$C_{10}H_{20}O_2$, M_r 172.27, d_{20}^{20} 0.948–0.955, n_D^{20} 1.455–1.460, is a colorless to pale yellow liquid with a fresh, soft and natural floral odor. It does not occur in nature. Cyclocondensation of 3-methyl-3-buten-1-ol with 3-methylbutanal on silica gel and alumina in the absence of solvents is proposed for synthesis [191].

It can be used in almost all perfume types to give elegant floral diffusion without changing the fragrance character. Its stability allows application in soap, toiletries, and household products.

Trade Name. Floral (Firmenich).

3-Pentyltetrahydro-2*H*-pyran-4-ol acetate [*18871-14-2*]

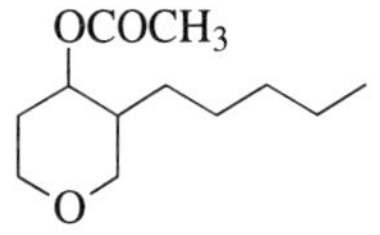

$C_{12}H_{22}O_3$, M_r 214.30, $bp_{0.13\,kPa}$ 102–103 °C, d_{25}^{25} 0.974–0.978, n_D^{20} 1.448–1.451, is a colorless to pale yellow liquid with a sweet-floral, fruity, slightly woody, jasmin-like odor. It is prepared by a Prins reaction of 1-octene with formaldehyde and acetic acid and is used in perfumes for various flower types, especially jasmin.
FCT 1992 (**30**) p. 5 S.

Trade Names. Jasmal (IFF), Jasmonyl LG (Giv.-Roure), Jasmophyll (H&R), Jasmopyrane (Quest).

4,6,6,7,8,8-Hexamethyl-1,3,4,6,7,8-hexahydrocyclopenta[g]benzopyran
[*1222-05-5*]

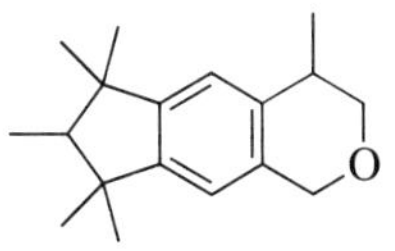

$C_{18}H_{26}O$, M_r 258.40, $bp_{1.1\,kPa}$ 129 °C, d_4^{20} 1.0054, n_D^{20} 1.5342, is a viscous liquid with a musklike odor. It is one of the most frequently used synthetic, artificial musk fragrances. The starting material for its synthesis is 1,1,2,3,3-pentamethylindane, which is prepared by cycloaddition of *tert*-amylene to α-methylstyrene. The pentamethylindane is hydroxyalkylated with propylene oxide in a Friedel–Crafts reaction using aluminum chloride as a catalyst (analogous to the synthesis of phenethyl alcohol from benzene and ethylene oxide (see p. 94). Ring closure of the resulting 1,1,2,3,3-pentamethyl-5-(β-hydroxyisopropyl)indane is accomplished with paraformaldehyde and a lower aliphatic alcohol via the acetal [192] or with paraformaldehyde and a carboxylic acid anhydride via the acylate [193].

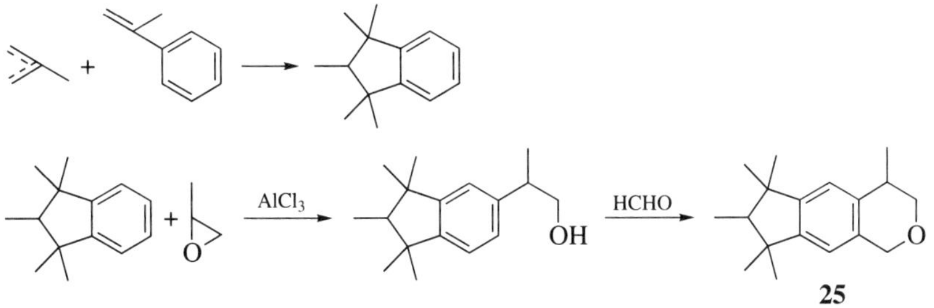

25

The commercial product is diluted with solvents (e.g., diethyl phthalate, isopropyl myristate, benzyl benzoate) to make it less viscous. It is alkali-stable and does not discolor in light. Therefore, it is a popular ingredient of perfume compositions for soap, detergents, and cosmetics and is used in large amounts.
FCT 1976 (**14**) p. 793.

Trade Names. Abbalide (BBA), Galaxolide (IFF), Musk 50 (Agan), Pearlide (Kao), Polalide (Polarome).

Phenylacetaldehyde glycerine acetal [*29895-73-6*], 2-benzyl-1,3-dioxolan-4-methanol

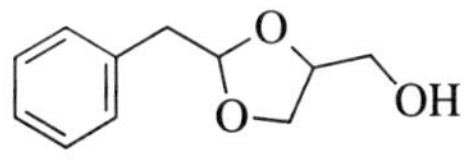

$C_{11}H_{14}O_3$, M_r 194.23, is not found in nature. It is a colorless slightly viscous liquid, d_4^{20} 1.154–1.162, n_D^{20} 1.529–1.534, with a tenacious honey, cyclamen, and rose note. It is prepared by acetalization of phenylacetaldehyde with glycerine.

It is used for tenacious compositions with green and broom character, e.g., for detergents.

FCT 1976 (**14**) p. 829.

Trade Name. Acetal CD (Giv.-Roure).

2,4,6-Trimethyl-4-phenyl-1,3-dioxane [*5182-36-5*]

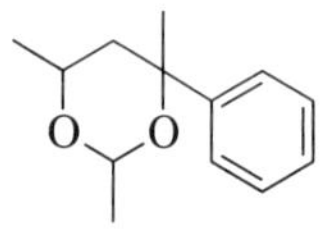

$C_{13}H_{18}O_2$, M_r 206.29, does not occur in nature. It is a colorless to pale yellow liquid, d_4^{20} 1.018–1.023, n_D^{20} 1.501–1.506, with a herbal-fresh odor, reminiscent of grapefruit. It is prepared by a Prins reaction of α-methylstyrene with acetaldehyde and used in perfume compositions for soap, detergents, and household products.

Trade Names. Floropal (H&R), Vertacetal (Dragoco).

4,4a,5,9b-Tetrahydroindeno[1,2-d]-*m*-dioxin [*18096-62-3*]

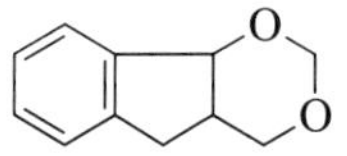

$C_{11}H_{12}O_2$, M_r 176.21, $bp_{1.3\,kPa}$ 134 °C, $d_{15.5}^{15.5}$ 1.159, n_D^{20} 1.559, forms crystals (*mp* 35–36 °C) with an indole-like odor. It is prepared by a Prins reaction from indene and formaldehyde in the presence of dilute sulfuric acid [194]. It is used in perfumes for soap and detergents.

Trade Names. Indoflor (H&R), Indolal (Dragoco), Indolarome (IFF).

The 2,4-dimethyl homologue [*27606-09-3*] is prepared by using acetaldehyde instead of formaldehyde. Its odor is reminiscent of geranium and magnolia [195].

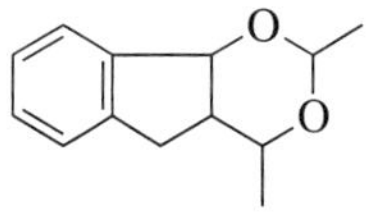

Trade Name. Magnolan (H&R).

Ethyl 2-methyl-1,3-dioxolane-2-acetate [*6413-10-1*], **2-methyl-1,3-dioxolane-2-acetic acid ethyl ester**

$CH_2COOCH_2CH_3$

$C_8H_{14}O_4$, M_r 174.20, is not found in nature. It is a colorless liquid, d_4^{20} 1.084–1.092, n_D^{20} 1.431–1.435, with a strong, fruity, apple-like, slightly green odor. It can be prepared by acetalization of ethyl acetoacetate with ethyleneglycol.

It is used in perfume bases for soap, toiletries, and detergents.
FCT 1988 (**26**) p. 315.

Trade Names. Applinal (Quest), Fructone (IFF), Jasmaprunat (H&R).

2.7.2 Lactones

Naturally occurring organoleptically important lactones are mainly saturated and unsaturated γ- and δ-lactones, and to a lesser extent macrocyclic lactones. The occurrence of these types of lactones reflects their ready formation from natural acyclic precursors.

The lactones are the intramolecular esters of the corresponding hydroxy fatty acids. They contribute to the aroma of butter and various fruits. 15-Pentadeca-nolide is responsible for the musklike odor of angelica root oil. Of the naturally occurring bicyclic lactones, phthalides are responsible for the odor of celery root oil, and coumarin for woodruff.

The macrocyclic esters hold a special position among the industrially produced lactone fragrance compounds. Like the well-known macrocyclic ketones, they have outstanding odor properties as musks. However, the lactones can be prepared more easily than the ketones, for example, by depolymerization of the corresponding linear polyesters. Since replacement of a methylene unit by oxygen affects the odor of these compounds very little, oxalactones with 15–17-membered rings are commercially produced in addition to 15-pentadecanolide. Several cyclic diesters prepared from long-chain α,ω-dicarboxylic acids and glycols are also valuable musk fragrances.

The γ-lactones described below can be prepared in good yield in a one-step process by radical addition of primary fatty alcohols to acrylic acid, using di-*tert*-butyl peroxide as a catalyst. A patent claims a high yield when the reaction is carried out in the presence of alkali phosphates or alkali sulfates [196].

$$H_3C(CH_2)_nCH_2\underset{OH}{\overset{}{\diagdown}} + \underset{HO^{\diagup}}{\overset{H_2C=CH}{\underset{CO}{|}}} \longrightarrow H_3C(CH_2)_n-HC\underset{O}{\overset{H_2C-CH_2}{\diagdown \ |}}CO$$

γ-Lactones

Because of the demand for natural γ- and δ-lactones in the flavor industry biosynthetic processes have been developed for their production [197].

γ-Octalactone [104-50-7], (n = 3)

$C_8H_{14}O_2$, M_r 142.20, $bp_{1.3\,kPa}$ 116–117 °C, d_4^{20} 0.977, n_D^{25} 1.4420, occurs as an aroma constituent in many processed and unprocessed foods. It is a pale yellow liquid with a fruity-coconut-like odor and is used both in aroma compositions and in heavy blossom perfumes.
FCT 1976 (14) p. 821.

γ-Nonalactone [104-61-0], (n = 4), so-called 'aldehyde C_{18}'

$C_9H_{16}O_2$ M_r 156.22, $bp_{1.7\,kPa}$ 136 °C, d_4^{20} 0.9676, n_D^{20} 1.446, occurs in many foods and is a pale yellow liquid with a coconut-like aroma. It has numerous applications, similar to those of γ-octalactone, in aroma compositions and perfumery.
FCT 1975 (13) p. 889.

   Trade Names. Abricolin (H&R), Prunolide (Giv.-Roure).

γ-Decalactone [706-14-9], (n = 5)

$C_{10}H_{18}O_2$, M_r 170.25, $bp_{2.3\,kPa}$ 156 °C, d_4^{21} 0.952, $n_D^{19.5}$ 1.4508, is present in a wide variety of foods and is an almost colorless liquid with an intensely fruity odor, reminiscent of peaches. It is used in perfumery for heavy, fruity flower odors and in aroma compositions, particularly peach flavors.
FCT 1976 (14) p. 741.

γ-Undecalactone [104-67-6], (n = 6), so-called 'aldehyde C_{14}'

$C_{11}H_{20}O_2$, M_r 184.28, $bp_{2.0\,kPa}$ 167–169 °C, d_4^{20} 0.944, n_D^{20} 1.4514, occurs in foods and is an almost colorless liquid with a peachlike odor. In addition to preparation by radical addition of l-octanol to acrylic acid, γ-undecalactone is also prepared by intramolecular cyclization of 10-undecylenic acid with 70–80% sulfuric acid with migration of the double bond.

   γ-Undecalactone has many applications in perfume and aroma compositions, similar to those of γ-decalactone.
FCT 1975 (13) p. 921.

   Trade Name. Peach pure (Giv.-Roure).

β-Methyl-γ-octalactone [39212-23-2], whiskey lactone

$C_9H_{16}O_2$, M_r 156.23, $bp_{0.5\,kPa}$ 96 °C, d_{20}^{20} 0.961–0.971, n_D^{20} 1.443–1.449, is found as its *cis-* and *trans*-isomers in whisky and in oakwood volatiles, the *cis*-isomer as the more important in sensory terms. It is a clear, almost colorless liquid with an intense, warm, sweet, coumarin-like odor.

A *cis–trans*-mixture can be prepared by radical addition reaction of pentanal with crotonic acid followed by reductive cyclization of the resulting γ-oxo acid with sodium boron hydride/sulfuric acid [198]. It is used in aroma compositions, e.g., for beverages.

Trade Name. PFW[®] Methyl octalactone (PFW).

4,5-Dimethyl-2(5*H*)-furanone [*28664-35-9*], **sotolone**

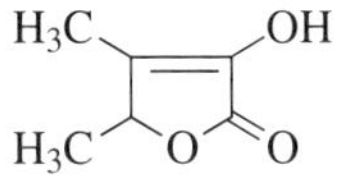

$C_6H_8O_3$, M_r 128.13, was found in, e.g., fenugreek, coffee, sake, and flor-sherry. Its aroma characteristic changes from caramel-like at low concentrations to curry-like at high concentrations. A method described for its preparation comprises condensation of ethyl propionate with diethyl oxalate and reaction of the intermediately formed diethyl oxalylpropionate with acetaldehyde. Acidic decarboxylation of the ethyl 4,5-dimethyl-2,3-dioxodihydrofuran-4-carboxylate gives the title compound [199].

It is used in food flavoring.

5-Ethyl-3-hydroxy-4-methyl-2(5)-furanone [*698-10-2*], **abhexone**

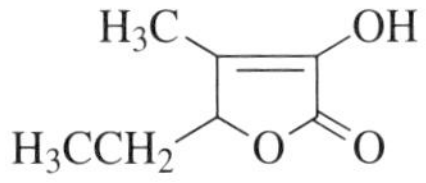

$C_7H_{10}O_3$, M_r 142.16, has been found in, e.g., lovage and roast coffee. It has a bouillon-like, coffee and lovage aroma, depending on its concentration. It can be synthesized by condensation of 2-oxobutanoic acid (from acrylic aldehyde with acetic anhydride and sodium cyanide, followed by reaction with methanol/hydrochloric acid) with methanol/sodium methylate and subsequent decarbomethoxylation [200]. It is used for aromatization of food.

δ-Decalactone [*705-86-2*]

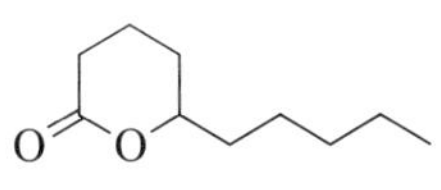

$C_{10}H_{18}O_2$, M_r 170.25, $bp_{3\,Pa}$ 117–120 °C, $d_4^{27.5}$ 0.9540, n_D^{26} 1.4537, is a flavor constituent of many types of fruit, cheese, and other dairy products. It is a colorless, viscous liquid with a creamy-coconut, peachlike aroma.

δ-Decalactone can be prepared by peracid oxidation of 2-pentylcyclopentanone. It is used in perfumes and for cream and butter flavorings.
FCT 1976 (**14**) p. 739.

Tetrahydro-6-(3-pentenyl)-2*H*-pyran-2-one [*32764-98-0*]

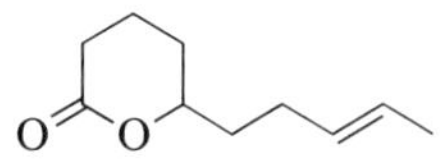

$C_{10}H_{16}O_2$, M_r 168.24, d_4^{20} 0.995–1.005, n_D^{20} 1.475–1.480, does not occur in nature. It is a colorless liquid with an oily-fruity odor with floral, petallike notes. For synthesis, 1,3-cyclohexanedione is reacted with crotyl bromide in the presence of potassium hydroxyde to give the 2-alkenyl-substituted 1,3-diketone. Ring cleavage with sodium hydroxyde leads to the unsaturated keto acid which is reduced with $NaBH_4$ under formation of the title compound [201].

It is used in compositions with fantasy notes for toiletries.

Trade Name. Jasmolactone (Firmenich).

15-Pentadecanolide [*106-02-5*], **15-hydroxypentadecanoic acid lactone**

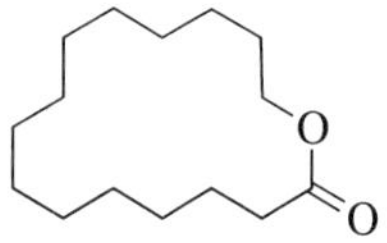

$C_{15}H_{28}O_2$, M_r 240.39, $bp_{1.3–1.4\,kPa}$ 169 °C, d_4^{40} 0.940, occurs in small quantities in, for example, angelica root oil. It forms colorless crystals (*mp* 37–38 °C) with a delicate, musklike odor.

Production.   The main industrial syntheses start from compounds produced from cyclododecatriene: either by ring expansion of cyclododecanone or by depolymerization of polyesters of 15-hydroxypentadecanoic acid (from 1,12-dodecanediol).

1. *Preparation by Ring Expansion of Cyclododecanone.* Radical addition of allyl alcohol to cyclododecanone, for example, with di-*tert*-butyl peroxide as a radical initiator, yields 2-(γ-hydroxypropyl)cyclododecanone. This is converted into 13-oxabicyclo[10.4.0]hexadec-1(12)-ene by acid-catalyzed dehydration [202]. Addition of hydrogen peroxide, in the presence of sulfuric acid, gives 12-hydroperoxy-13-oxabicyclo[10.4.0]hexadecane. Cleavage of the peroxide by

heating in xylene gives 15-pentadecanolide as well as a small amount of 15-pentadec-11(and 12)-enolide and 12-hydroxy-15-penta-decanolide [203].

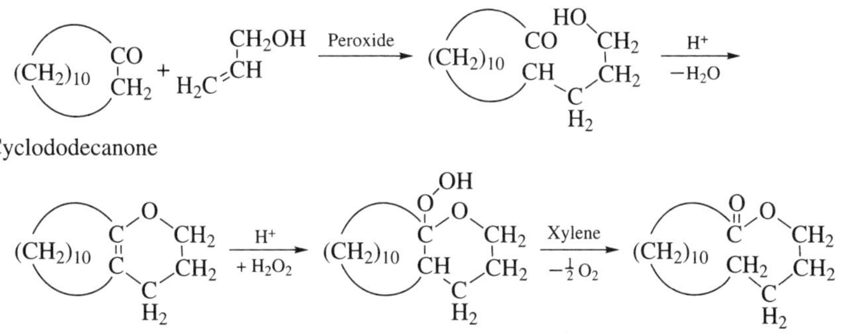

Cyclododecanone

In another process, (**27**) is reacted with isopropyl nitrite to give 12-oxo-15-pentadecanolide, which is hydrogenated to the corresponding hydroxy compound. The hydroxy group is derivatized and the reaction product pyrolyzed to the unsaturated lactone. Hydrogenation of the latter yields 15-pentadeca-nolide [204].

2. *Preparation from Polyesters of 15-Hydroxypentadecanoic Acid.* In a Japanese process, the required ω-hydroxy acid is prepared from 1,12-dodecanediol in several steps. The diol is added to methyl acrylate in a radical reaction, using di-*tert*-butyl peroxide as a catalyst. The free hydroxyl group in the resulting ω-hydroxy-γ-pentadecalactone is acetylated with acetic anhydride, and the resulting ω-acetoxy-γ-pentadecalactone is converted into 15-hydroxypentade-canoic acid by hydrogenolysis and hydrolysis [205].

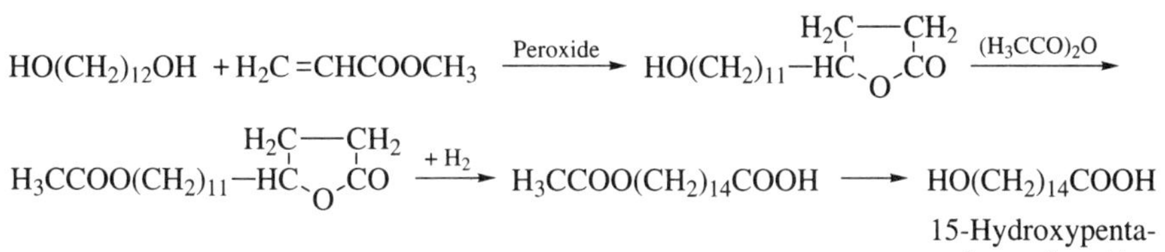

The polyester of 15-hydroxypentadecanoic acid is prepared by customary methods and is cleaved under high vacuum in the presence of transesterifica-tion catalysts.

Uses. 15-Pentadecanolide is a highly valuable fragrance material that is used in fairly large amounts in fine fragrances as a fixative with a delicate musk odor. FCT 1975 (**13**) p. 787.

Trade Names. Cyclopentadecanolid (H&R), Exaltolide (Firmenich), Pentalide (Soda Aromatic), Thibetolide (Giv.-Route).

9-Hexadecen-16-olide [*28645-51-4*], oxacycloheptadec-10-en-2-one

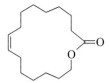

$C_{16}H_{28}O_2$, M_r 252.40, is not reported as being found in nature. It is a colorless to slightly yellow liquid, d_4^{20} 0.949–0.957, n_D^{20} 1.477–1.482, with an intense and powerful musk odor. It is prepared by treating aleuritic acid (9,10,16-trihydroxy-hexadecanoic acid) with trimethyl orthoformate to give a dioxolane derivative. Reaction with acetic anhydride yields ω-acetoxy 9-*trans*-hexadecenoic acid methyl ester. This is lactonized with potassium hydroxide to give the title compound [206].

It is used in perfumery for fine fragrances, highly appreciated for its diffusion and fixative properties.

Trade Name. Ambrettolide (Giv.-Roure), Ambrettolide (IFF).

12-Oxa-16-hexadecanolide [*6707-60-4*], 16-hydroxy-12-oxahexadecanoic acid lactone

$$O \overset{\diagup (CH_2)_m}{\underset{\diagdown (CH_2)_n}{\overset{\overset{\textstyle CO}{|}}{O}}}$$

($m = 10$, $n = 4$) $C_{15}H_{28}O_3$, M_r 256.38, does not occur in nature. Its odor is comparable to that of 15-pentadecanolide, but less intense. It is prepared by reacting methyl 11-bromoundecanoate with the monosodium salt of 1,4-butanediol. The resulting methyl 16-hydroxy-12-oxapalmitate is condensed to the corresponding polyester, which is subsequently depolymerized.

$$HO(CH_2)_4ONa + Br(CH_2)_{10}COOCH_3 \longrightarrow HO(CH_2)_4O(CH_2)_{10}COOCH_3$$

$$\longrightarrow \text{Polyester} \longrightarrow O \overset{\diagup (CH_2)_{10}}{\underset{\diagdown (CH_2)_4}{\overset{\overset{\textstyle CO}{|}}{O}}}$$

12-Oxa-16-hexa-
decanolide

12-Oxa-16-hexadecanolide as well as the stronger smelling 11-oxa ($m = 9$, $n = 5$) [*3391-83-1*], and 10-oxa ($m = 8$, $n = 6$) [*1725-01-5*] isomers which are obtained in the same way from the corresponding hydroxy-oxa acids, are used as substitutes for 15-pentadecanolide in fine fragrances.

FCT 1982 (**20**) p. 789: 12-Oxa-.
    1982 (**20**) p. 787: 11-Oxa-.
    1992 (**30**) p. 99 S: 10-Oxa-.

Trade Names. 12-Oxa-16-hexadecanolide = Cervolide (Quest), Musk 781 (IFF); 11-Oxa-16-hexadecanolide = Musk R 1 (Quest); 10-Oxa-16-hexadecanolide = Oxalide (Takasago).

1,12-Dodecanedioic acid ethylene ester [*54982-83-1*], (*n* = 10)

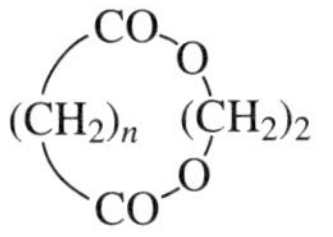

$C_{14}H_{24}O_4$, M_r 256.34, *mp* 18 °C, $bp_{2.7\,kPa}$ 139–141 °C, d_4^{60} 1.0303, n_D^{20} 1.4588, is a synthetic musk that is prepared by thermal depolymerization of the polyester obtained from 1,12-dodecanedioic acid and ethylene glycol in the presence of a catalyst (e.g., stannous salts of aliphatic monocarboxylic acids) [207].

The compound is used in perfumery as a musk fragrance, but is not as long-lasting as the following homologous compound, ethylene brassylate.

Trade Name. Arova 16 (Hüls).

1,13-Tridecanedioic acid ethylene ester [*105-95-3*], **ethylene brassylate** (*n* = 11)
$C_{15}H_{26}O_4$, M_r 270.37, $bp_{1.3\,kPa}$ 140 °C, d_4^{20} 1.0180, n_D^{20} 1.4702, is an artificial fragrance compound, with a sweetish, slightly fatty, musk odor. Like the previous compound, the ester is obtained by depolymerization of the corresponding polyester. Brassylic acid (1,13-tridecanedioic acid) is prepared by ozonolysis of erucic acid [208]:

$$H_3C(CH_2)_7CH{=}CH(CH_2)_{11}COOH \xrightarrow{O_3} H_3C(CH_2)_7COOH + HOOC(CH_2)_{11}COOH$$

     Erucic acid   Brassylic acid

Ethylene brassylate is used in large amounts in perfumery as a fixative and for rounding off sweet-flowery odor notes.

FCT 1975 (**13**) p. 91

Trade Name. Musk T (Takasago).

Coumarin [*91-64-5*], **2*H*-1-benzopyran-2-one**

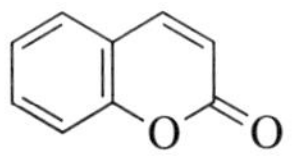

$C_9H_6O_2$, M_r 146.15, $bp_{1.33\,kPa}$ 153.9 °C, d_4^{20} 0.935, occurs widely in nature and determines, for example, the odor of woodruff. It forms white crystals (*mp* 70.6 °C) with a haylike, spicy odor. When treated with dilute alkali, coumarin is hydrolized to the corresponding coumarinic acid salt (*cis*-2-hydroxycinnamic acid). Heating with concentrated alkali or with sodium ethanolate in ethanol results in the formation of *o*-coumaric acid salts (*trans*-2-hydroxycinnamic acid). 3,4-Dihydro-coumarin is obtained by catalytic hydrogenation, for example, with Raney nickel as a catalyst; octahydrocoumarin is obtained if hydrogenation is carried out at high temperature (200–250 °C).

Production.   Coumarin is currently produced by Perkin synthesis from salicyl-aldehyde. In the presence of sodium acetate, salicylaldehyde reacts with acetic anhydride to produce coumarin and acetic acid. The reaction is carried out in the liquid phase at elevated temperature.

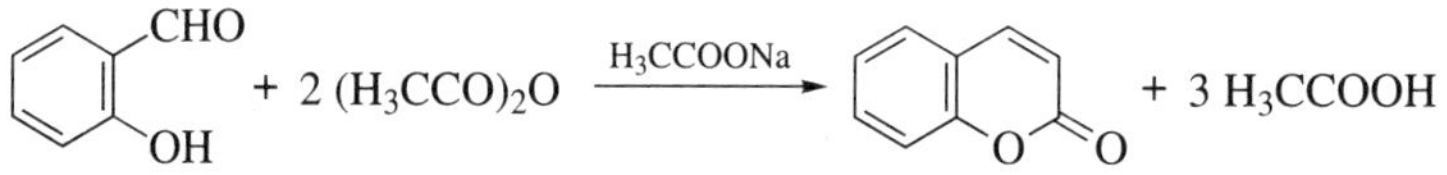

In a special process, the sodium acetate catalyst is retained in the reactor by a built-in filter and is reused [209].

Since the odor of coumarin is relatively weak, strong-smelling byproducts (e.g., vinylphenol) must be removed. Many purification methods have been reported and patented.

Uses.   Coumarin is one of the most widely used fragrance compounds. It is used in fine fragrances as well as in soap perfumes for spicy green notes. It is also used in galvanization as a brightener.

FCT 1974 (**12**) p. 385.

Dihydrocoumarin [*119-84-6*], **3,4-dihydro-2*H*-benzopyran-2-one**

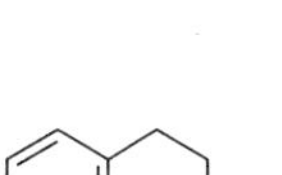

$C_9H_8O_2$ M_r 148.16, $bp_{2.3\,kPa}$ 145 °C, n_D^{25} 1.5528, occurs in a few essential oils and forms colorless crystals (*mp* 24 °C) with a sweet-herbal odor. Dihydrocoumarin is prepared by hydrogenation of coumarin, for example, in the presence of a Raney nickel catalyst. Another process employs the vapor-phase dehydrogenation of hexahydrocoumarin in the presence of Pd or Pt–Al_2O_3 catalysts [210]. Hexahydro-coumarin is prepared by cyanoethylation of cyclohexanone and hydrolysis of the nitrile group, followed by ring closure to the lactone [211].

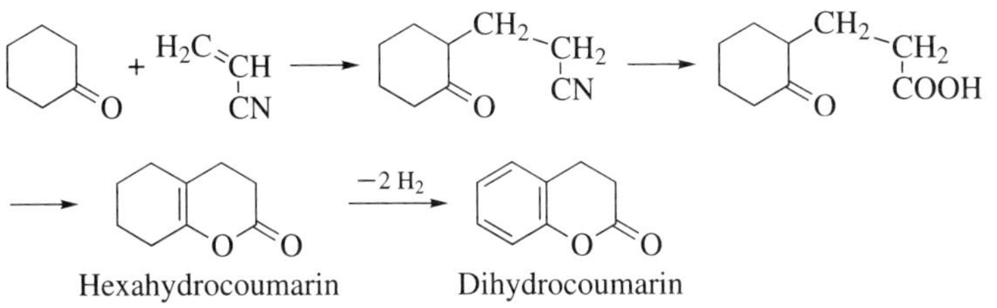

Hexahydrocoumarin   Dihydrocoumarin

Dihydrocoumarin is used in woodruff-type flavor compositions.
FCT 1974 (**12**) p. 521.

2.7.3 Glycidates

A number of glycidates are important intermediates in the synthesis of fragrance compounds. A few glycidates are fragrance compounds in themselves. They are prepared either by epoxidation of the corresponding acrylates or by condensa-tion of aldehydes or ketones with α-chloro substituted fatty acid esters (Darzens reaction).

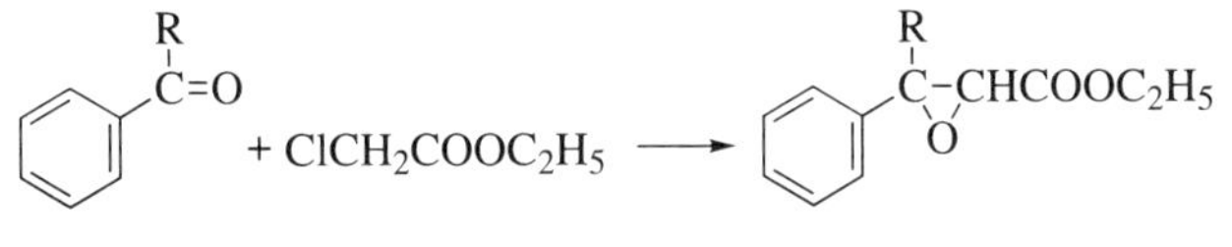

Ethyl 3-phenylglycidate [*121-39-1*], **so-called 'aldehyde C$_{16}$ special'**

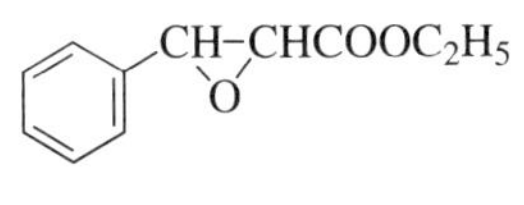

$C_{11}H_{12}O_3$, M_r 192.21, $bp_{0.04\,kPa}$ 104 °C, d^{20} 1.1023, n_D^{30} 1.5095, is a colorless liquid with a strawberry-like odor; it is not known to occur in nature.

It is prepared by treating ethyl cinnamate with peracetic acid [212] or by conden-sation of benzaldehyde with ethyl chloroacetate (in the above Darzens reaction, R = H). The glycidate is used as a long-lasting fragrance compound for creating harmonic, fruity notes in household and fine fragrances.
FCT 1975 (**13**) p. 101

Ethyl 3-methyl-3-phenylglycidate [*77-83-8*], **so-called 'aldehyde C₁₆'**

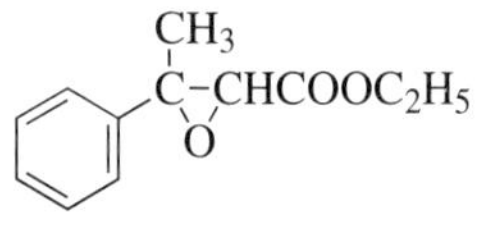

$C_{12}H_{14}O_3$, M_r 206.24, $bp_{2.4\,kPa}$ 153–155 °C, d_{25}^{25} 1.506–1.513, occurs in two optically active pairs of *cis* and *trans* isomers; each isomer has a characteristic odor [213]. The commercial product is a racemic mixture of all four isomers and has a strong, sweetish, strawberry odor. The *cis* : *trans* ratio obtained in the Darzens condensation of acetophenone (R = CH₃) and ethyl chloroacetate depends on the base used in the reaction.

   The glycidate is used in household perfumery for fruity notes.
FCT 1975 (**13**) p. 95.

   Trade Name. Strawberry pure (Giv.-Roure).

2.7.4 Miscellaneous Compounds

2-Furylmethanethiol [*98-02-2*], **furfuryl mercaptan**

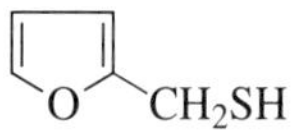

C_5H_6OS, M_r 114.16, $bp_{101.3\,kPa}$ 160 °C, d_4^{20} 1.1319, n_D^{20} 1.5329, is an important constituent of the aroma of roasted coffee. It is a liquid with an unpleasant odor, which becomes like coffee when diluted.

   Furfuryl mercaptan is prepared from furfuryl alcohol, thiourea, and hydrogen chloride. The resulting *S*-furfurylisothiouronium chloride is cleaved with sodium hydroxide to give furfuryl mercaptan.

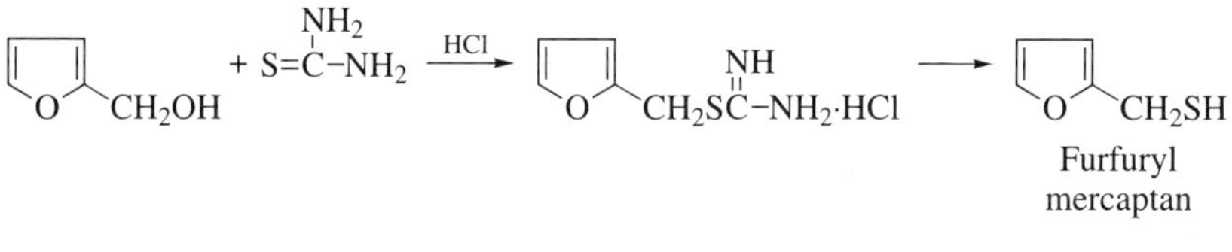

The thiol is used in coffee aromas.

2.8 N- and N,S-Heterocycles

Many nitrogen- and sulfur-containing heterocycles have been identified in the aroma fractions of foods [214]. In roasted products (e.g., coffee) and heat-treated foods (e.g., baked bread or fried meat), these heterocycles are formed from reducing sugars and simple or sulfur-containing amino acids by means of Maillard reactions [215], [216]. Their odor threshold values are often extremely low and even minute amounts may significantly contribute to the aroma quality of many products [217], [218]. Therefore, N- and N,S-heterocyclic fragrance and flavor substances are produced in far smaller quantities than most of the products previously described.

Pyrroles, indoles, pyridines, quinolines, and pyrazines are examples of N-hetero-cycles that are produced as fragrance and flavor compounds. Thiazoles and dithiazines are examples of nitrogen- and sulfur-containing heterocycles. These heterocyclic compounds are mainly used in aroma compositions, exceptions are indoles and quinolines, which are important fragrance substances.

Representatives of the above-mentioned classes are as follows:

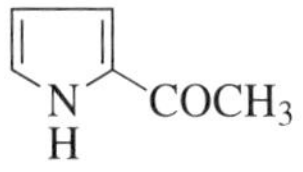

2-acetylpyrrole [*1072-83-9*]: roast odor

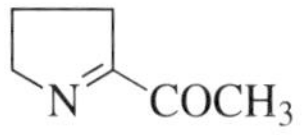

2-acetyl-3,4-dihydro-5*H*-pyrrole [*85213-22-5*]: charac-teristic odor of white bread crust

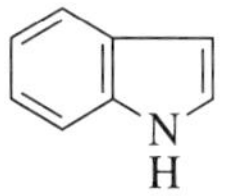

indole [*120-72-9*]: fecal odor, floral in high dilution

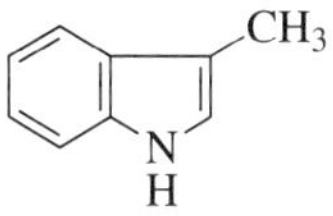

3-methylindole, skatole [*83-34-1*]: indole-like odor

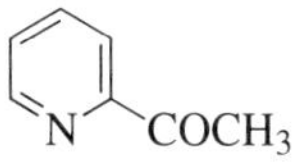

2-acetylpyridine [*1122-62-9*]: roast odor

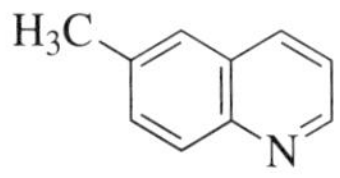

6-methylquinoline [*91-62-3*]: blossom odor, sweet-animalic upon dilution

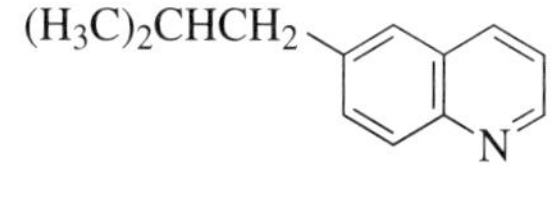

6-isobutylquinoline [*68141-26-4*]: mossy-earthy odor

2-acetylpyrazine [*22047-25-2*]: popcorn-like odor

2-methoxy-3-isopropylpyrazine [*25773-40-4*]: green pea odor

2-methoxy-3-isobutylpyrazine [*24683-00-9*]: green-pepper odor

2,3-dimethylpyrazine [*5910-89-4*] and its 2,5-[*123-32-0*] and 2,6-[*108-50-9*] isomers: roast odor, reminiscent of nuts

trimethylpyrazine [*14667-55-1*]: roast odor, reminiscent of coffee and cocoa

mixture of 3-ethyl-2,5-dimethylpyrazine [*13360-65-1*] and 2-ethyl-3,5-dimethylpyrazine [*13925-07-0*]: roast odor, reminiscent of nuts

5-methyl-6,7-dihydro[5*H*]cyclopentapyrazine [*23747-48-0*]: nutty, roast odor

2,5-dimethylthiazole [*4175-66-0*]: meat-like odor

2-isobutylthiazole [*18640-74-9*]: tomato odor

4-methyl-5-thiazolethanol [*137-00-8*]: meaty, roast odor

2-acetyl-2-thiazoline [*29926-41-8*]: cooked beef odor

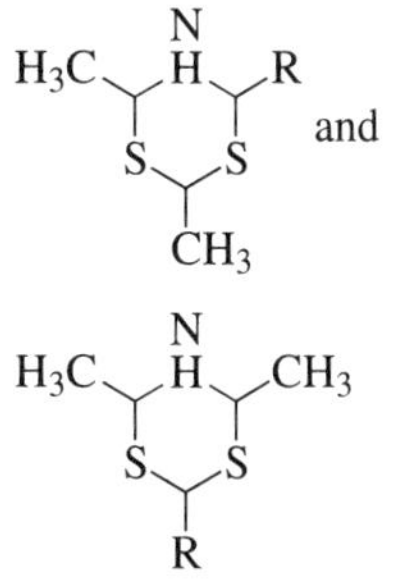

alkyldimethyl-1,3,5-dithiazines: roast odor

R = Methyl, isopropyl, isobutyl, 2-butyl

3 Natural Raw Materials in the Flavor and Fragrance Industry

3.1 Introduction

Although synthetic flavor and fragrance materials are produced on an industrial scale, naturally occurring raw materials continue to be essential, important ingredients in the manufacture of flavor and fragrance compositions for several reasons. First, the composition and organoleptic nature of natural products are often too complex to be reproduced by a combination of synthetic fragrance substances. Second, the characteristic flavor and fragrance substances of a particular product often cannot be synthesized at a competitive price. Third, the use of natural materials in the production of certain flavor compositions is compulsory. Moreover, an increasing demand for perfumes based on natural materials has been observed recently.

Currently, raw materials for the flavor and fragrance industry are obtained from more than 250 different plant species, but only a handful of products originate from animals.

Raw materials are isolated from various parts of plants, e.g., blossoms, buds, fruit, peel, seeds, leaves, bark, wood, roots, or from resinous exudates. Different parts of the same plant may yield products with different compositions. For instance, steam distillation of the bark of the cinnamon tree gives cinnamon bark oil, which contains mainly cinnamaldehyde, whereas cinnamon leaf oil obtained from the leaves of the tree contains eugenol as its major constituent.

The quality of natural products depends considerably on their geographic origin, even if they are isolated from the same plant species. This may be partly due to variations in cultivation conditions, such as soil structure and climate, but also results from the fact that different varieties of the same plant species are cultivated in different areas. Thus, more than 500 natural raw materials are available for the creation of perfumes and flavors.

The flavor and fragrance industry has expanded so much that the plants required to supply the raw materials are now grown on a very large scale. Examples are the peppermint and spearmint plantations in the United States, the lavandin plantations in southern France and the cornmint plantations in China and India.

The economic importance of the cultivation of aromatic plants has led to the systematic breeding of new varieties, which are obtained either by alternation of generations or by vegetative means in an attempt to improve yield, oil quality, and resistance to disease and insects.

The production of some essential oils has decreased to low levels or even been discontinued due to competition from synthetic products. Nevertheless, the worldwide production of flavor and fragrance materials of natural origin has increased recently due to breeding successes, but their total market share has decreased. Annual worldwide sales currently amount to 700–880 million US $ [219, 220]. The total amount of annually produced essential oils is estimated at 45.000 t [221]. More details of the economic importance of the individual essential oils are given in references [219–222]

Raw materials derived from intensive agricultural cultivation are usually relatively inexpensive. However, the prices of some natural materials may exceed $ 1000 per kilogram because cultivation and harvesting of these plants are tedious and product yields are very low. Examples of extremely valuable ingredients of fragrance and flavor creations include rose oil, jasmine absolute, tuberose absolute, orris root oil, ambrette seed oil, angelica root oil, and orange flower oil [220].

3.2 Isolation of Natural Fragrance and Flavor Concentrates

Three main methods are used to concentrate plant flavor and fragrance substances:

1. distillation

2. mechanical separation ('pressing')

3. solvent extraction

The qualitative and quantitative composition and, thus, the organoleptic properties of the product depend on the isolation procedure. For example, an extract contains large amounts of nonvolatile components that are not found in essential oils obtained by distillation. Since these components markedly influence odor development (complexing and fixing), the two products may have completely different organoleptic properties, even though the compositions of their odorous volatile constituents are comparable.

In addition, the distillation of essential oils at elevated temperature results in the transformation of thermolabile substances, and some typical components are only released from their precursors in the plants under distillation conditions.

Solvent extraction is generally applied in the separation of heat-labile plant materials or if an essential oil can be obtained only in very low yield (e.g., from blossoms). It is also used if the nonvolatile components are desired for their fixative properties (e.g., in the preparation of resinoids from exudates).

3.2.1 Essential Oils

Production.   Essential oils are obtained from plant materials by distillation with water or steam. After condensation of the vapor phase, the oil separates from the aqueous phase and is removed. The yield of essential oil, based on the starting plant material, generally ranges from a few tenths of 1% to a few percent. The apparatus used in the production of natural fragrance concentrates is described in [223].

Essential oils consist of volatile, lipophilic substances that are mainly hydrocarbons or monofunctional compounds derived from the metabolism of mono- and sesquiterpenes, phenylpropanoids, amino acids (lower mass aliphatic compounds), and fatty acids (long-chain aliphatic compounds). Unlike fatty oils, essential oils do not leave a grease stain when dabbed on filter paper.

Essential oils are to be distinguished from the so-called distillates which are ethanol-containing products that are obtained from plant materials by distillation with ethanol or with ethanol–water mixtures. *Essence oils* are defined as essential oils that separate from the aqueous phase in the distillation receiver during the distillative concentration of fruit juices (usually citrus juices).

Citrus peel oils are a special type of essential oil. They are isolated by pressing the peel to release the volatile substances stored in the pericarp in small oil glands. The resulting products are termed essential oils because they consist largely of highly volatile terpene hydrocarbons. However, they also contain small amounts of nonvolatile compounds, such as dyes, waxes, and furocoumarines.

Uses.   Most essential oils are used directly as starting materials in the production of flavor and fragrance compositions. However, some essential oils are fractionated or concentrated by distillation, partitioning, or adsorption. Substances that are important for the desired characteristic odor and taste are thus concentrated, and other components, which possess either an unpleasant or very faint odor or are unsuitable for the application in question, are removed.

Individual compounds can be isolated from essential oils containing one or only a few major components by distillation or crystallization. Examples are eugenol from clove oil, menthol from cornmint oil, citronellal from *Eucalyptus citriodora* oil and citral from *Litsea cubeba* oil. These compounds are used as such or serve as starting materials for the synthesis of derivatives, which are also used as flavor and fragrance substances. However, the importance of some of these oils has decreased substantially because of the development of selective synthetic processes for their components.

Although essential oils or their fractions are mixtures of many substances, these oils are occasionally converted as a whole into derivatives. Examples of such derivatives are vetiveryl acetate from vetiver oil, guaiyl acetate from guaiac wood oil, and acetyl cedrene from cedarwood terpenes. These products are also employed as fragrance substances.

3.2.2 Extracts

Extracts of fragrance and flavor substances obtained from plants are termed pomades, concretes, absolutes, resinoids, or tinctures according to their method of preparation.

Pomades consist of fats that contain fragrance substances and are produced by the hot or cold enfleurage of flowers. Hot enfleurage is the oldest known procedure for preserving plant fragrance compounds. In this method, flowers (or other parts of a plant) are directly immersed in liquid or molten wax.

In cold enfleurage, the volatile components released by flowers into their surroundings are absorbed with fats over a longer period of time. This industrial procedure was developed in southern France in the 19th century for the production of high-grade flower concentrates. It involves the application of fresh flowers to a fat layer, consisting of a mixture of specially refined lard and beef tallow, which is spread on a glass plate in a closed container. This method, however, has been almost totally replaced by the less tedious technique of solvent extraction.

Concretes are prepared by extracting fresh plant material with nonpolar solvents (e.g., toluene, hexane, petroleum ether). On evaporation, the resulting residue contains not only volatile fragrance materials, but also a large proportion of non-volatile substances including waxy compounds. For this reason, concretes (like pomades) are not completely soluble in alcohol and, thus, find limited use as perfume ingredients. However, they can be employed in the scenting of soaps.

Concretes, which are actually intermediate products (see below), are prepared mainly from flowers (rose, jasmine, tuberose, jonquil, ylang-ylang, mimosa, boronia, etc.), but also from other plant materials (lavender, lavandin, geranium, clary sage, violet leaves, oak moss, etc.). A yield of ca. 0.3% based on the starting flower material, is obtained in the production of jasmine concrete.

Absolutes are prepared by taking up concretes in ethanol. Compounds that precipitate on cooling are then removed by filtration. After evaporation of the ethanol, a wax-free residue called an absolute is left behind. Absolutes are completely soluble in ethanol and, therefore, can be freely used as perfume ingredients. They are usually formed in a yield of ca. 50%, based on the concrete as starting material.

In rare cases, absolutes can be obtained directly by extracting the plant material with alcohol (e.g., tonka absolute).

Resinoids are prepared by extracting plant exudates (balsams, oleo gum resins, natural oleo resins, and resinous products) with solvents such as methanol, ethanol, or toluene. Yields range from 50 to 95 %. The products are usually highly viscous and are sometimes diluted (e.g., with phthalates or benzyl benzoate) to improve their flow and processing properties.

Resinoids mainly consist of nonvolatile, resinous compounds and are primarily used for their excellent fixative properties.

The resinoids described above should be distinguished from prepared oleoresins (e.g., pepper, ginger, and vanilla oleoresins), which are concentrates prepared from spices by solvent extraction. The solvent that is used depends on the spice; currently, these products are often obtained by extraction with supercritical carbon dioxide. Pepper and ginger oleoresins contain not only volatile aroma compounds, but also substances responsible for pungency.

Tinctures are alcoholic solutions that are prepared by treating natural raw materials with ethanol or ethanol–water mixtures. They can also be obtained by dissolving other extracts in these solvents. Tinctures are sometimes called *infusions*.

3.3 Survey of Natural Raw Materials

The following survey of the most important, well-known raw materials used in the flavor and fragrance industry is by no means complete; the materials are listed in alphabetical order. Physical standards for essential oils are described as specified by the International Organization for Standardization (ISO), the Association Française de Normalisation (AFNOR), or the Essential Oil Association of the United States (EOA). Gas chromatograms are widely used for analysis and quality control, but have not been included due to lack of space. Further details are given in the literature, e.g., [224], and in ISO and AFNOR specifications, which now include gas chromatograms.

The enumeration of the components of the individual products has been limited to the main constituents and the odor determining compounds; further information is available in the literature, e.g., [214], [225–230]. Physical data for extracts or concentrates consisting largely of nonvolatile material are not given because the composition of these products varies widely according to the isolation and manufacturing procedure used.

The botanical names of plants are cited in accordance with the International Code of Botanical Nomenclature (ICBN) as described, for example, [231].

Allium oils are obtained from garlic and onion (Liliaceae). Their quality is assessed on the basis of their odor and aroma rather than their physical and chemical properties. The EOA specifications given below are, therefore, of limited value only.

1.   *Garlic oil* is obtained by steam distillation of crushed bulbs of the common garlic, *Allium sativum* L.; it is a clear, reddish-orange liquid, with a strong, pungent, characteristic garlic odor.

d_{25}^{25} 1.040–1.090, n_D^{20} 1.5590–1.5790 [232].

Diallyl disulfide [*2179-57-9*] is an essential odor component of garlic oil [233].

$CH_2 = CHCH_2SSCH_2CH = CH_2$
[*8000-78-0*].

2.   *Onion oil* is obtained by steam distillation of the crushed bulbs of the common onion, *Allium cepa* L. It is an amber-yellow to amber liquid with a strongly pungent, lasting, characteristic onion odor.

d_{25}^{25} 1.050–1.135; n_D^{20} 1.5495–1.5695 [234].

Aliphatic sulfur compounds, in particular disulfides such as methyl propyl disulfide, dipropyl disulfide, and especially *cis*- and *trans*-propenyl propenyl disulfide are mainly responsible for the typical odor of onion oil [235].

Garlic and onion oil are used in seasoning mixtures for the food industry [*8002-72-0*].

Allspice oil, see Piment Oils.

Ambergris (ambra), see Animal Secretions.

Ambrette seed oil is obtained by steam distillation of the dried, crushed seeds of *Ambelmoschus moschatus* Medik. (*Hibiscus abelmoschus* L., Malvaceae), a flowering shrub growing in tropical areas. Due to its content of long-chain fatty acids, the crude product is a waxy mass and, therefore, also called 'Ambrette beurre.' Removal of the fatty acids with alkali gives a clear yellow to amber liquid with the strong, musky odor of ambrettolide.

d_{25}^{25} 0.898–0.920; n_D^{20} 1.4680–1.4850; $\alpha_D^{20} - 2°30'$to $+3'$; acid number: max. 3; saponification number: 140–200 [236].

The constituents responsible for the musk odor of the oil are (*Z*)-7-hexadecen-16-olide [*123-69-3*] (ambrettolide) and (*Z*)-5-tetradecen-14-olide [*63958-52-1*]:

```
        (CH2)5                (CH2)3
   HCCO          HCCO
   HC      O           HC      O
      (CH2)8               (CH2)8
```
Ambrettolide   5-Tetradecen-14-olide

Other components are acyclic aliphatic esters and terpenes, such as farnesol and farnesyl acetate [237–239]. Ambrette seed oil is one of the most expensive essential oils and, thus, is used mainly in fine fragrances and in alcoholic beverages. FCT 1975 (**13**) p. 705; [*8015-62-1*].

Amyris oil is obtained by steam distillation of the wood from the tree *Amyris balsamifera* L. (Rutaceae), which grows in the Caribbean area and around the Gulf of Mexico. It is a pale yellow to amber, slightly viscous liquid with a mild wood odor.

d_{20}^{20} 0.946–0.978; n_D^{20} 1.505–1.510; α_D^{20} + 10° to +60°; solubility: 1 vol in 1 vol of 90% ethanol at 20 °C; solutions sometimes become opalescent on dilution; acid number: max. 3.0; ester number (after acetylation): 180–198 [240].

The oil is sometimes incorrectly called West Indian sandalwood oil. However, its composition and odor are different from those of the oils obtained from sandalwood species. The major components of amyris oil are sesquiterpenoids such as elemol [639-99-6], β-eudesmol [473-15-4], and *epi-γ*-eudesmol [15051-81-7] [241–244].

Elemol   β-Eudesmol   *epi-γ*-Eudesmol

Amyris oil is used in perfume compositions, mainly as a fixative [8015-65-4].

Angelica oil is prepared from *Angelica* roots or seeds.

1. *Angelica root oil* is obtained by steam distillation of the dried roots of *Angelica archangelica* L. [*Archangelica officinalis* (Moench) Hoffm.], a plant occurring predominantly in Europe (Apiaceae). The oil is a pale yellow to deep amber liquid with a green, herbaceous, peppery, musklike odor and a bittersweet taste.

d_{25}^{25} 0.850–0.880, occasionally up to 0.930 for oils from stored roots; n_D^{20} 1.4735–1.4870; α_D^{20} 0° to +46°; acid number: max. 7; ester number: 10–65; solubility: 1 vol in at least 1 vol of 90% ethanol, often with turbidity [245].

2. *Angelica seed oil* is similarly obtained from fresh seeds of the plant. It is a light yellow liquid with an odor that is sweeter and more delicate than that of the root oil.

d_{25}^{25} 0.853–0.876; n_D^{20} 1.4800–1.4880; α_D^{20} + 4° to +16°; acid number: max. 3; ester number: 14–32; solubility: 1 vol in at least 4 vol 90% ethanol, often with considerable turbidity [246].

Angelica root oil contains ca. 90% terpenoids and sesquiterpenoids. α-Pinene, 3-carene, limonene, and β-phellandrene are the major components (together ca. 60%). In addition, the oil contains a large number of oxygen-containing compounds, of which the macrolides 15-pentadecanolide and 13-tridecanolide are essential odor components [247–253]. The two angelica oils are used mainly in the alcoholic beverage industry, and in very low dosages also in fine fragrances. FCT 1974 (**12**) p. 821; [8015-64-3].

Animal secretions are of minor commercial importance because some of the animal species from which they are obtained are virtually extinct, and the killing quota for others has been sharply reduced. Few odoriferous secretions of mammals have actually been shown to possess pheromone-like properties linked with reproduction. However, the odor of animal secretions is known to be important for communication and behavior of a particular species. Most of the products described below contain strong-smelling compounds with relatively high molecular masses. Therefore, they are used as long-lasting fragrance complexes. Most of their odoriferous constituents are now produced synthetically and are used for the same purposes. Therefore, these fragrance materials are mentioned only for reasons of history.

1.   *Ambergris (ambra)* is a secretion of the sperm whale *Physeter catodon* (*P. macrocephalus* L.), that possibly results from a pathological condition. Ambergris has a lower density than water and washes ashore along the ocean coasts. The major quantity is obtained from killed animals, but only a low percentage contain ambergris in their intestines.

Fresh ambergris is almost black, but it turns light gray and develops a pleasant odor when exposed to light and seawater over a period of time. The major components of ambergris are epicoprosterol and the odorless triterpene alcohol, ambrein [*473-03-0*].

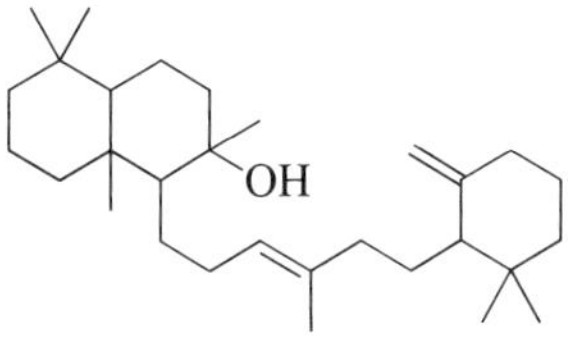

Ambrein

Ambrein is the likely precursor of a number of strongly odoriferous mono-, bi-, and tricyclic compounds that are formed by autoxidation or photooxidation [254, 255]. Examples are as follows:

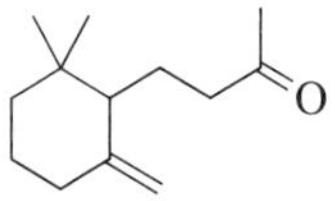

γ-Dihydroionone
[*13720-12-2*]
(odor: tobacco)

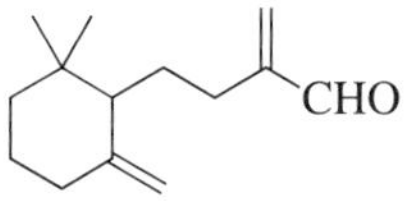

2-Methylene-4-(2,2-dimethyl-
6-methylenecyclohexyl)butanal
[*72892-63-8*]
(odor: seawater)

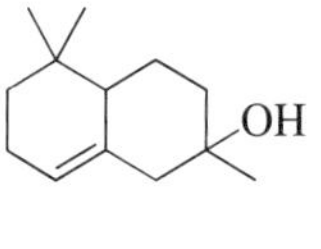

α-Ambrinol
[*41199-19-3*]
(odor: moldy, animal, fecal)

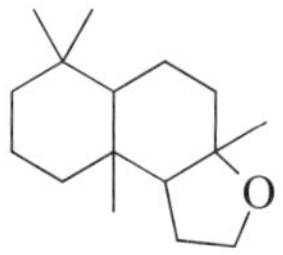

3a,6,6,9a-Tetramethyldodeca-
hydronaphtho[2,1-b]furan
[*6790-58-5*]
(odor: ambergris)

Together, these compounds largely represent the odor of ambergris. Ambergris is applied as a 3% tincture in 90% ethanol, which is matured by standing over a period of time with occasional shaking. It was used in fine fragrances. FCT 1976 (**14**) p. 675.

2. *Beeswax absolute* is obtained by alcohol extraction of beeswax. The yield is generally less than 1%. The yellowish-brown viscous product has a mild, honey-like odor and high tenacity; it is used almost exclusively in fine fragrances. FCT 1976 (**14**) p. 691.

3. *Castoreum* is an unpleasantly sharp-smelling, oily substance secreted by special glands of beavers, *Castor fiber* L. (Castoridae), living in Canada, Alaska, and Siberia. Both sexes secrete the substance, which accumulates in an abdominal pouch, also called castoreum. Dilute castoreum (e.g., as a tincture in ethanol) smells pleasantly of birch tar and musk and is slightly fruity.

Castoreum is a by-product of the fur industry. The beaver pouches are dried in the air or over a wood fire, the color of their contents then changes from yellow to dark brown, and the consistency from a butter-like to resinous character.

In addition to alcoholic tinctures, castoreum is available in the form of resinoids, which are prepared by extracting dried, comminuted pouches with suitable solvents.

The intense, for the Siberian beaver leathery, odor of castoreum is caused largely by phenolic compounds (e.g., 4-alkylphenols and catechol derivatives [256]), which beavers take in with their food and excrete into their abdominal pouch. Castoreum was used mainly in fine fragrances for its characteristic, long-lasting odor, particularly for delicate leather nuances. FCT 1973 (**11**) p. 1061; [*8023-83-4*].

4. *Civet* is a glandular secretion produced by both sexes of the civet cat (Viverridae). Two species are known: *Civetticitis civetta* that inhabits Ethiopia, and *Viverra zibetha* that is found in India and southeast Asia.

The animals are kept in cages, and the fresh secretion is taken from the pouches at regular (about one-week) intervals. Civet is almost liquid with a light yellow color. It darkens when exposed to light and takes on a consistency like salve. In dilutions (e.g., as an alcoholic tincture), civet has a pleasant, sweetish odor. A resinoid prepared by extraction with acetone is a dark brown-red mass [257]. Civetone (see p. 84) is the main odoriferous constituent of civet (2.5–3.4%). Civet contains other macrocyclic ketones such as cyclohexa- and cycloheptadecanone and 6-*cis*-cycloheptadecenone. Traces of indole and skatole contribute to the animal note [258].

Civet has a distinctly different odor from musk and was formerly a versatile ingredient of fine fragrances. FCT 1974 (**12**) p. 863; [*68991-27-5*].

5. *Musk* is secreted exclusively by the male animals of *Moschus moschiferus*, a wild deer living in the mountains of Nepal, Tibet, and Mongolia. The light yellow

secretion with a salve consistency accumulates in an abdominal pouch and probably serves to attract females. When the pouch is dried, the secretion solidifies to form a brittle, brown mass with a characteristic odor. Since several *Moschus* species occur, large variations exist in the quality and specifications of musk. Hunting of the animals has been prohibited; therefore, only small quantities of musk are occasionally offered at extremely high prices [259]. In the USA and Europe, natural musk is no longer used as a fragrance ingredient.

Muscone and related macrocyclic ketones are responsible for the odor of musk [260, 261]. Like other animal secretions, musk was preferentially used as an alcoholic tincture in fine fragrances. FCT 1983 (**21**) p. 865; [*68991-41-3*].

Anise oil is obtained by steam distillation of the fruits of *Pimpinella anisum* L. (Apiaceae). It is a colorless to pale yellow liquid or crystalline mass with a powerful, sweet odor, characteristic of anethole.

d_{20}^{20} 0.980–0.990; n_D^{20} 1.552–1.559; α_D^{20} −2° to +2°; solubility: 1 vol in 3 vol of 90% ethanol at 20 °C; *fp* 15–19.5 °C [262].

The main component of anise oil is *trans*-anethole, which may be present at a concentration >95% and which determines the melting point of the oil [263–266].

The oil was formerly produced in many countries, mainly in eastern Europe, but has now been replaced, to a large extent, by the less expensive star anise and fennel oils which also contain a high percentage of anethole.

Anise oil is used for flavoring foods, beverages, and oral care products. FCT 1973 (**11**) p. 865; [*8007-70-3*].

Artemisia oil (Armoise oil) is obtained by steam distillation of the herb *Artemisia herba-alba* Asso (Asteraceae) that grows in Morocco. It is a light yellow to yellow liquid with a light herbaceous odor characteristic of thujone.

d_{25}^{25} 0.917–0.935; n_D^{20} 1.4600–1.4720; α_D^{20} −25° to −8°.

The major components of artemisia oil (Marrakesh-type) are the ketones camphor (40%) and α- and β-thujone (see p. 211) (35% and 5% respectively) [267–270]. Since *Artemisia herba-alba* exists as various chemotypes, the composition of the oil may vary widely. For example, each of the above-mentioned ketones may occur in a concentration over 50% or less than 10%.

Artemisia oil is used in fairly large amounts in fine fragrances (e.g., for chypre notes). FCT 1975 (**13**) p. 719; [*8022-37-5*].

Basilicum

Basil oil is available in several types that differ in their major components; the most important are described.

1. *Basic oil, methylchavicol-type (Réunion type, exotic type)* is obtained by steam distillation of the flowering tops or whole plants of *Ocimum basilicum* L. (Lamiaceae). This oil is produced mainly in Réunion, the Comores, Madagascar, and the Seychelles. It is a light yellow liquid with a fresh, green, spicy odor characteristic of methylchavicol (estragole) [*140-67-0*].

d_{20}^{20}0.948–0.970; n_D^{20} 1.5100–1.5200; α_D^{20} −1° to +2°; solubility: 1 vol in max. 8 vol. 80% ethanol; content by GC: methylchavicol 75–85%; linalool 0.5–3% [271].

$$CH_3O-\langle\bigcirc\rangle-CH_2CH=CH_2$$

Methylchavicol (estragole)

Annual production (ca. 10 t) is used predominantly for seasoning foods.

2. *Basil oil, linalool-type* (*European type*; *mediterranean type*) is produced mainly in the mediterranean area (France, Egypt). It is light yellow to amber-colored oil with typical fresh-spicy odor.
d_{20}^{20} 0.895–0.920; n_D^{20} 1.4750–1.4950; α_D^{20} −2° to −14°; content by GC: linalool 45–62%; methylchavicol trace to 30%; eugenol 2–15% [272].
It is used for food flavoring and in perfumery. For further constituents of these two basil oil types see [273–279].

3. *Indian Basil* oil is produced exclusively in India. It contains ca. 70% methylchavicol and 25% linalool. It is used mainly for the isolation of the pure compounds. Methylchavicol is used as starting material for the production of anethole.
FCT 1973 (**11**) p. 867; [*8015-73-4*].

Lorbeerbaum

Bay oil is obtained by steam distillation of the leaves of *Pimenta racemosa* (Miller) Moore (Myrtaceae). It is a dark brown liquid with a strong, spicy, clovelike odor.
d_{20}^{20} 0.943–0.984; n_D^{20} 1.505–1.517; phenols content: min. 50% [280].
Evergreen bay trees or bay-rum trees, which are up to 12 m high, grow wild and are also cultivated in northern South America and in the West Indies. The main cultivation area is the island of Dominica.
The major components of the oil are myrcene, eugenol, and chavicol [281–283]. The phenol content is determined largely by the latter two compounds.
Total production of bay oil is ca. 50 t/yr. It has antiseptic properties because of its high phenol content and is, therefore, a classical ingredient for perfuming after-shave lotions. FCT 1973 (**11**) p. 869; [*8006-78-8*].

Benzoe resinoids
1. *Benzoe Siam resinoid* is obtained by solvent extraction of the resin from *Styrax tonkinensis* (Pierre) Craib ex Hartwich trees (Styracaceae). The wild growing *Styrax* tree is widespread in Thailand, Laos, Cambodia, and Vietnam.
Benzoe Siam resinoid is a reddish to light brown, viscous liquid with a long-lasting, chocolate-like, sweet, balsamic odor. It is used in perfumery for balsamic nuances and as a fixative [*9000-72-0*].

2. *Benzoe Sumatra resinoid* is obtained by solvent extraction of the resin from *Styrax benzoin* Dryand, a tree growing predominantly on the island of Sumatra.

Benzoe Sumatra resinoid is a dark brown viscous liquid with a warm, powdery, sweet-balsamic odor. Its main volatile, odor-determining components are derivatives of benzoic and cinnamic acids and vanillin [284].

Benzoe Sumatra resinoid is used in perfumery, mainly as a fixative with a warm, balsamic note [9000-73-1]. FCT 1973 (**11**) p. 871.

Bergamot oil see Citrus Oils.

Bitter almond oil (free from hydrogen cyanide) contains benzaldehyde as its main component. Benzaldehyde does not occur as such in the plant, but is formed, together with hydrogen cyanide, by the hydrolytic cleavage of the glycoside amygdalin.

Amygdalin is present in bitter almonds, the seeds of *Prunus amygdalus* Batsch var. *amara* (DC.) Focke, and ripe apricot kernels, *Prunus armeniaca* L. (Rosaceae). The press cake, which remains after removal of the fatty oils, is macerated with water and left to stand for several hours, after which the 'essential oil' is separated by steam distillation. The crude oil contains 2–4% hydrogen cyanide, which is removed by washing with alkaline solutions of iron(II) salts. Subsequent redistillation yields an oil free from hydrogen cyanide. It is a colorless to slightly yellow liquid with an intense, almond-like, cherry aroma and a slightly astringent, mild taste.

d_{25}^{25} 1.025–1.065; n_D^{20} 1.5350–1.5550; acid value: max. 8; solubility: 1 vol in max. 6 vol 50% ethanol. HCN content: <0.01%; benzaldehyde content by GC: min. 98% [285].

Bitter almond oil is used almost exclusively in natural aroma compositions.

Blackcurrant absolute (Bourgeons de cassis absolute, cassis absolute) is obtained by solvent extraction via the concrete obtained from the dormant buds of the blackcurrant bush *Ribes nigrum* L. (Saxifragaceae). The yield is ca. 3%. Blackcurrant absolute is a dark green paste with the characteristic, powerful, penetrating odor of blackcurrants. The typical 'catty' note is caused by a sulfurous trace constituent, 4-methoxy-2-methyl-2-butenethiol [80324-91-0] [286–288].

Blackcurrant absolute is used in fine fragrances and in fruity food flavors.

Bois de rose oil, see Rosewood Oil.

Buchu leaf oils are obtained by steam distillation of *Barosma betulina* (Bergius) Bartl. and Wendl. and *B. crenulata* (L.) Hook. (Rutaceae) leaves. The oils are dark yellow to brown liquids with a characteristic minty-fruity odor, reminiscent of blackcurrant.

d_{25}^{25} 0.912–0.956, n_D^{20} 1.474–1.488; α_D^{20} −36° to −8°; acid number: 1–5; ester number: 20–85.

The bushes grow wild and are cultivated in South Africa. The major components of the oils are (+)-limonene (ca. 10%) and other cyclic terpenoids that are structurally related to menthone. However, the constituents responsible for the characteristic blackcurrant odor are *trans-p*-menthane-8-thiol-3-one [*34352-05-1*] and its *S*-acetate derivative [*57074-34-7*] which are two of the small number of naturally occurring sulfur-containing terpenoids known to date [289, 290].

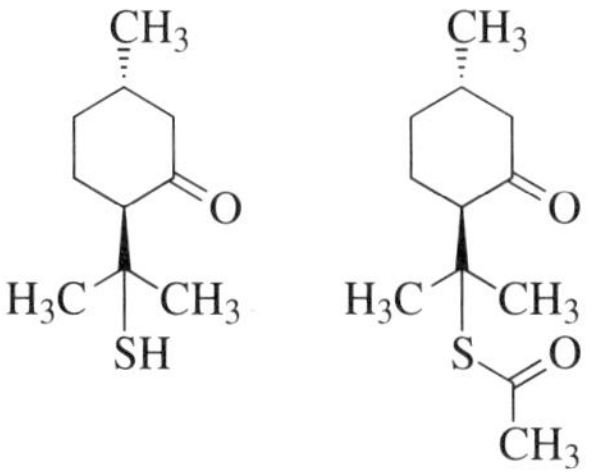

Buchu oil is used as a flavor ingredient (e.g., in fruit aromas) and in perfumery in chypre bases and in certain types of eau de cologne; only very small amounts are employed because of its intensity [*68650-46-4*].

Calamus oil (sweet flag oil) is obtained by steam distillation of fresh or unpeeled, dried roots of *Acorus calamus* L. (Araceae). It is a yellow to medium brown, moderately viscous liquid with a pleasant, spicy, aromatic odor.

The plant occurs in polyploid varieties and the corresponding essential oils differ predominantly in their content of β-asarone (*cis*-isoasarone) [*5273-86-9*]:

diploid (American)	0%
triploid (European)	0–10%
tetraploid (East Asian)	up to 96%

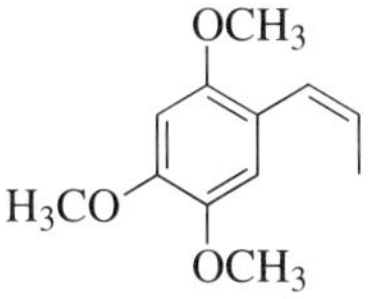

β-Asarone

The following data are typical for European and Indian oils, respectively:
d_{25}^{25} 0.940–0.980 and 1.060–1.080; n_D^{20} 1.5010–1.5160 and 1.5500–1.5525; α_D^{20} +5° to +35° and −2° to +6.5°; acid number: max. 4; ester number: 3–20; solubility: 1 vol in 5 vol of 90% ethanol; solutions may be turbid [291]. Main constituents of European oil are sesquiterpene ketones (shyobunone, preisocalamenediol and acorenone), but its characteristic odor is formed by a number of trace constituents, especially unsaturated aldehydes like Z,Z-4,7-decadienal [*22644-09-3*] [292–294].

Calamus oil is used in perfumery for spicy-herbaceous notes; small quantities are also employed in the alcoholic beverage industry. However, use is legally restricted because of the potential toxicity of β-asarone. For determination of β-asarone content in calamus oil see [295]. FCT 1977 (**15**) p. 623; [*8015-79-0*].

Camphor oil is obtained by steam distillation of the wood of the camphor tree *Cinnamomum camphora* Sieb. (Lauraceae) growing in China, Taiwan, and Japan. The main constituent of the crude oil is camphor (ca. 50%), which can be separated by crystallization on cooling and subsequent centrifugation. Fractionation of the mother liquor gives two oils:

1. *White camphor oil* is the first distillation fraction (ca. 20% of the crude camphor oil). It is a colorless or almost colorless liquid with a cineole-like odor.
d_{25}^{25} 0.855–0.875; n_D^{20} 1.4670–1.4720; $[\alpha]_D$ +16° to +28°; solubility: 1 vol in 1 vol of 95% ethanol; solutions usually become cloudy on further dilution [296].
In addition to monoterpene hydrocarbons, this oil contains up to 35% 1,8-cineole.

2. *Brown camphor oil* is a fraction with a boiling point higher than that of camphor (ca. 20%). It is a pale yellow to brown liquid with the characteristic odor of sassafras oil.
d_{25}^{25} 1.064–1.075; n_D^{20} 1.5100–1.5500; α_D^{20} 0° to +3°; *fp* 6 °C; solubility: 1 vol in 2 vol of 90 % ethanol [297].
The oil contains more than 80% safrole and, like Brasilian sassafras oil is therefore used as a raw material for the production of heliotropin (piperonal, see p. 132) via isosafrole.
Camphor oils with a high safrole content can also be obtained by steam distillation from other Cinnamomum species (see sassafras oils).
The production of natural camphor and camphor oils was formerly several thousand of tons per year, but has declined as a result of the production of synthetic camphor. The same is true for the distillation of linalool-containing camphor oils (Ho oil, Ho leaf oil), which are derived from other varieties of the camphor tree.
China still produces 500 t per year of camphor oil [298]; [*8008-51-3*].

Cananga oil, see Ylang-Ylang and Cananga Oils.

Kümmel

Caraway oil is obtained by steam distillation of crushed caraway seeds from *Carvum carvi* L. (Apiaceae). It is a colorless to yellow liquid with a characteristic caraway odor and a mild-spicy taste.
d_{20}^{20} 0.905–0.920; n_D^{20} 1.4840–1.4890; α_D^{20} +67° to +80°; solubility: 1 vol in max. 5 vol. 80% ethanol [299].
The major constituents of common caraway oil are (+)-limonene and (+)-carvone, which together may make up more than 95% of the oil. (+)-Carvone is the essential odor component.
Caraway oil is used primarily for flavoring foods and alcoholic beverages, but also for the production of (+)-carvone. FCT 1973 (**11**) p. 1051; [*8000-42-8*].

Cardamom oil is obtained by steam distillation of the seeds of *Elettaria cardamomum* (L.) Maton (Zingiberaceae). It is a colorless or very pale yellow liquid with an aromatic, penetrating, slightly camphoraceous odor and a persistent, pungent, strongly aromatic taste.

d_{20}^{20} 0.919–0.936; n_D^{20} 1.4620–1.4680; α_D^{20} +22° to +41°; solubility: 1 vol in max. 5 vol. 70% ethanol [300].

The major components of cardamom oil are 1,8-cineole and α-terpinyl acetate (ca. 30% each). Trace constituents like unsaturated aliphatic aldehydes may be important for the typical aroma [301–307]. It is produced from cultivated or wild plants in the mountainous regions of southern India, Sri Lanka, Indonesia, and Guatemala.

Cardamom oil is used primarily for seasoning foods, alcoholic beverages, and in small dosages also occasionally in perfumery. FCT 1974 (**12**) p. 837; [*8000-66-6*].

Carrot seed oil is obtained by steam distillation of the crushed seeds of *Daucus carota* L. (Apiaceae). It is a light yellow to amber-yellow liquid with a pleasant, aromatic odor.

d_{20}^{20}0.900–0.945; n_D^{20}1.4800–1.4930; α_D^{20} −30° to −4°; solubility: 1 vol in max. 2 vol of 90% ethanol; solutions in up to 10 vol of ethanol are clear to opalescent [308].

The main constituent of carrot seed oil is carotol [*465-28-1*], which may be present in over 50% concentration [309–313].

Carotol

Carrot seed oil is used in the alcoholic beverage industry, in food flavors, and in perfume compositions. FCT 1976 (**14**) p. 705; [*8015-88-1*].

Castoreum, see Animal Secretions.

Cedar Oils. Several different conifer species are called cedars and the corresponding oils vary considerably in composition. The following cedar oils are commercially important:

1. *Cedar leaf oil* (Thuja oil) is produced by steam distillation of fresh leaves and branch ends of the tree *Thuja occidentalis* L. (Cupressaceae). It is a colorless to yellow liquid with a powerful, herbaceous odor, characteristic of thujone, see p. 211, [314–319].

d_{25}^{25} 0.910–0.920; n_D^{20} 1.4560–1.4590; α_D^{20} −14° to −10°; ketone content (calculated as thujone): min. 60%; solubility: 1 vol in 3 vol of 70% ethanol at 25 °C [320].

The oil is produced in the northern states of the United States and in Canada. It is used in perfumery for dry nuances in citrus and woody compositions. FCT 1974 (**12**) p. 843; [*8007-20-3*].

2.   *Chinese cedarwood oil* is similar in composition to Texas cedarwood oil (see below). Chinese cedarwood oil is obtained by steam distillation of *Chamaecyparis funebris* (Endl.) France (*Cupressus funebris* Endl., Cupressaceae), which is a weeping cypress, indigenous to China. Commercial Chinese cedarwood oil is a colorless to slightly yellow oil with an odor more smoke-like than the American oils.

d_{20}^{20} 0.955–0.966; n_D^{20} 1.5000–1.5080; α_D^{20} −35° to −20°; solubility: 1 vol in max. 5 vol. 95% ethanol; cedrol content by GC: min. 10% [321]. Crude Chinese cedarwood oil contains a higher quantity of cedrol which is separated by crystallization.

3.   *Texas cedarwood oil* is produced by steam distillation of chopped wood of the Texas cedar, *Juniperus mexicana* Schiede (Cupressaceae). It is a brown to reddish-brown, viscous liquid that may partially solidify at room temperature. It has a characteristic cedarwood odor.

d_{20}^{20} 0.954–0.967; n_D^{20} 1.5050–1.5080; α_D^{20} −50° to −35°; total alcohol content (calculated as cedrol): 35–48%; solubility: 1 vol in 5 vol of 90% ethanol at 20 °C [322].

For uses, see Virginia cedarwood oil.

4.   *Virginia cedarwood oil* is produced by steam distillation of sawdust, finely chipped waste wood from the manufacture of cedarwood products, or from stumps and logs of the red cedar; *Juniperus virginiana* L. (Cupressaceae). It is a light yellow to pale brown, viscous liquid with a characteristic cedarwood odor. The oil sometimes solidifies at room temperature.

d_{20}^{20} 0.941–0.970; n_D^{20} 1.5020–1.5080; α_D^{20} −38° to −14°; total alcohol content (calculated as cedrol): 20–46%; solubility: 1 vol in max. 5 vol of 95% ethanol [323].

While the Texas cedar grows in Mexico and other Central American countries, the Virginia cedar grows exclusively in the Southeast of the United States. Both cedar oils are used mainly for perfuming soaps and other products, as well as a starting material for the isolation of cedrol [77-53-2]. Further fragrance substances such as cedryl acetate (see p. 71) and cedryl methyl ether (see p. 58) are produced from cedrol. The oils contain other sesquiterpenes, particularly α-cedrene and *cis*-thujopsene [324–326] which can be converted to valuable fragrance materials by acetylation (cedryl methyl ketone, see p. 67). The worldwide annual production amounts to 1500–2000 t. FCT 1974 (**12**) p. 845, 1976 (**14**) 711; [*8000-27-9*].

(+)-Cedrol

Celery seed oil is obtained by steam distillation of the crushed, ripe seeds of field-grown celery, *Apium graveolens* L. (Apiaceae). It is an almost colorless to brownish-yellow liquid with a characteristic, pervasive, sweet-spicy, long-lasting odor.

d_{20}^{20} 0.875–0.908; n_D^{20} 1.4780–1.4900; α_D^{20} +48° to +78°; solubility: 1 vol in no more than 10 vol of 90% ethanol at 20 °C; saponification number: 30–70 [327].

Major mono- and sesquiterpene hydrocarbons in the oil are (+)-limonene (up to 50% or more) and β-selinene [*17066-67-01*] Its typical, long-lasting odor is caused primarily by two lactones, 3-butylphthalide [*6066-49-5*] and sedanenolide [*62006-39-7*] which each occur at a concentration of ca.1% [328–334].

Celery seed oil is used chiefly for flavoring foods, although small quantities are also used in perfumery. FCT 1974 (**12**) 849; [*8015-90-5*].

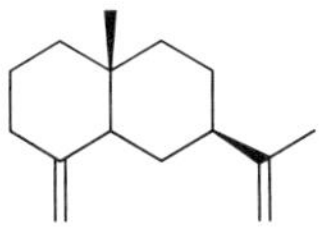

(+)-β-Selinene

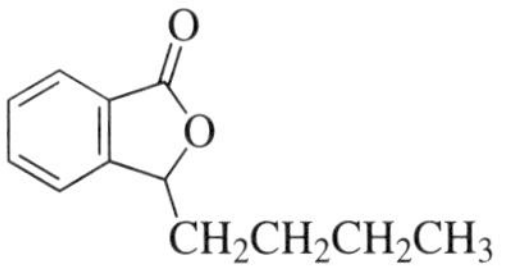

3-Butylphthalide

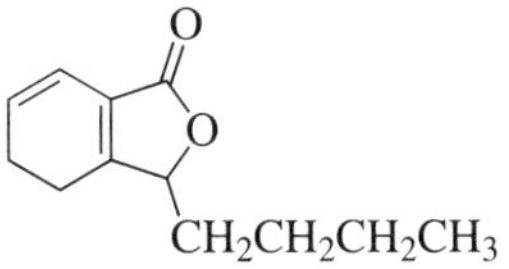

CH₂CH₂CH₂CH₃

Sedanenolide (3-butyl-
4,5-dihydrophthalide)

CH₂CH₂CH₂CH₃

Chamomile oils are available in two types:

1. *Blue chamomile oil* (German chamomile oil) is obtained by steam distillation of the flowers and stalks of *Chamomilla recutita* (L.) Rauschert (*Matricaria recutita*. L., *M. chamomilla* L. p.p.-Asteraceae), which is grown mainly in central and eastern Europe, Egypt, and Argentina. It is a deep blue or bluish-green liquid that turns green and, finally, brown when exposed to light and air. The oil has a strong, characteristic odor and a bitter-aromatic taste. Chamazulene [*529-05-5*] is responsible for its blue color. Chamazulene and (−)-α-bisabolol [*23089-26-1*] contribute to the anti-inflammatory properties of blue chamomile oil [335, 336].

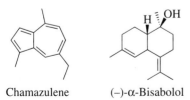

Chamazulene   (−)-α-Bisabolol

d_{25}^{25} 0.910–0.950; acid number: 5–50; ester number: 0–40; ester number (after acetylation): 65–155; solubility: solutions in 95% ethanol usually remain turbid [337]. Blue chamomile oil is used preferentially in pharmaceutical preparations, but also in small quantities in flavoring of alcoholic beverages. FCT 1974 (**12**) p. 853, [*8002-66-2*].

2.   *Roman chamomile oil* (English chamomile oil) is produced from the dried flowers of *Chamaemelum nobile* (L.) All. (*Anthemis nobilis* L.), which are cultivated primarily in Belgium, but also in England, France, and Hungary. The oil is distilled in France and England. It is a light blue or light greenish-blue liquid with a strong aromatic odor characteristic of the flowers.

d_{20}^{20} 0.900–0.920; n_D^{20} 1.4380–1.4460; acid value: max. 8, ester value. 250–340; solubility: 1 vol in max. 0.6 vol. 90% ethanol [338].

The constituents of Roman chamomile oil include esters of angelic and tiglic acid. Main constituents are isobutyl and isoamyl esters of angelic acid [339–344]. Roman chamomile oil is used in very low dosages in flavoring of alcoholic beverages and in fine fragrances. FCT 1974 (**12**) p. 853, [*8002-66-2*].

Cinnamon oils

1.   *Cassia oil* (Chinese cinnamon oil) is obtained by steam distillation of the leaves, twigs, and bark of *Cinnamomum aromaticum* Nees (*C. cassia* Bl., Lauraceae). It is a reddish-brown liquid with a sweet-spicy, cinnamon-like odor.

d_{20}^{20} 1.052–1.070; n_D^{20} 1.6000–1.6140; solubility: 1 vol in 3 vol of 70% ethanol at 20 °C; acid number: max. 15; content of carbonyl compounds (calculated as cinnamaldehyde): min 80% [345].

In contrast to cinnamon bark oil, cassia oil contains a considerable amount of 2-methoxycinnamaldehyde (3–15%) in addition to its main constituent, cinnamaldehyde (min. 70%) [346–350]. Cassia oil is used predominantly in flavoring soft drinks (cola-type). Annually a few hundred tons are are produced. FCT 1975 (**13**) p. 109; [*8007-80-5*].

2.   *Cinnamon leaf oil* is produced by steam distillation of the leaves of the cinnamon tree, *Cinnamomum zeylanicum* Bl. (*C. verum* J.S. Presl). The main countries in which the oil is produced are Sri Lanka, the Seychelles, southern India, Madagascar, and the Comoro Islands. It is a reddish-brown to dark brown liquid with a characteristic spicy odor, reminiscent of clove buds.

Specifications of cinnamon leaf oil from Sri Lanka are d_{20}^{20} 1.037–1.053; n_D^{20} 1.5300–1.5400; α_D^{20} −2.5° to +2°; solubility: 1 vol in 2 vol of 70% aqueous ethanol at 20 °C; phenol content: 75–85%; content of carbonyl compounds: 5% [351].

The main component of cinnamon leaf oil is eugenol [352–354]. The oil is used as such in spicy oriental perfumes, for flavoring sweets, alcoholic beverages or as a source of high-grade eugenol.

Annually, some hundred tons are produced. FCT 1975 (**13**) p. 749; [*8015-96-1*].

3.   *Sri Lanka cinnamon bark oil* is obtained by steam distillation of the dried bark of the cinnamon tree. It is a yellow liquid with the odor and burned-spicy taste of cinnamon. Main constituent is cinnamaldehyde [350, 355, 356].

d_{25}^{25}1.010–1.030; n_D^{20} 1.5730–1.5910; α_D^{20} −2° to 0°; aldehyde content (calculated as cinnamaldehyde): 55–78%; solubility: 1 vol in at least 3 vol of 70% ethanol [357].

The oil is used predominantly in flavor compositions. FCT 1975 (**13**) p. 111; [*8015-91-6*].

Cistus oil, see Labdanum Oil.

Citronella oil, see Cymbopogon Oils.

Citrus oils comprise both essential oils obtained from the peels of citrus fruits and essence oils obtained by concentrating citrus juice (see Section 3.2.1).

Oils isolated from other parts of citrus plants (blossoms and leaves) are not classified as citrus oils because the former show marked differences in composition and organoleptic properties, they are described in other sections (see Neroli Oil, and Petitgrain Oils).

Production of Citrus Peel Oils. Apart from distilled lime oil, citrus peel oils are produced by pressing. Pressing of the peels for oil is often combined with juice production.

In the first industrial production process, the citrus fruit was cut into halves and the juice was then pressed out. Further pressing of the peel liberated the desired oil. This mechanical procedure, which is still used, is known as cold pressing. Currently, citrus peel oils are also produced by other methods. For instance, the outer peel of the whole fruit (albedo) is rasped or punctured before juice extraction to release the oil. The oil is then rinsed off with water and is subsequently separated from the resulting emulsion by centrifugation. The so-called 'sfumatrice' process belongs to the type of first mentioned methods, the so-called 'pelatrice' process to the second type. In Italy, both methods are used to produce cold-pressed lemon oil. Occasionally, inferior qualities of some kinds of citrus oils are also obtained by steam distillation of the oil which remains in the sponge-like inner peel (flavedo) after the cold-pressing process (e.g. lemon and bergamot oils).

Production of Citrus Essence Oils. Distillative concentration of citrus juices yields essence oils, which separate from the aqueous phase in the receiver when the distillate condenses. The composition of essence oils is similar to that of peel oils, but the essence oils usually contain larger quantities of aliphatic ethyl esters (e.g., ethyl butyrate in orange essence oil). Thus, their aroma resembles that of a particular juice more than that of peel oils.

Citrus oils contain up to 95% monoterpene hydrocarbons (usually limonene, but others as well, e.g., lemon oil also contains α-terpinene and β-pinene). The important aroma-determining components of citrus oils are functionalized terpenes and aliphatic compounds (predominantly carbonyl compounds and esters), present only in relatively low concentrations [358]. Thus, several methods are employed to concentrate citrus oils on an industrial scale. The monoterpene hydrocarbon content is decreased by distillation, liquid–liquid partitioning between two immiscible solvents, or absorption on a carrier such as silica gel. The deterpenized concentrates are marketed under the name 'Citrus oil *x*-fold,' depending on the concentration factor.

1. *Bergamot oil, Italian* is obtained by pressing peels from the unripe fruits of *Citrus aurantium* L. subsp. *bergamia* (Risso et Poit.) Engl. (Rutaceae). It is a green to greenish-yellow liquid, which sometimes contains a deposit. The oil has a pleasant, fresh, sweet, fruity odor.

d_{20}^{20} 0.876–0.884; n_D^{20} 1.4650–1.4700; α_D^{20} +15° to +32°; evaporation residue: 4.5–6.4%; solubility: 1 vol in max. 1 vol. 85% ethanol; acid value: max. 2; ester value: 86–129, corresponding to an ester content of 30–45%, calculated as linalyl acetate; bergaptene content by HPLC: 0.18–0.38% [359].

Bergamot is grown in the Italian province of Calabria, in Brazil, and in the Ivory Coast. The quality of the oil is determined by its ester content and varies with the species. Annual production is 100–200 t/yr.

Linalyl acetate, linalool (10%), and citral (1%) are important components of bergamot oil. Its terpene content (25–50%) is relatively low for a citrus oil [360–370].

Use of the untreated oil in cosmetics is limited by the photosensitizing properties of bergaptene (a furocoumarin) present in the oil. Bergamot oil is used in flavoring food ('Earl Grey tea'); furocoumarin-free bergamot oil is used in many perfumes and is an important ingredient in eau de cologne. FCT 1973 (**11**), p. 1031, p. 1035; [*8007-75-8*], [*68648-33-9*] (furocoumarin-free).

2. *Grapefruit oil* is obtained by cold pressing of the outer peels of the fruits of *Citrus paradisi* Macfad. (Rutaceae). It is a greenish-yellow liquid, with an odor resembling that of sweet orange oil, but more herbaceous and bitter.

d_{20}^{20} 0.852–0.860; n_D^{20} 1.4740–1.4790; α_D^{20} +91° to +96°; content of carbonyl compounds (calculated as decanal): min. 1.0%; evaporation residue: min. 5.0%, max. 10.0% [371].

Grapefruit oil is produced mainly in the United States and its composition varies with the species. White grapefruit oil obtained from the varieties Marsh seedless and Duncan that are grown in Florida, contains ca. 90% terpene hydrocarbons (mainly (+)-limonene), 0.5% alcohols (mainly linalool), 1.8% aldehydes (mainly octanal and decanal), 0.5% ketones (mainly nootkatone), 0.3% esters (mainly octyl and decyl acetate), and 7.5 % nonvolatile components [370, 372–379]. Worldwide production is ca. 200 t/yr.

The sesquiterpene ketone nootkatone is primarily responsible for the characteristic aroma of grapefruit oil.

Grapefruit oil is used mainly for flavoring fruit beverages. FCT 1974 (**12**) p. 723; [*8016-20-4*].

3. *Lemon oil* is obtained by pressing peel from the fruits of *Citrus limon* (L.) Burm. (Rutaceae). It is a pale yellow to pale greenish-yellow liquid with a characteristic lemon peel odor.

d_{20}^{20} 0.852–0.858; n_D^{20} 1.4740–1.4760; α_D^{20} +57° to +65°; evaporation residue: 1.6–3.6%; acid number: max. 1.4%; content of carbonyl compounds (calculated as citral): 3.0–5.0% [380].

The composition of lemon oils depends on the variety of lemon and the country of origin. Their main components are terpenes – (+)-limonene (ca. 65%), β-pinene, and γ-terpinene (8–10% each) being the most important. The characteristic odor of lemon oil differs from that of other citrus oils and is largely due to neral and geranial. The content of these compounds in Italian lemon oil generally exceeds 3% [369, 381–394].

Annual worldwide production is ca. 3000 t, most of which originates from the United States, Italy, and Argentina.

Lemon oil is used in many food flavors. Because of its fresh odor, relatively large quantities are also employed in eau de cologne and other perfumery products. FCT 1974 (**12**) 725; [*8008-56-8*].

4. *Lime oil* may be either pressed or distilled, but the distilled oil is produced on a much larger scale.

Two varieties of limes are of importance in the commercial production of lime oils: the West Indian (Mexican or Key) lime, *Citrus aurantifolia* (Christm.) Swingle and the Persian (Tahiti) lime, *C. latifolia* Tanaka. The former has small fruits with many seeds, and the latter bears larger, seedless fruits. West Indian limes are grown primarily in Mexico, the West Indies, and Peru; Persian limes are cultivated in Florida and Brazil.

Persian lime oils contain lower concentrations of the typical organoleptic components than the West Indian lime oils and are thus comparatively mild and flat [395]. Consequently, West Indian lime oils are more popular.

Pressed lime oils are obtained by rasping and puncturing as described on p. 146. Other methods are also employed for West Indian limes: the whole fruits may be chopped and the separated oil–juice emulsion subsequently centrifuged [396], [397].

Pressed lime oil is a yellow to greenish-yellow liquid with a strong, characteristic odor, reminiscent of lemon.

d_{20}^{20} 0.874–0.882, n_D^{20} 1.482–1.486, α_D^{20} +35° to +40°; content of carbonyl compounds (calculated as citral): 4.5–9.0%; evaporation residue: 8.0–13.5% [398].

The composition [399–406] and uses of pressed lime oil are similar to those of pressed lemon oil. FCT 1974 (**12**) p. 731, 1993 (**31**), p. 331; [*90063-52-08*].

Distilled lime oils are produced by steam distillation of an oil–juice emulsion that is obtained by chopping the whole fruit. The acid present in the juice acts on the oil released from the peel and changes its characteristics. The original components are modified to form a series of new compounds.

Distilled (Mexican) lime oil is a colorless to pale yellow liquid, with a characteristic odor, which differs from that of the fresh fruit and the cold-pressed oil.

d_{20}^{20} 0.856–0.865; n_D^{20} 1.474–1.478; α_D^{20} +34° to +45°; evaporation residue: max. 2.5%; content of carbonyl compounds (calculated as citral): max. 1.5% [407].

Acid-catalyzed cyclization and dehydration of citral and linalool give rise to several compounds that occur at comparatively high concentrations and contribute to the typical aroma of distilled lime oil (e.g., 1,4-cineole [*470-67-0*], 1,8-cineole [*470-82-6*], 2,2,6-trimethyl-6-vinyltetrahydropyran [*7392-19-0*], and 2-(2-buten-2-yl)-5,5-dimethyltetrahydrofuran [*7416-35-5*]) [406, 408–412].

Annual worldwide production of distilled West Indian lime oil is ca. 600 t. The oil is used primarily in soft drinks of cola-type. FCT 1974 (**12**) p. 729; [*8008-26-2*].

5. *Mandarin oil* is obtained by cold pressing the peel of mandarin oranges, the fruits of *Citrus reticulata* Blanco (Rutaceae). The oil is a greenish-yellow to reddish-orange liquid, depending on the degree of ripeness of the fruit, with a pale blue fluorescence and a characteristic odor, reminiscent of mandarin peel.

d_{20}^{20} 0.848–0.855; n_D^{20} 1.474–1.478; α_D^{20} +64° to +75°; evaporation residue: 1.6–4.0%; content of carbonyl compounds (calculated as decanal): 0.4–1.2% [413].

The main components are limonene (ca. 65–75%) and γ-terpinene (ca. 20%). The characteristic feature of mandarin oil is its content of α-sinensal, methyl *N*-methylanthranilate (0.3–0.6%, which is responsible for fluorescence), and long-chain unsaturated aliphatic aldehydes [369, 381, 414–421].

Annual production of mandarin oil in Italy is ca. 100 t; smaller quantities are produced in Spain, Brazil, and Argentina.

Mandarin oil is used to enrich the bouquet of flavor compositions containing sweet orange oils as the main component. It is also used in liqueurs and perfumery. FCT 1992 (**30 suppl.**) p. 69 S; [*8008-31-9*].

6. *Bitter orange oil* is obtained by pressing fresh peel from the fruits of *Citrus aurantium* L. subsp. *aurantium* (Rutaceae). It is produced mainly in Mediterranean countries and is a pale yellow to yellowish-brown liquid with a slightly mandarin-like odor and a somewhat bitter aroma.

d_{20}^{20} 0.840–0.861; n_D^{20} 1.4720–1.4760; α_D^{20} +88° to +98°; evaporation residue: 2.2–5%; aldehyde content, calculated as decanal: 0.5–2.9% solubility: 1 vol in max. 8 vol. 90% ethanol [422].

The composition of bitter orange oil differs from that of sweet orange oil; i.e., its aldehyde content is lower and its ester content is higher [369, 370, 394, 423–428].

Worldwide production of bitter orange oil is much lower than that of other pressed peel oils. Bitter orange oil is predominantly used for flavoring alcoholic beverages (liquers). FCT 1974 (**12**) 735; [*68916-04-1*].

7. *Sweet orange oil* is obtained from the peel of the fruits of *Citrus sinensis* (L.) Osbeck. It is a yellow to reddish-yellow liquid with the characteristic odor of orange peel and may become cloudy when chilled. Sweet orange oil is often produced in combination with orange juice (e.g., in the United States, Brazil, Israel, and Italy).

d_{20}^{20} 0.842–0.850; n_D^{20} 1.4700–1.4760; α_D^{20} +94° to +99°; evaporation residue: 1–5%; aldehyde content, calculated as decanal: 0.9–3.1% solubility: 1 vol in max. 8 vol. 90% ethanol [429].

The oils have a high terpene hydrocarbon content (>90%, mainly (+)-limonene), but their content of oxygen-containing compounds differs and affects their quality. Important for aroma are aldehydes, mainly decanal and citral, and aliphatic and terpenoid esters. The sesquiterpene aldehydes α-sinensal [*17909-77-2*] and β-sinensal [*6066-88-8*], which contribute particularly to the special sweet orange aroma, also occur in other citrus oils, although in lower concentration [370, 394, 421, 430–438].

α-Sinensal   β-Sinensal

Worldwide production of cold-pressed orange oil is nearly 20 000 t/a. Its main uses are the flavoring of beverages and confectioneries and perfuming eau de cologne, soaps and household products. FCT 1974 (**12**) 733; [*8008-57-9*].

Civet, see Animal Secretions.

Clary sage oil, see Sage Oils.

Clove oils are produced from the 15–20 m high clove tree *Syzigium aromaticum* (L.). Merr. et Perry [*Eugenia caryophyllus* (Spreng.) Bullock and Harrison].

1.   *Clove bud oil* is obtained in 15–20% yield by steam distillation of the dried flower buds. Clove bud oil, like the leaf oil, is a yellow to brown, sometimes slightly viscous liquid. It turns dark purple-brown on contact with iron. The oil has the spicy odor characteristic of eugenol.
d_{20}^{20} 1.044–1.057; n_D^{20} 1.5280–1.5380; α_D^{20} −1.5° to 0°; solubility: 1 vol in 2 vol of 70% ethanol at 20 °C; phenol content: min. 85%, max. 93%; content by GC: eugenol 75–85%, caryophyllene 2–7%, eugenyl acetate 8–15% [439]. FCT 1975 (**13**) p. 761; [*8000-34-8*].

2.   *Clove leaf oil* is obtained in 2–3% yield by steam distillation of the leaves of the above-mentioned.
d_{20}^{20} 1.039–1.051; n_D^{20} 1.5310–1.5350; solubility: 1 vol in 2 vol of 70% ethanol at 20 °C; phenol content: min. 78%; content by GC: eugenol 80–92%, caryophyllene 4–17%, eugenyl acetate 0.2–4% [440]. FCT 1975 (**13**) p. 765; [*8015-97-2*].

3.   *Clove stem oil* is obtained in ca. 5% yield by steam distillation of the dried flower stems. It is a yellow to light brown oil with a sweet-spicy, slightly woody odor similar to that of bud oil but without the fresh-fruity top-note.
d_{20}^{20} 1.043–1.063; n_D^{20} 1.5280–1.5380; α_D^{20} −1.5° to 0°; solubility: 1 vol in max. 2 vol. 70% ethanol; content by GC: eugenol 83–92%, caryophyllene 4–12%, eugenyl acetate 0.5–4% [441]. FCT 1978 (**16**) p. 695; [*8015-98-3*].

The main component of all clove oils is eugenol (up to 80%, sometimes more), which is responsible for their odor and antiseptic properties. Other major constituents are eugenyl acetate and caryophyllene [442–449]. Clove bud oil has a higher acetate content and a more delicate odor than the other oils, therefore it is much more expensive. Leaf oil is produced and used in the largest quantities. The composition of clove stem oil resembles that of bud oil but with a lower content of eugenyl acetate.

The most important countries that produce clove oils are Madagascar, Tanzania, and Indonesia. Smaller quantities are produced in other tropical areas (e.g., Malaysia and Sri Lanka). Worldwide production of clove oils is more than 2000 t/yr, of which Indonesia produces about half.

The oils are used in many perfume and aroma compositions, because of their spicy clove odor. A small amount is used as an antiseptic, mainly in dentistry. The leaf oil, in particular, is also used as a raw material for the production of eugenol, which is the starting material for further commercially important fragrance compounds, such as isoeugenol and eugenyl esters.

Copaiba (balsam) oils are obtained by steam distillation of the exudate (balsam) from the trunk of several species of *Copaifera* L. (Fabaceae), a genus of trees growing in the Amazon basin. They are colorless to light yellow liquids with the characteristic odor of the corresponding balsams and an aromatic, slightly bitter, pungent taste.

d_{25}^{25} 0.880–0.907; n_D^{20} 1.4930–1.5000; α_D^{20} −33° to −7°; solubility: 1 vol in 5–10 vol of 95% ethanol [450].

The oils consist primarily of sesquiterpene hydrocarbons; their main component is caryophyllene (>50%).

Copaiba balsam oils and balsams are used mainly as fixatives in soap perfumes. FCT 1973 (**11**) p. 1075, 1976 (**14**) p. 687; [*8013-97-6*].

Coriander oil is obtained by steam distillation of ripe fruits of *Coriandrum sativum* L. (Apiaceae). It is an almost colorless to pale yellow liquid with a characteristic odor, reminiscent of linalool.

d_{20}^{20} 0.862–0.878; n_D^{20} 1.4620–1.4700; α_D^{20} +7° to +13°; acid number: max. 3.0; content of linalool by GC: 65–78% [451].

The main component of coriander oil is (+)-linalool (60–80%) [452–460]. Mono- and polyunsaturated fatty aldehydes, although minor components, contribute to the characteristic aroma of the oil because of their powerful odor. In contrast to the seed oil, coriander leaf oil contains these aldehydes as main constituents, e.g. 2-decenal and 2-dodecenal.

Coriander is mainly cultivated in Eastern Europe. World-wide oil production is 50–100 t/yr. Coriander oil is no longer important as a raw material for the production of linalool and its derivatives. However, it is still used extensively in seasoning mixtures and in perfume compositions. FCT 1973 (**11**) 1077; [*8008-52-4*].

Cornmint oil, see Mint Oils.

Cymbopogon oils are produced from several aromatic grasses that belong to the genus *Cymbopogon* Spreng. (Poaceae). The oils are obtained by steam distillation of the aerial parts of the plants. The following oils are of commercial interest:

1.   *Citronella oil* is available in two types:

(a)   *Sri Lanka (Ceylon) citronella oil* is produced by steam distillation of fresh or partly dried leaves and stems of the grass species *Cymbopogon nardus* (L.) W. Wats. – 'lenabatu' – cultivated in Sri Lanka. It is a pale yellow to brownish liquid with a fresh, grassy, camphoraceous odor.

d_{20}^{20} 0.894–0.910; n_D^{20} 1.479–1.487; α_D^{20} −22° to −12°; solubility: 1 vol in 2 vol of 80% ethanol at 20 °C; ester number (after acetylation): 157–200; carbonyl number: 18–55, corresponding to 5–15% carbonyl compounds (calculated as citronellal) [461].

Sri Lanka (Ceylon) oil is less valuable than Java oil and is used almost exclusively for perfuming toilet soaps, washing powders, and household products.

(b)   *Java citronella oil* is obtained by steam distillation of fresh or partially dried stems and leaves of *Cymbopogon winterianus* Jowitt – 'mahapengiri' – which is grown in Southeast Asia, India, China and Indonesia, as well as in Central and South America. It is a pale yellow to pale brown liquid with a slight, sweet, flowery, roselike odor with the strong citrus note of citronellal.

d_{20}^{20} 0.880–0.895; n_D^{20} 1.466–1.473; α_D^{20} −5° to 0°; solubility: 1 vol in 2 vol of 80% ethanol at 20 °C, opalescence is sometimes observed when ethanol is continuously added; ester number (after acetylation): min. 250, corresponding to 85% acetylizable compounds (calculated as geraniol, this percentage includes citronellal, since it is converted quantitatively into isopulegyl acetate under the acetylation conditions); carbonyl number: min. 127, corresponding to 35% carbonyl compounds (calculated as citronellal) [462].

Java citronella oil may contain up to 97% acetylizable compounds and up to 45% carbonyl compounds, depending on the time of harvesting. Is is used extensively not only in perfumery, but also as one of the most important raw materials for the production of citronellal. In addition, a fraction with a high geraniol content is obtained from the oil. Both citronellal and the geraniol fraction are starting materials for the synthesis of a large number of other fragrance compounds. The oil produced in Taiwan and in Java contains, in addition to the major components citronellal and geraniol, 11–15% citronellol, 3–8% geranyl acetate, 2–4% citronellyl acetate, and many other minor components [463, 464].

Annual worldwide production was reported to be >5000 t in 1971 and is now ca. 2000 t. Main producers are Taiwan, China, and Java. FCT 1973 (**11**) p. 1067; [*8000-29-1*].

2.   *Lemongrass oil* is available in two types, which are produced by steam distillation.

(a)   *West Indian or Guatemala lemongrass oil* is obtained from *Cymbopogon citratus* (DC.) Stapf in Central and South America, as well as in a number of African and East Asian countries. It is a pale yellow to orange-yellow liquid with a lemon-like odor, characteristic of citral.

d_{20}^{20} 0.872–0.897; n_D^{20} 1.483–1.489; α_D^{20} −3° to +1°; content of carbonyl compounds (calculated as citral): min. 75%; solubility: freshly distilled oil is soluble in 70% ethanol at 20 °C, but solubility diminishes on storage and the oil may become insoluble in 90% ethanol. Residues remain after vacuum distillation of oils stored for longer times due to the high molecular mass products formed by polymerization of myrcene. The oil contains up to 20% myrcene [465].

(b) *Indian lemongrass oil* is obtained from the so-called Indian variety of lemongrass, *Cymbopogon flexuosus* (Nees ex Steud.) W. Wats. The oil is produced mainly in India. It is a pale yellow to brownish-yellow oil.

d_{20}^{20} 0.855–0.905; n_D^{20} 1.4830–1.4890; α_D^{20} −3° to +1°; solubility: 1 vol in max. 3 vol 70% ethanol; content of carbonyl compounds (calculated as citral): min. 73% [466].

The two oils were formerly the main source of natural citral, obtained as a ca. 4:1 mixture of geranial and neral by distillation [463]. However, lemongrass oil has declined commercial importance due to the competitive synthesis of citral and isolation of natural citral from Litsea cubeba oil. Nevertheless, more than 1000 t/yr are still produced. In addition to being processed into citral, it is used to some extent for perfuming soap and household products. FCT 1976 (**14**) p. 455; [*8007-02-1*].

3. *Palmarosa oil* is obtained by steam distillation of wild or cultivated *Cymbopogon martinii* (Rox.) W. Wats. var. *motia*, collected when in blossom. It is a pale yellow liquid with a characteristic roselike odor and a grassy note.

d_{20}^{20} 0.880–0.894; n_D^{20} 1.4710–1.4780; α_D^{20} +1.4° to +3°; solubility: 1 vol in 2 vol of 70% ethanol at 20 °C; ester number (after acetylation): 255–280, corresponding to a total alcohol content of 80–95%, free alcohol content (calculated as geraniol): 72–94% [467]. High-grade palmarosa oil may contain up to 95% geraniol and its esters [463]; it is produced in smaller quantities than other oils obtained from aromatic grasses. Annual worldwide production (India, Guatemala, Indonesia) exceeds 100 t [468]. Palmarosa oil the starting material for geraniol and geranyl esters of high odor quality, but it is so used for perfuming soaps and cosmetics. FCT 1974 (**12**) 947; [*8014-19-5*].

4. *Gingergrass oil* is produced in India from the *sofia* variety of *Cymbogon martinii* and is less important than palmarosa oil.

Cypress oil is produced by steam distillation of terminal branches of *Cupressus sempervirens* L. (Cupressaceae). It is a liquid with a woodlike odor that has an ambergris note.

d_{20}^{20} 0.863–0.885; n_D^{20} 1.4680–1.4780; α_D^{20} +15° to +30°; solubility: 1 vol in 8 vol of 90% ethanol [469].

This oil is produced exclusively in southern France and Algeria; its major components are α-pinene and 3-carene. Degradation products of higher terpenoids are responsible for the typical ambergris note [470–473]. FCT 1978 (**16**) 699; [*8013-86-3*].

Davana oil is obtained by steam distillation of the herb *Artemisia pallens* Wall. (Asteraceae), grown in south India. It is an orange-brown liquid with a sweet tealike odor reminiscent of dried fruits. The composition of the oil is very complex; its main components are furanoid sesquiterpenes [474–476]. It is used predominantly for aroma compositions. FCT 1976 (**14**) p. 737; [*8016-03-3*].

Dill oil is obtained from the dill plant, *Anethum graveolens* L. (Apiaceae), in two different forms:

1. *Dill weed oil*, which is the most important, is obtained by steam distillation from dill weed (herb) before the fruits become mature. Its main constituents are α-phellandrene (10–20%), limonene (30–40%), carvone (30–40%) and the so-called (+)-dill ether [*74410-10-9*] (up to 10%) [477–485]. The latter is responsible for the typical organoleptic properties of the dill plant and, thus, of dill weed oil.

(+)-Dill ether

2. *Dill seed oil* is prepared by steam distillation of the crushed ripe fruits of the dill plant. Its main components are limonene (up to 40%) and (+)-carvone (up to 60%) [486]. In contrast to the weed oil, this oil has a typical caraway odor and taste which is characteristic of (+)-carvone.

Indian dill oil is obtained by steam distillation of the seeds of a closely related plant, *Anethum sowa* Roxb. It contains also larger quantities of limonene and carvone [486].

Dill oils are used primarily for seasonings in the pickling and canning industries. FCT 1976 (**14**) p. 747; [*8006-75-5*].

Elemi oil, **Elemi resinoid** are obtained from exuded gum resin of *Canarium luzonicum* (Miqu.) A. Gray (Burseraceae), a tree growing in the Phillipines. The resin is extracted with a solvent to form the resinoid, which is a yellow to orange mass of high viscosity. The oil is produced by steam distillation of the gum resin and is a colorless to light yellow liquid.

d_{20}^{20} 0.850–0.910; n_D^{20} 1.4720–1.4900; α_D^{20} +44° to +85° [487].

The major components of elemi oil are limonene (30–70%), α-phellandrene (10–24%), and the sesquiterpene alcohol elemol (1–25%) [488–491]. Both the resinoid and the oil have a fresh, citrus-like, peppery odor and are used predominantly in soap perfumes. (FCT 1976 (**14**) p. 755; [*8023-89-0*] (oil), [*9000-74-2*] (resin).

Estragon oil, see Tarragon Oil.

Eucalyptus oils are produced from plants belonging to the genus *Eucalyptus* (Myrtaceae), which includes ca. 500 species in Australia, the country of origin, alone. Correct botanical classification was possible only by determining the chemical composition of the essential oils obtained from the leaves. At present, few of these oils are commercially important.

1. *Cineole-rich Eucalyptus oils*
    (a) *Australian eucalyptus oil* is obtained by steam distillation of the foliage of certain Eucalyptus species indigenous to Australia, mainly from *Eucalyptus fruticetorum* F.v. Muell. (*E. polybractea* R. T. Bak.)
    d_{20}^{20} 0.918–0.928; n_D^{20} 1.458–1.465; α_D^{20} −2° to +2°; solubility: 1 vol in 3 vol of 70% ethanol at 20 °C; 1,8-cineole content: 80–85% [492].
    The minor components of this oil differ from those of *E. globulus* oil. Despite its high cineole content, annual production of Australian eucalyptus oil is only 50 t.
    (b) *Eucalyptus globulus oil* is produced by steam distillation of the leaves of *Eucalyptus globulus* Labill. It is an almost colorless to pale yellow liquid with a fresh odor, characteristic of cineole. The crude oil contains ca. 65% 1,8-cineole and more than 15% α-pinene [493–499]. Rectified qualities have a cineole content of 70–75% or 80–85%. The respective specifications of these three types are as follows:
    d_{20}^{20} 0.905–0.925/0.904–0.920/0.906–0.920; n_D^{20} 1.457–1.470/1.460–1.468/1.458–1.465; α_D^{20} +2° to +8°/0° to +10°/0° to +5°; solubility: 1 vol in max. 7/5/4 vol 70% ethanol [500].
    The oil is produced mainly in Spain and Portugal, where the wood is used in the cellulose industry, and in China. Worldwide production exceeds 1500 t/yr.
    Eucalyptus oils with a high cineole content are used for cineole production. The oils and cineole itself are used primarily in pharmaceutical preparations. Fairly large quantities are also used in perfumery, e.g., to imitate the odor of cineole-containing essential oils and flavoring of food (sweets) and oral care products. FCT 1975 (**13**) p. 107; [*8000-48-4*].

2. *Eucalyptus citriodora oil* is obtained by steam distillation of leaves and terminal branches of *Eucalyptus citriodora* Hook. It is an almost colorless, pale yellow, or greenish-yellow liquid with a citronellal-like odor.
    d_{20}^{20} 0.860–0.870; n_D^{20} 1.4500–1.4560; α_D^{20} −1° to +3°; solubility: 1 vol in max. 2 vol of 80% ethanol; content of carbonyl compounds calculated as citronellal: min. 70%; content of citronellal by GC: min. 75% [501].
    In addition to the main component, citronellal, the oil contains citronellol and isopulegol (5–10% each) [502–504].
    Young *E. citriodora* trees that are grown exclusively for essential oil production are cut back to a height of 1–1.50 m and develop into shrubs. The leaves can be harvested throughout the year; more than 200 kg of oil can be obtained per hectare.
    The major producer is Brazil with over 500 t/yr, but considerable quantities are also produced in other countries (e.g., South Africa and India). *Eucalyptus citriodora* oil is a starting material for the manufacture of citronellal and products

derived from it. It is also used in perfumery for the same purposes as citronellal. FCT 1988 (**26**) p. 323; [*85203-56-1*].

3.   *Eucalyptus dives oil* is obtained by steam distillation of fresh leaves of *Eucalyptus dives, Schau*, piperitone-type, grown in Australia and South Africa.

In addition to (−)-piperitone the oil contains 15–25% α-phellandrene [505]. The oil was previously used as a starting material in the manufacture of (−)-menthol, but has lost much of its significance. Annual worldwide production has dropped to 50 t [*85203-58-3*].

Fennel oil is commercially available in sweet and bitter types that are obtained from two varieties of a subspecies of common fennel, *Foeniculum vulgare* Mill., subsp. *capillaceum* Gilib. (Apiaceae). The sweet oil is obtained by steam distillation of the ripe fruits (seeds) of the *dulce* variety, the bitter oil from the aerial parts of the *vulgare* variety before the fruits become mature.

1.   *Sweet fennel oil* is colorless to pale yellow liquid with a clean sweet-aromatic odor and a sweet warm-spicy taste.

d_{20}^{20} 0.961–0.977; n_{D}^{20} 1.5280–1.5430; α_{D}^{20} +11° to +24°; *fp* 3–11.5 °C; solubility: 1 vol in 5 vol of 80% ethanol.

Main constituent is anethole (up to 80% and more) [486, 506–513].

2.   *Bitter fennel oil* is colorless to pale yellow liquid with a camphoraceous sweet-spicy odor and a slightly bitter sweet camphoraceous taste. Main constituents are monoterpene hydrocarbons (e.g. α-pinene, α-phellandrene and limonene), fenchone (10–20%) and anethole (30–40%). With increasing maturity of the fruits, the content of monoterpenes decreases. The ripe fruits yield an oil with 50–70% anethole and 10–25% fenchone [486, 506–513].

Fennel oils (anise liqueurs) are used in flavoring oral care products and alcoholic beverages and in pharmaceutical preparations. FCT 1976 (**17**) p. 529; [*8006-84-6*].

Fir needle oils, see Pinaceae Oils.

Galbanum oil and **galbanum resinoid** are produced from the gumlike exudate of *Ferula galbaniflua* Boissier and Buhse (Apiaceae) growing in northern Iran and *F. rubricaulis* Boissier growing in southern Iran.

The gum is collected from a cut in the upper part of the uncovered roots. The annual yield of gum is ca. 100 t.

The oil is produced by steam distillation and is a yellow liquid with a green, slightly spicy odor.

d_{20}^{20} 0.867–0.890; n_{D}^{20} 1.4780–1.4850; α_{D}^{20} +7° to +15°; acid value: max. 2 [514].

In addition to 75% monoterpene hydrocarbons and ca. 10% sesquiterpene hydrocarbons, galbanum oil contains a fairly large number of terpene and sesquiterpene alcohols and their acetates. Minor components, with entirely different structures and low odor threshold values, contribute strongly to the characteristic odor [515–519]. Examples are as follows:

(*E,Z*)-1,3,5-Undecatriene
[*51447-08-6*]

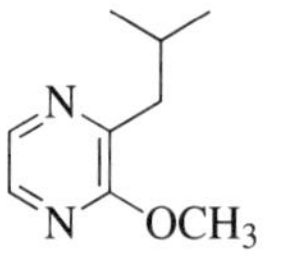

2-Methoxy-3-isobutylpyrazine
[*24683-00-9*]

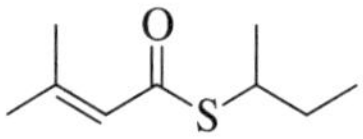

S-sec-Butyl 3-methyl-2-butenethioate
[*34322-09-3*]

Galbanum oil is used for creating green top notes in perfume compositions.

Galbanum resinoid is produced by extraction of the gum with a nonpolar solvent. It is used for the same purposes as the oil and has excellent fixative properties. FCT 1978 (**16**) p. 765 (oil), 1992 (**30**) p. 395 (resin); [*8023-91-4*].

Geranium oil is obtained by steam distillation of the flowering herb *Pelargonium graveolens* l'Heritier ex Aiton, *P. roseum* Willdenow, and other nondefined hybrids that have developed into different ecotypes in different geographical regions. The oil is an amber to greenish-yellow liquid with the characteristic roselike odor of the plant.

The main cultivation areas are Réunion and Madagascar (Bourbon type), Egypt (North African type), and China. The Bourbon quality is more valuable and thus more expensive. Annual worldwide production is some hundred tons. Specifications of geranium oils [520] are given in Table 3.

The composition of the Bourbon oil differs quantitatively as well as qualitatively from that of North African oil. However, they both contain an unusually high percentage of (−)-citronellol, isomenthone, formates, and tiglates, which are rarely found in essential oils. The two types of oil can be differentiated by two characteristic minor constituents: the Bourbon type contains (−)-6,9-guaiadiene [*36577-33-0*] and the African type contains 10-*epi*-γ-eudesmol [*15051-81-7*] see p. 167 [521–531].

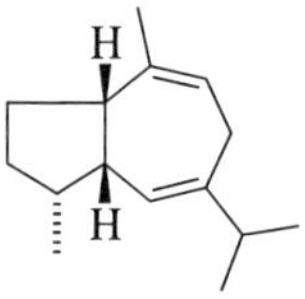

(−)-6,9-Guaiadiene

Table 3. Specifications of geranium oils

Parameter	Origin of geranium oil		
	Bourbon	North Africa	China
d_{20}^{20}	0.884–0.892	0.883–0.905	0.882–0.892
n_D^{20}	1.4610–1.4700	1.4610–1.4770	1.4600–1.4720
α_D^{20}	−14° to −10°	−14° to −8°	−14° to −8°
Solubility in 70% ethanol, vol:vol	1:3	1:3	1:5
Acid value	max. 10	max. 10	max. 10
Ester value	53–76	31–80	55–75
Ester value (acetylated)	205–239	192–240	215–232
Content of carbonyl compounds, calc. as isomenthone [%]	16	16	10

The qualitative composition of Chinese oil is very similar to that of Bourbon oil, but it contains more citronellol (ca. 40%) and lower amounts of linalool (ca. 4%) and geraniol (ca. 8%).

Geranium oil is one of the most important natural raw materials in the fragrance industry. It shows a broad variety of application possibilities. FCT 1974 (**12**) p. 883, 1975 (**13**), p. 451.

Ginger oil and **ginger oleoresin** are produced from the ginger plant *Zingiber officinale* (L.) Rosc. (Zingiberaceae).

Ginger oil is produced by steam distillation of dried, ground rhizomes. It is a light yellow to yellow liquid with the aromatic, persistent odor of ginger, but lacking the pungency usually associated with ginger. The citrus note of ginger oil is produced by citral.

d_{25}^{25} 0.871–0.882; n_D^{20} 1.4880–1.4940; α_D^{20} −45° to −28°; saponification number: max. 20; soluble in ethanol, solutions are usually turbid [532].

The major components of the oil are β-sesquiphellandrene [*20307-83-9*] and zingiberene [*495-60-3*] [533–540].

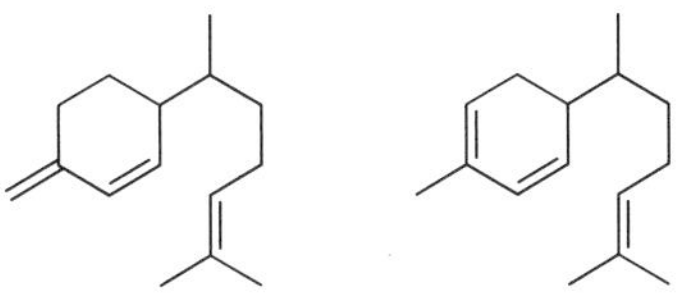

β-Sesquiphellandrene   Zingiberene

Ginger oleoresin is prepared by extracting ginger rhizomes with acetone or alcohol. The product contains the essential oil along with the substances responsible for the pungency of ginger. These compounds are substituted phenols of the following structure:

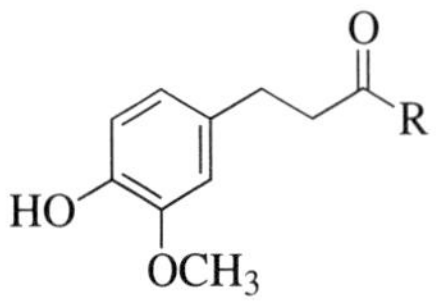

Gingerols: R = CH$_2$CH(CH$_2$)$_n$CH$_3$; n = 1–4, 8, 10
                         |
                        OH
Shogaols: R = CH=CH(CH$_2$)$_n$CH$_3$; n = 2, 4, 6, 8

   Main cultivation areas for ginger are India, Malaysia, Taiwan, Australia and the Fiji Islands. Ginger concentrates are used in large amounts in beverages (e.g., ginger ale), as well as in baked goods and confectioneries. FCT 1974 (**12**) p. 901; [*8007-08-7*].

   Grapefruit oil, see Citrus Oils.

   Green cognac oil, see Lie de Vin Oil.

   Guaiac wood oil is obtained by steam distillation of ground wood and sawdust from the tree *Bulnesia sarmienti* Lorentz (Zygophyllaceae), which is up to 20 m high and grows wild in the Gran Chaco region of Paraguay and in Argentina. The oil is a dark yellowish, viscous liquid with a mild, pleasant odor reminiscent of that of tea roses and faintly of violets. The oil solidifies at room temperature to a yellow-white to light amber colored mass (*mp* 40–50 °C).
   d_{25}^{25} 0.9600–0.975; n_D^{20} 1.5020–1.5070; α_D^{20} −12° to −3°; solubility: 1 vol in at least 7 vol of 70% ethanol; solutions are sometimes slightly turbid or opalescent; total alcohol content (calculated as guaiol): min. 85% [541].
   The main constituents of the oil are the sesquiterpene alcohols guaiol [*489-86-1*] and bulnesol [*22451-73-6*] [542]. The oil may be used as a starting material for the synthesis of guaiazulene, which has anti-inflammatory properties.

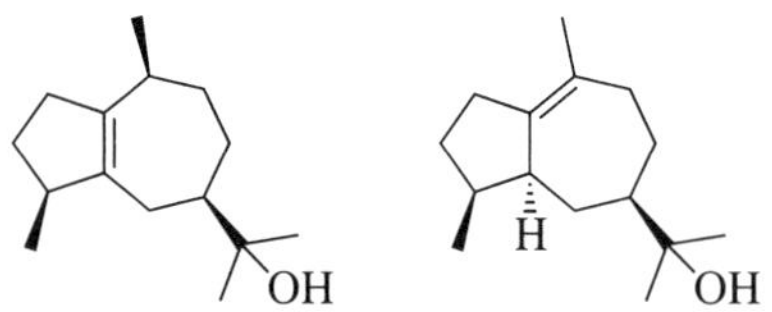

Guaiol   Bulnesol

   Guaiac wood oil is used extensively in perfume compositions for its excellent fixative properties (see also guajyl acetate, p. 71). FCT 1974 (**12**) p. 905.

Gurjun balsam oil is produced by steam distillation of balsams obtained from several *Dipterocarpus* species (Dipterocarpaceae); the trees grow in South and East Asia. The oil is a yellow, slightly viscous liquid with a weak, woodlike odor.

d_{15} 0.918–0.930; n_D^{20} 1.5010–1.5050; α_D^{20} −130° to −35°; solubility: 1 vol in 10 vol of ethanol.

It consists almost entirely of sesquiterpene hydrocarbons, and its main component (>60%) is α-gurjunene [489-40-7] [543]. Other qualities of gurjun balsam oil containing Calarene [17334-55-3] and α-Copaene [38565-25-5] as main constituents are also found on the market.

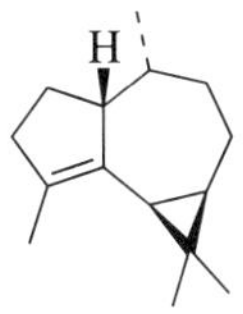

α-Gurjunene 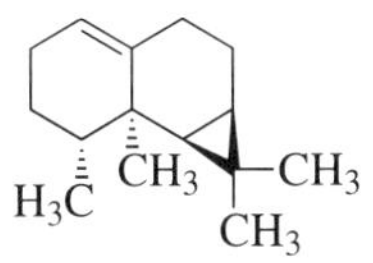   Calarene 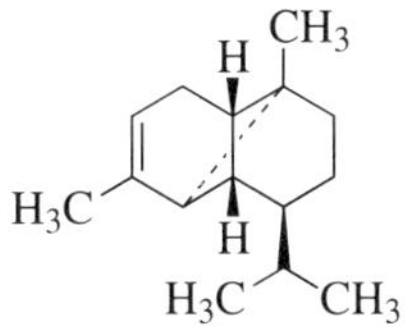   α-Copaene

Gurjun balsam oil and gurjun balsams are used for their good fixative properties, e.g., in soap perfumes and serve also as a starting material for the production of guaiazulene. FCT 1976 (**14**) p. 789, p. 791; [8030-55-5].

Japanese Mint Oil, see Mint Oils.

Jasmin absolute is obtained by solvent extraction, via the concrete, from the flowers of *Jasminum grandiflorum* L. (Oleaceae), cultivated in Egypt, Italy, Morocco, and India, and of *J. sambac* (L.) Ait. from China and India. The concrete is usually a brown to dark brown waxy mass, with a characteristic jasmin odor.

mp 48–51 °C; ester number: 70–125.

The absolute is generally a reddish-brown liquid with a delicate jasmin odor; the color deepens on storage.

n_D^{20} 1.4780–1.4920; acid number: 8–14; ester number: 120–220.

When extracted two or three times with hexane one ton of jasmin blossoms yields ca. 2.5–3 kg of concrete. Extraction of the concrete with ethanol gives a ca. 60% yield of the absolute.

The main volatile component of jasmin oil is benzyl acetate. However, minor components such as indole [120-72-9], *cis*-jasmone [488-10-8], and methyl jasmonate [1211-29-6] contribute strongly to the typical jasmin fragrance [544–549].

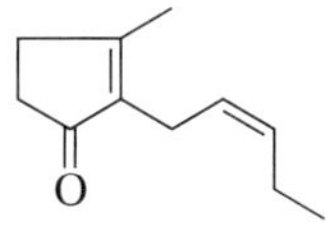

cis-Jasmone 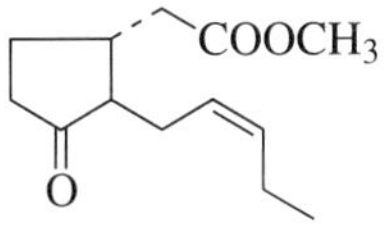   Methyl jasmonate

Annual worldwide production of jasmin concrete is ca. 10 t. The absolute is one of the most valuable blossom fragrances used in fine fragrances. FCT 1976 (**14**) p. 331; [*8022-96-6*].

Wacholder

Nadelholz

Juniper berry oil is obtained by steam distillation of ripe fruits of *Juniperus communis* L. var. *erecta* Pursh. (Cupressaceae). It is a colorless, pale green or yellowish liquid with a characteristic, conifer-like odor, and an aromatic-bitter taste.

d_{20}^{20} 0.857–0.872; n_D^{20} 1.4710–1.4830; α_D^{20} −15° to 0°; solubility: 1 vol in max. 10 vol of 95% ethanol [550].

The slightly turpentine-like odor and the relatively low solubility of the oil are caused by its high content of α-pinene and other monoterpene hydrocarbons. The main oxygen-containing component is 1-terpinen-4-ol [551–556].

The oil is used to a limited extent in perfumery for creating fresh, dry effects and as an aroma ingredient in alcoholic beverages of the gin type. FCT 1976 (**14**) p. 333; [*8012-91-7*].

Labdanum absolute and **Labdanum oil** are obtained from labdanum gum, which is exuded when twigs of *Cistus ladaniferus* L. (Cistaceae) are boiled in water.

Labdanum absolute is produced by extraction of the gum with alcohol (e.g., methanol or ethanol). So-called colorless absolutes are obtained from absolutes by extraction with, for example, hexane.

Steam distillation of the gum yields *labdanum oil* which is a golden yellow, viscous liquid that quickly turns dark brown on standing:

d_{25}^{25} 0.905–0.993; n_D^{20} 1.4920–1.5070; α_D^{20} +0°15′ to +7°, often difficult to determine due to its dark color; solubility: 1 vol in 0.5 vol of 90% ethanol, solutions frequently opalescent to turbid, paraffins may separate upon further dilution; acid number: 18–86; ester number: 31–86 [557].

Other odoriferous materials are derived from the leaves and young twigs of *Cistus ladaniferus*. Cistus oil is obtained by steam distillation; solvent extraction yields cistus concrete. Cistus oil, in contrast to labdanum oil, consists mainly of monoterpene hydrocarbons.

Cistus and labdanum products come from Mediterranean countries, mainly Spain, where the *Cistus* shrub grows abundantly. They are used widely in perfumery, giving perfume compositions a warm, balsamic tonality with a touch of ambergris. The typical odor originates from a number of compounds which are formed by oxidative degradation of diterpenes with labdane skeleton, which are the main constituents of labdanum gum [558–565]. The resinoids and absolutes are excellent natural fixatives. FCT 1976 (**14**) p. 335; [*8016-26-0*].

Lorbeer

Laurel leaf oil is obtained by steam distillation of leaves from *Laurus nobilis* L. (Lauraceae), an evergreen tree cultivated primarily in Mediterranean countries. The oil is a light yellow to yellow liquid with an aromatic, spicy odor.

d_{25}^{25} 0.905–0.929; n_D^{20} 1.4650–1.4700; α_D^{20} −19° to −10°; solubility: 1 vol in at least 1 vol of 80% ethanol; acid number: max. 3; saponification number: 15–45; saponification number (after acetylation): 36–85 [566].

The main component of the oil is 1,8-cineole (30–70%); other important components are linalool (ca. 10%) and eugenol [567–572].

Laurel leaf oil is used extensively in the food industry, e.g., for seasoning meat products and soups. FCT 1976 (**14**) p. 337; [*8002-41-3*].

Lavandula products comprise the following oils and extracts:

1.   *Lavender oil* is produced by steam distillation of freshly cut, flowering tops of *Lavandula angustifolia* Mill. (Lamiaceae). It is a pale yellow, amber-tinged liquid with a fresh, sweet, floral, herbaceous odor on a woody balsamic base.

d_{20}^{20} 0.880–0.890; n_D^{20} 1.4580–1.4640; α_D^{20} −11.5° to −7°; solubility: 1 vol in max. 2 vol of 75% ethanol; acid value: max. 1; ester value 102.5–165, corresponding to an ester content calculated as linalyl acetate: 38–58%; content of linalyl acetate by GC: 25–45% (specification for French oil) [573].

True French lavender grows in the Haute Provence at an altitude of 600–1500 m. The plants are grown from seeds of the wild lavender ('population' lavender). Lavender oil is produced in a yield of 10–25 kg/ha. It has the following typical composition (%): *cis*-ocimene (5–9), *trans*-ocimene (3–5), 1,8-cineole (<1), camphor (<0.4), linalool (30–35), linalyl acetate (30–40), 1-terpinen-4-ol (3–4), and lavandulyl acetate (3–4) [574–583].

Cloned varieties of lavender (e.g., *mailette*) yield more oil per hectare and can be grown at lower altitudes; however, they produce a poorer quality oil. Bulgarian lavender oil is similar in composition to that obtained from French population lavender, but has a higher linalyl acetate content (45–50%).

Lavender oils of a special quality are also produced in several other areas throughout the world (e.g., Tasmania, China and Russia). Worldwide annual production is estimated at 200 t. FCT 1976 (**14**) p. 451; [*8000-28-0*].

2.   *Spanish spike lavender oil* is produced by steam distillation of the flowering tops of spike, *Lavandula latifolia* Medik. It is an almost colorless to pale greenish-yellow liquid with a characteristic, rough odor slightly like cineole and camphor.

d_{20}^{20} 0.894–0.917; n_D^{20} 1.4620–1.4680; α_D^{20} −7° to +2°; solubility: 1 vol in 3 vol of 70% ethanol; solutions may become opalescent on dilution; acid number: max. 1; ester number: 3–14; ester number (after acetylation): 130–200 [584].

The main components of Spanish spike lavender oil are linalool (25–50%), 1,8-cineole (20–35%), and camphor (8–20%) [585–591].

Spike plants grow wild in the entire Mediterranean area and prefer warmer, lower-lying regions than lavender and lavandin. Oil is primarily produced from plants cultivated in Spain; in comparison to lavender and lavandin oils, only small quantities are produced annually. FCT 1976 (**14**) p. 453; [*8016-78-2*].

3.   *Lavandin oil* is obtained by steam distillation of freshly cut flowering tops of lavandin, which is a hybrid of lavender and spike (*Lavandula angustifolia* Mill. × *Lavandula latifolia* Medik.). It is a pale yellow to amber liquid with a lavender-like and a slightly camphoraceous note.

d_{20}^{20} 0.885–0.897; n_D^{20} 1.459–1.466; α_D^{20} −7° to −2°; solubility: 1 vol in 4 vol of 70% ethanol at 20°C; acid number: max. 1.0; ester number: 80–105, corresponding to an ester content of 28–37%, calculated as linalyl acetate (lavandin oil *abrial*) [592]. d_{20}^{20} 0.891–0.899; n_D^{20} 1.4580–1.4620; α_D^{20} −7° to −3.5°; solubility: 1 vol in max. 3 vol of 70% ethanol; acid value: max. 1; ester value 100–137, corresponding to an ester content, calculated as linalyl acetate: 35–48%; (lavandin oil *grosso*) [593].

Lavandin plants are sterile and can be propagated only by using cuttings. The oils from the most important varieties, *abrial* and *grosso*, contain linalool and linalyl acetate as major constituents as well as 6–11% 1,8-cineole and more than 4–8% camphor [594–601]. A third variety is called *super* because its oil contains a high concentration of linalyl acetate (35–47%), and, thus, resembles lavender oil most closely.

Although lavender oil is more valuable than lavandin oil as a fragrance raw material, lavandin plants are more commonly cultivated because they give a higher yield of oil (ca. 50–100 kg/ha) and are hardier than lavender plants.

Cultivation in southern France is no longer restricted to the traditional lavender regions, but also now includes the Languedoc. Approximately 650–800 t of oil are produced annually in this area.

All three oils are used primarily in soap perfumes; considerable quantities are also employed in eau de cologne and in bath products. FCT 1976 (**14**) p. 447; [*8022-15-9*].

4.   *Lavender and lavandin extracts* are also commercially important and are produced in southern France by solvent extraction of flowering lavander and lavandin herbs. Production of lavandin concrete is higher than that of lavender. Extraction of the pastelike concretes with ethanol, followed by evaporation, yields absolutes. These extracts differ from the essential oils in being more soluble and in having a green color and a longer-lasting odor with a haylike, spicy note. They are also used in eau de cologne and fine fragrances, sometimes after decoloration (removal of chlorophyll with activated charcoal). FCT 1992 (**30 suppl.**) p. 65.

Lemon oil, see Citrus Oils.

Lemongrass oil, see Cymbopogon Oils.

Lie de vin oil (green cognac oil or wine lees oil) is obtained by steam distillation of the yeast and other sediments (lees) formed in wine. It is a green to bluish-green liquid with a characteristic cognac aroma.

d_{25}^{25} 0.864–0.870; n_D^{20} 1.4275–1.4295; α_D −1° to +2°; acid number: 32–70; ester number: 200–245; solubility: 1 vol in at least 2 vol of 80% ethanol [602].

Lie de vin oil consists mainly of the ethyl and isoamyl esters of fatty acids, formed during fermentation [603]. It is used mostly in aroma compositions; only very small amounts are employed in perfume compositions.

Lime oil, see Citrus Oils.

***Litsea cubeba* oil** is produced by steam distillation of the fruits of *Litsea cubeba* C. H. Persoon (Lauraceae) growing in East Asia. It is a pale yellow liquid with a fresh odor, reminiscent of citral.

d_{20}^{20} 0.880–0.892; n_D^{20} 1.4800–1.4870; α_D^{20} −1° to +10°; carbonyl content (calculated as citral): min. 74%; solubility: 1 vol in 3 vol of 70% ethanol at 20°C [604].

The oil is used mainly for the production of citral as a starting material for many other fragrance materials [605–607]; smaller quantities are employed for perfuming household products. The main producer is China. Worldwide annual production is ca. 1000 t. FCT 1982 (**20**) p. 731; [*68855-99-5*].

Lovage oils are obtained by steam distillation of either the leaves (leaf oil) or the fresh roots (root oil) of the perennial plant *Levisticum officinale* Koch (Apiaceae). Both oils are yellow-greenish-brown to dark brown liquids with a strong, characteristic, aromatic odor and taste.

d_{20}^{20} 1.020–1.060; n_D^{20} 1.5350–1.5550; α_D^{20} −1° to +5°; solubility: 1 vol in max. 1.5 vol of 85% ethanol; acid value: max. 2–16; ester value 170–260 [608].

The main constituent of the leaf oil is α-terpinyl acetate, while the main constituent and odor-determining component of the root oil is ligustilide [*4431-01-0*] (3-butylidene-4,5-dihydrophthalide) [609–612].

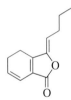

Ligustilide

The oils are very expensive. They are mainly used in the alcoholic beverage industry and for flavoring of tobacco; small amounts are also used in perfumery. FCT 1978 (**16**) p. 813; [*8016-31-7*].

Mace oil, see Nutmeg Oil.

Mandarin oil, see Citrus Oils.

Marjoram oil (sweet marjoram oil) is obtained by steam distillation of the herb *Origanum majorana* L. (*Majorana hortensis* Moench, Lamiaceae). The plants are cultivated in France, Hungary, Egypt, and Tunisia. The oil is a yellow to greenish-yellow liquid with a characteristic earthy-spicy odor.

d_{25}^{25} 0.890–0.906; n_D^{20} 1.4700–1.4750; α_D^{20} +14° to +24°; acid number: max. 2.5; saponification number: 23–40; saponification number (after acetylation) 68–86; solubility: 1 vol in 2 vol of 80% ethanol [613].

The main constituent of the oil is 1-terpinen-4-ol (>20%), which with (+)-*cis*-sabinene hydrate [*15537-55-0*] (3–18%) is responsible for the characteristic flavor and fragrance of marjoram oil [614–620].

(+)-*cis*-Sabinene hydrate

Sweet marjoram oil is used mainly for flavoring foods, but also in smaller amounts in perfumery for spicy shadings in men's fragrances. FCT 1976 (**14**) p. 469; [*8015-01-8*].

Spanish marjoram oil is distilled from *Thymus mastichina* L., a wild plant growing in Spain. The oil is a slightly yellow liquid with a fresh camphoraceous odor. Its main constituent is 1,8-cineole (up to 65%).

Mimosa absolute is obtained from the blossoms of the mimosa trees *Acacia dealbata* Link and *A. decurrens* Willd. var. *mollis* (Fabaceae), which grow in the forests of southern France and in Morocco.

The absolute is a yellowish brown liquid with a slightly green, flowery odor and extremely high tenacity.

Annual production is only some few tons [621]. Mimosa absolute is used mainly in fine fragrances as a flowery fixative. FCT 1975 (**13**) p. 873; [*8031-03-6*].

Mint oils are listed in order of the quantities produced.

1.   *Cornmint oil, Japanese mint oil* is produced by steam distillation of the flowering herb *Mentha arvensis* var. *piperascens* Malinv. The crude oil contains ca. 70% (−)-menthol, which can be isolated by crystallization at low temperature. The residual dementholized oil is a colorless to yellow liquid with a characteristic minty odor; typical specifications are given in Table 4 [622].

Dementholized oil still contains ca. 35–45% (−)-menthol; (−)-menthone and (+)-isomenthone (ca. 30% together) are other major components [623–626]. Dementholized *Mentha arvensis oil* was formerly produced primarily in Brazil; maximum production was 3000 t (together with 3000 t of (−)-menthol) in 1973. Currently, cornmint oil is produced mainly in China (several thousand tons per year), and in India which has become a second major cornmint oil producer since ca. 1990 and produced ca. 4000 t in 1994.

Dementholized cornmint oil tastes more bitter and stringent than peppermint oil. Thus, the former is used as a cheaper substitute for the latter and for the production of (−)-menthol. FCT 1975 (**13**) p. 771; [*68917-18-0*].

Table 4. Specifications of dementholized cornmint oils

Parameter	Origin of cornmint oil	
	China	India
d_{20}^{20}	0.890–0.908	0.890–0.910
n_D^{20}	1.4570–1.4650	1.4570–1.4650
α_D^{20}	$-24°$ to $-15°$	$-22°$ to $-13°$
Acid value	max. 1	max. 1
Ester value	8–20	8–25
Content of ester, calc. as menthyl acetate [%]	3–7	3–9

2. *Peppermint oil* is produced by steam distillation of the flowering herb *Mentha piperita* L. It is an almost colorless to pale greenish-yellow liquid with a characteristic peppermint odor.

d_{20}^{20} 0.903–0.912; n_D^{20} 1.460–1.464; α_D^{20} $-28°$ to $-17°$; solubility: 1 vol in 5 vol of 70% ethanol at 20°C; ester number: 14–29; ester number (after acetylation): 157–193; carbonyl number: 68–115 [627].

As in cornmint oil, the main component of peppermint oil is (−)-menthol; it also contains (−)-menthone (ca. 20%) and (−)-menthyl acetate (ca. 10%). However, peppermint oil, unlike cornmint oil, has a high content of (+)-menthofuran [*17957-94-7*] (2–3%, sometimes far higher) [628–633]

(+)-Menthofuran

The leading producer of peppermint oil is the United States, where annual production from the variety *M. piperita* L. var. *vulgaris* Sole (Black Mitcham) is ca. 3000 t. Cultivation areas are located in the Midwest states, Idaho, Oregon, and Washington. The oils differ in quality and are named according to their geographic origin (e.g., Midwest, Idaho, Madras, Willamette, Yakima). Production of European oil has decreased significantly, despite its high quality. Peppermint oil is used mainly for flavoring toothpaste, other oral hygiene products, and chewing gum. Smaller quantities are used for flavoring confectioneries. Due to its high price, peppermint oil is not used for the production of menthol [*8006-90-4*].

3. *Spearmint oils* are obtained by steam distillation of the flowering herbs of *Mentha spicata* L. (native spearmint) and *Mentha cardiaca* Ger. (Scotch spearmint). They are colorless to yellow-green liquids with a fresh, caraway-minty odor.

Table 5. Characteristic components of Scotch and native spearmint oils

Compound	Content of compound in spearmint oil, %	
	Scotch	Native
Limonene	15	9
3-Octanol	2	1
Menthone	1	<0.1
Dihydrocarvone	1	2.5
Sabinene hydrate	0.1	1.5
Dihydrocarvyl acetate	<0.1	0.6
cis-Carvyl acetate	0.1	0.6

d_{20}^{20} 0.920–0.937; n_D^{20} 1.485–1.491; α_D^{20} −60° to −45°; solubility: 1 vol in 1 vol of 80% ethanol at 20 °C; carbonyl number: min. 200, corresponding to a carvone content of 55% [634].

The compositions of Scotch and native spearmint oils differ as shown in Table 5 [486, 635–637]. The main producer of spearmint oil is the United States, primarily the state of Washington. Smaller amounts are also produced in some Midwest states. Total annual production is nearly 2000 t, 30–50% being of the native type and 50–70% of the Scotch type. Other minor cultivation areas for spearmint oils exist in China and India where oils with up to 80% carvone are produced.

By far the most spearmint oil is used for flavoring toothpaste and chewing gum. Smaller quantities are used in other oral care products and in pharmaceutical preparations. FCT 1976 (**16**) p. 871; [*8008-79-5*].

Musk, see Animal Secretions.

Myrrh oil and **myrrh resinoids** are produced from the air-dried gum of *Commiphora myrrha* (Nees) Engl. var. *molmol* Engl., *C. abyssinica* (Berg) Engl. and *C. schimperi* (Berg) Engl. (Burseraceae), shrubs that grow in Northeast Africa and Arabia.

Myrrh resinoids are obtained in 20–45% yield by extracting the gum with suitable solvents (e.g., toluene, hexane). They are waxlike, brown-yellow to red-brown masses with a balsamic odor and an aromatic bitter taste. Specifications depend on the solvents used.

Acid number: 20–35; ester number: 165–200; carbonyl number: 20–75 [638].

Myrrh oil is obtained from the gum by steam distillation; it is a light brown or green liquid with the characteristic odor of the gum.

d_{25}^{25} 0.995–1.014; n_D^{20} 1.5190–1.5275; α_D^{20} −83° to −60°; acid number: 2–13; saponification number: 9–35; solubility: 1 vol in 7–10 vol of 90% ethanol; solutions are occasionally opalescent or turbid [639].

Typical aroma-determining compounds of the myrrh plant are furanosesqui-terpenoids such as lindestrene [*2221-88-7*] [640, 641]. FCT 1976 (**14**) p. 621, 1992 (**30 suppl.**) p. 91S; [*8016-37-3*].

Lindestrene

Neroli oil and **orange flower absolute** are obtained from the blossoms of the bitter orange tree, *Citrus aurantium* L. subsp. *aurantium*, which is grown in France, Italy, and North Africa. *Neroli oil* is produced by steam distillation and is a pale yellow to amber-colored liquid, with a slight blue fluorescence and a characteristic sweet, spicy-bitter odor of orange blossom.

d_{20}^{20} 0.866–0.876; n_D^{20} 1.470–1.474; α_D^{20} +6° to +11°; solubility: 1 vol in 3.5 vol of 85% ethanol at 20 °C; acid number: max. 2.0; ester number: 28–50 (North African quality) [642].

After separation of neroli oil, the aqueous layer of the steam distillate, known as orange blossom water, is extracted with suitable solvents (e.g., petroleum ether). Evaporation gives *orange flower water absolute* (absolue de l'eau de fleurs d'oranger), which is a dark brown-red liquid. It contains less terpene hydrocarbons and correspondingly more polar compounds than neroli oil.

Orange flower absolute is obtained from the blossoms by solvent extraction via the concrete. It is a dark brown liquid with a warm, spicy-bitter odor.

The main volatile constituent of all three products is linalool. Their typical flavor is created by a number of nitrogen-containing trace constituents, such as indole and derivatives of anthranilic acid [428, 643, 644].

Neroli oil and the related products are some of the most expensive natural raw materials and are produced only in small quantities (a few tons per year). They are used in fine fragrances; neroli oil, for example, is one of the classical components of eau de cologne. FCT 1976 (**14**) p. 813, 1982 (**20**) p. 785; [*8016-38-4*].

Nutmeg (mace) oil is obtained by steam distillation of the seeds (nutmeg) and/or the seed-coverings (mace), which are obtained from the fruits of *Myristica fragrans* Houtt. (Myristicaceae). The tree grows in Indonesia and in the West Indies and becomes 15–20 m high.

Nutmeg/mace oils are colorless to pale yellow liquids with a pleasant spicy odor. The physical constants and odor vary with the origin. Specifications of Indonesian and (in parentheses) West Indian nutmeg oils are as follows:

d_{20}^{20} 0.885–0.905 (0.862–0.882); n_D^{20} 1.4750–1.4850 (1.4720–1.4760); α_D^{20} +8° to +18° (+25° to +40°); solubility (both oils): 1 vol in 5 vol of 90% ethanol at 20 °C (solutions sometimes opalescent), for freshly distilled oils, 1 vol in 3–4 vol [645].

Indonesian oils contain a higher percentage of higher boiling components than do West Indian oils (produced mainly in Grenada).

The oils contain ca. 90% terpene hydrocarbons, mainly sabinene (14–29%) and α- (15–28%) and β-pinene (13–18%). Major oxygen-containing constituents are 1-terpinen-4-ol and phenol ether derivatives like myristicine (5–12% in Indonesian oil) [646–650].

Nutmeg oil is used mainly in food flavorings (soft drinks of the cola-type) and to a lesser extent in perfumery. More than 1000 t/a is produced in Indonesia alone. FCT 1976 (**14**) p. 631, 1979 (**17**) p. 851; [*8007-12-3*].

Oakmoss absolute and **tree moss absolute** are obtained from tree lichens. *Oakmoss absolute* is derived from *Evernia prunastri* (L.) Arch. (Usnaceae), a lichen growing on oak trees. The lichen is first extracted with nonpolar solvents to give a concrete. The waxes are then removed by precipitation with ethanol, leaving an absolute.

The concretes are green to brown waxy pastes; the absolutes are more or less viscous liquids with an earthy, mossy, woody odor and a slight phenolic, leather note.

Resorcinol derivatives, e.g., orcinol [*504-15-4*], β-orcinol [*488-87-9*], their monomethyl ethers, and methyl 3,6-dimethylresorcylate [*4707-47-5*] (see p. 137) are mainly responsible for the characteristic earthy-moss-like odor of the oakmoss products [651–655].

Orcinol   β-Orcinol   Methyl 3,6-dimethylresorcylate

Tree moss concretes and absolutes are prepared from *Evernia furfuracea* Fr., a lichen growing on conifer bark. Their odors are different from those of the correponding oakmoss products.

Large quantities of oakmoss (mousse de chêne) and tree moss (mousse d'arbre) are collected annually in the Balkan countries (ca. 5000 t), France (ca. 2500 t), and Morocco (ca. 1000 t) [656].

The extracts and absolutes are used in perfumery for nuances and as a fixative to give compositions a dry, sweet base note, e.g, in fougère and chypre perfumes [*900-50-4*], [*68606-93-9*], [*68917-10-2*].

Olibanum oil and **olibanum resinoid** are obtained from frankincense, which is a gum resin collected from the bark of the tree *Boswellia carterii* Birdw. or *B. frereana* Birdw. (Burseraceae) growing in Arabia and Somalia. The resinoid is produced by solvent extraction, and steam distillation gives the oil, which is a pale yellow, slightly viscous liquid with a balsamic odor and a faint lemon note.

d_{25}^{25} 0.862–0.889; n_D^{20} 1.4650–1.4820; α_D^{20} −15° to +35°; solubility: 1 vol in 6 vol of 90% ethanol, solutions occasionally opalescent; acid number: max. 4.0; ester number: 4–40 [657].

Various qualities are commercially available. Their compositions may vary considerably because they are prepared from the resins of different *Boswellia* species. Main constituents of the oil are monoterpene hydrocarbons [658–661].

Olibanum oil and resinoid are used in oriental type perfumes, the resinoid especially for its good fixative properties. FCT 1978 (**16**) p. 835, p. 837; [*8016-36-2*].

Opopanax oil and **opopanax resinoid** are obtained from the resin of *Commiphora erythraea* Engl. var. *glabrescens* Engl., a tree growing in Somalia (Burseraceae). The resinoid is prepared by solvent extraction, and steam distillation of the resin gives the essential oil, which is a yellow to greenish-yellow liquid with a warm, sweet, balsamic odor.

d_{25}^{25} 0.865–0.932; n_D^{20} 1.488–1.504; α_D^{20} −32° to −9°; acid number: max. 4; saponification number: 4–20; solubility: 1 vol in 10 vol of 90% ethanol; solutions are occasionally turbid [662]. Sesquiterpene hydrocarbons like α-santalene, (*E*)-α-bergamotene and (*Z*)-α-bisabolene make up the main constituents of opopanax oil [663].

Opopanax oil and resinoid are used in perfume compositions with oriental characteristics, the resinoid also for its fixative properties, [*8021-36-1*].

Orange flower absolute, see Neroli Oil.

Orange oils, see Citrus Oils.

Origanum oils are produced from several species of the flowering herb of mediterranean *Origanum* (Lamiaceae) mainly *O. vulgare* L. ssp. *hirtum* [664]. Spanish origanum oil is derived from *Coridothymus capitatus* Rchb., syn. *Thymbra capitata* (L.) Cav. The oils differ mainly in their content of carvacrol and thymol, which are major constituents [665–672]. Oil with a high carvacrol content [*499-75-2*] is a yellowish-red to dark brown liquid with a spicy, herbaceous odor, reminiscent of thyme. The color quickly turns to black when in contact with iron.

d_{25}^{25} 0.935–0.960; n_D^{20} 1.5020–1.5080; α_D^{20} −2° to +3°; phenol content: 60–75%; solubility: 1 vol in 2 vol of 70% ethanol; solutions may become turbid when further diluted [673].

Origanum oils are used in perfume compositions for herbal-spicy, leathery notes and in seasoning mixtures. FCT 1974 (**12**) p. 945; [*8007-11-2*].

Orris root oil is obtained by steam distillation of the rhizomes of the sweet iris, *Iris pallida* Lam. (Iridaceae) cultivated in the Italian province of Tuscany or *Iris germanica* L. cultivated in Morocco. Prior to distillation the rhizomes are stored for several years and are then ground. The steam distillate is a light yellow to brown-yellow solid mass with a violet-like odor. The solid liquefies to a yellow to yellow-brown liquid at 38–50 °C.

Acid number: 175–235, corresponding to an acid content of 71–95% (calculated as myristic acid); ester number: 4–35; ketone content (calculated as irone): 9–20%; soluble with ethanol in all proportions at 50 °C [674].

Because of its high content of myristic and other fatty acids, the steam distillate is a waxy mass that resembles a concrete and is sold under this name or as *Beurre d'Iris* (Orris butter). In addition, a neutral product, obtained after removal of the acids with alkali, is marketed under the name *orris oil absolute* or as *orris oil 10-fold*.

Main constituents of the oils are *cis-γ*-irone (usually 30–40%) and *cis-α*-irone (usually 20–30%) which are also responsible for the typical odor. *I. pallida* oil contains the dextrorotatory enantiomers while *I. germanica* oil contains the laevorotatory enantiomers [675, 676].

(–)-*cis*-γ-Irone   (–)-*cis*-α-Irone
[*89888-04-0*]   [*89888-03-9*]

Fresh orris roots do not contain any irones. These compounds are formed by an oxidative degradation process from higher molecular precursors, the so-called iridales, only when the dried orris roots are stored for a longer period. To avoid such a time-consuming procedure, microbiological processes for the oxidative degradation have been developed recently [677].

The oil is very expensive and is used in perfumery and in flavor compositions. FCT 1975 (**13**) p. 895; [*8002-73-1*].

Palmarosa oil, see Cymbopogon Oils.

Parsley oils are produced from the plant *Petroselinum crispum* (Mill.) Nym. ex A. W. Hill (P. *sativum* Hoffm.), (Apiaceae):

1. *Parsley leaf oil* is produced by steam distillation of the aerial parts of the plant, including the immature seeds. It is a yellow to light-brown liquid with the typical odor of the herb.

d_{25}^{25} 0.908–0.940; n_D^{20} 1.5030–1.5300; α_D^{20} −9° to +1°; acid number: max. 2; soluble in 95% ethanol; solutions may be opalescent [678].

The oil consists mainly of monoterpene hydrocarbons. Its main constituent is 1,3,8-menthatriene [*18368-95-1*], which is important for the aroma of parsley leaves [679–688].

2. *Parsley seed oil* is obtained by steam distillation of ripe fruits of parsley. It is an almost colorless to amber-yellow liquid whose dry odor is characteristic of the crushed fruit, but different from that of the green parts of the plant.

d_{20}^{20} 1.043–1.083; n_D^{20} 1.5110–1.5220; α_D^{20} −11° to −4°; acid number: max. 6; ester number: max. 11; solubility: 1 vol in at least 5 vol of 80% ethanol, solutions are sometimes slightly turbid [689].

Characteristic components of parsley seed oil are myristicin [607-91-0], apiol [523-80-8], and 2,3,4,5-tetramethoxyallylbenzene [15361-99-6] [679–688].

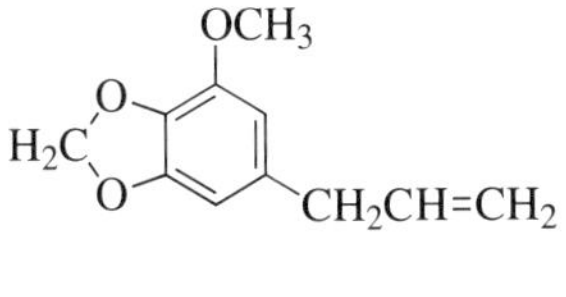

Myristicin 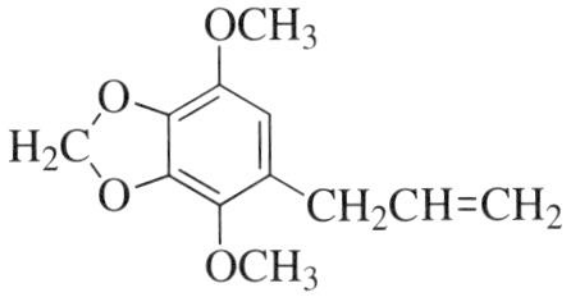   Apiol

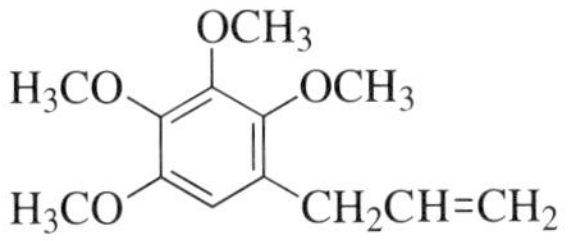

2,3,4,5-Tetramethoxyallylbenzene

Parsley oils are used in the food industry in seasonings, for example, for meat and sauces. FCT 1983 (21) p. 871, 1975 (13) p. 897; [8000-68-8].

Patchouli oil is produced by steam distillation of the dried leaves of *Pogostemon cablin* (Blanco) Benth. (Lamiaceae). It is a reddish-brown to greenish-brown, more or less viscous liquid with a characteristic, slightly camphoraceous, woody balsamic odor.

d_{20}^{20} 0.955–0.983; n_D^{20} 1.5050–1.5120; α_D^{20} −66° to −40° (in exceptional cases to −32°); solubility: 1 vol in 10 vol of 90% ethanol at 20 °C; acid number: max. 4.0; ester number: max. 10 [690].

The patchouli shrub is cultivated primarily in Indonesia. It grows to 1 m, but during harvesting is cut back to 10–15 cm. The oil yield from preferably sun-dried leaves is ca. 2 %.

Although the sesquiterpene alcohol (−)-patchoulol [5986-55-0] is the main component of patchouli oil (30–40%), this compound contributes less to the characteristic odor of the oil than norpatchoulenol [41429-52-1] present only at a concentration of 0.3–0.4% [261, 691–695]

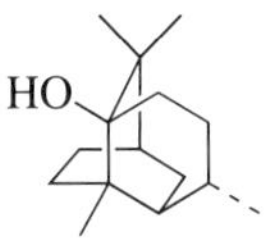

Patchoulol 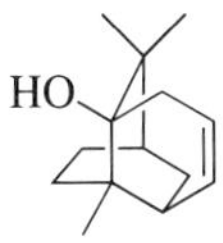   Norpatchoulenol

Worldwide production exceeds 500 t/a. Patchouli oil is very tenacious and is used in perfumery for oriental and masculine notes. FCT 1982 (20) p. 291; [8014-09-3].

Pepper oil and **pepper oleoresin** are obtained from the black pepper *Piper nigrum* L. (Piperaceae). Pepper oil is produced by steam distillation of whole or crushed fruits. The oil is an almost colorless to bluish-green liquid with a characteristic odor, reminiscent of pepper, but without the pungency of the spice.

d_{20}^{20} 0.870–0.890; α_D^{20} −16° to +4°; solubility: 1 vol in 3 vol of 95% ethanol; ester number: max. 11 [696].

The oil contains mainly monoterpene hydrocarbons (ca. 80%), its main constituent being sabinene [*3387-41-5*] (20–25%) [697–700].

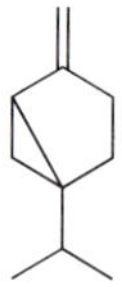

Sabinene

Pepper oleoresin is produced by solvent extraction and, unlike the oil, contains the principal pungent compound, piperine [*94-62-2*], along with some essential oil.

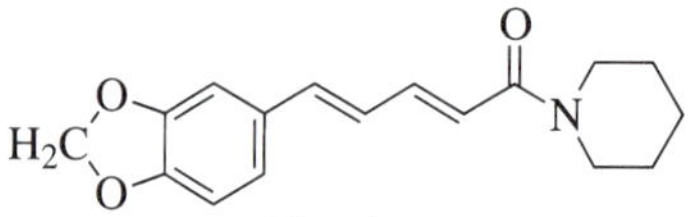

Piperine

Pepper products are used for flavoring foods; pepper oil is also employed to a small extent in perfume compositions. FCT 1978 (**16**) p. 651; [*8006-82-4*].

Peppermint oil, see Mint Oils.

Peru balsam oil is obtained by distillation of the balsam of the tree *Myroxylon pereira* (Royle) Klotzsch (Fabaceae), called *peru balsam*. It is a yellow to pale brown, slightly viscous liquid, which may form crystals. The oil has a rather sweet, balsamic odor.

d_{25}^{25} 1.095–1.110; n_D^{20} 1.5670–1.5790; α_D^{20} −1° to +2°; acid number: 30–60; ester number: 200–225; solubility: 1 vol in at least 0.5 vol of 90% ethanol [701].

Peru balsam is produced almost exclusively in El Salvador. The oil yield from the balsam is ca. 50%. Major components are benzyl benzoate and benzyl cinnamate. Nerolidol and vanillin contribute to the odor.

The oil is used for its excellent fixative properties in perfumes for soap and cosmetics. Use of Peru balsam itself is banned because it is allergenic. FCT 1974 (**12**) p. 951, p. 953; [*8007-00-9*].

Petitgrain oils are obtained by steam distillation of the leaves of citrus trees. The oils derived from the bitter orange tree are the most important. Other petitgrain oils (mandarinier, citronnier, and bergamottier) are less important. Petitgrain oil mandarinier is a source of natural methyl *N*-methylanthranilate, which is present at a concentration of more than 50%. Petitgrain oils are essential constituents of eau de cologne.

Petitgrain oil Paraguay is obtained from an acclimatized variety of the bitter orange tree, *Citrus aurantium* L. subsp. *aurantium*, which is grown in Paraguay. It is pale yellow liquid with a characteristic, strong, pleasant odor, reminiscent of linalool and linalyl acetate.

d_{20}^{20} 0.884–0.892; n_D^{20} 1.457–1.465; α_D^{20} −3.5° to 0°; solubility: 1 vol in 4 vol of 70% ethanol at 20 °C; acid number: max. 2; ester number: 110–170, corresponding to an ester content of 39–60% (calculated as linalyl acetate) [702]. Main constituents are linalool and linalyl acetate. A number of trace constituents contribute essentially to the odor [428, 703–706].

Annual production is ca. 200 t.

Petitgrain oil bigarade is derived from the bitter orange tree *Citrus aurantium* L. subsp. *aurantium*, grown in France, Italy, Spain, and North Africa.
FCT 1992 (**30 suppl.**) p. 101S; [*8014-17-3*].

Pimento oils (allspice oil) are derived from pimento fruits and leaves. Pimento berry oil is obtained by steam distillation of the dried, fully grown, unripe, berry-like fruits of the pimento shrub, *Pimenta dioica* (L.) Merrill. (Myrtaceae), growing in the islands of the West Indies and Central America. It is a pale yellow to brown liquid with a spicy odor, reminiscent of eugenol.

d_{20}^{20} 1.027–1.048; n_D^{20} 1.5250–1.5400; α_D^{20} −5° to 0°; solubility: 1 vol in 2 vol of 70% ethanol at 20 °C; phenol content: min. 65% [707]

The major components of pimento berry oil are eugenol (up to 75%), 1,8-cineole, and caryophyllene [708–712]. Distillation of the leaves gives an oil that has an even higher content of eugenol (80–90%). Annual production of leaf oil is ca. 50 t, which exceeds that of the berry oil.

Pimento oils, like the spice itself, are used mainly in the food industry, as well as in perfume compositions for creating spicy, clovelike notes. FCT 1974 (**12**) p. 971, 1979 (**17**) p. 381; [*8006-77-7*].

Pinaceae needle oils from Pinaceae species contain (−)-bornyl acetate as their main odoriferous component. Other main constituents are monoterpene hydrocarbons such as α- and β-pinene, limonene, 3-carene, and α- and β-phellandrene [713–718]. The oils are used in perfumes for soap, bath products, and air fresheners and in pharmaceutical preparations.

1. *European silver fir oil* is produced in central Europe by steam distillation of needle-bearing twigs of *Abies alba* Mill. It is a colorless to pale yellow liquid with a pleasant odor of freshly cut fir needles.

d_{25}^{25} 0.867–0.878; n_D^{20} 1.4700–1.4750; α_D^{20} −67° to −34°; solubility: 1 vol in 7 vol of 90% ethanol; ester content (calculated as bornyl acetate): 4–10% [719].

Silver fir cone oil (templin oil) obtained from the cones of *Abies alba* Mill. has similar properties.

2. *Siberian fir needle oil* is obtained by steam distillation of needles and twigs of *Abies sibirica* Ledeb., the Siberian silver fir. It is an almost colorless or pale yellow liquid with a characteristic fir odor.

d^{25} 0.898–0.912; n_D^{20} 1.4685–1.4730; α_D^{20} −45° to −33°; solubility: 1 vol in 1 vol of 90% ethanol, solutions may become hazy when further diluted; ester content (calculated as bornyl acetate): 32–44% [720]. FCT 1975 (**13**) p. 450; [*8021-29-22*].

3.   *Canadian fir needle oil* (balsam fir oil) is produced in Canada and in several northern states of the United States from needles and twigs of *Abies balsamea* (L.) Mill. It is an almost colorless to pale yellow liquid with a pleasant balsamic odor.
d_{25}^{25} 0.872–0.878; n_D^{20} 1.4730–1.4760; α_D^{20} −24° to −19°; solubility: 1 vol in 4 vol of 90% ethanol (turbidity may occur), ester content (calculated as bornyl acetate): 8–16% [721].

4.   *Pine-needle oil* is produced primarily in Austria (Tyrol), and Eastern and Southeastern Europe by steam distillation of the needles of the Norwegian or Scotch pine, *Pinus sylvestris* L. or *Pinus nigra* Arnold. It is a colorless or yellowish liquid with an aromatic, turpentine-like odor.
d^{25} 0.857–0.885; n_D^{20} 1.4730–1.4785; α_D^{20} −4° to +10°; solubility: 1 vol in 6 vol of 90% ethanol, sometimes slightly opalescent; ester content (calculated as bornyl acetate): 1.5–5% [722]. FCT 1976 (**14**) p. 845; [*8023-99-2*].

5.   *Dwarf pine-needle oil* is obtained in Austria (Tyrol) and other mountainous areas of central and southeast Europe by steam distillation of fresh needles and twig tips of *Pinus mugo* Turra subsp. *mugo* Zenan and subsp. *pumilio* (Haenke) Franco. It is a colorless liquid with a pleasant, balsamic odor.
d_{25}^{25} 0.853–0.871; n_D^{20} 1.475–1.480; α_D^{25} −16° to −3°; solubility: 1 vol in 10 vol of 90% ethanol; bornyl acetate content is similar to that of pine needle oil. FCT 1976 (**14**) p. 843; [*8000-26-8*].

6.   *Spruce and hemlock oils* are produced in Canada and the Northeast of the United States by steam distillation of needles and twigs from *Picea mariana* (Mill.) B.S.P. (black spruce), *Picea glauca* (Moench) Voss (white spruce), *Tsuga canadensis* (L.) Carr. (Hemlock spruce), and related species. They are very pale to light yellow liquids with a pleasant odor reminiscent of pine needles.
d_{25}^{25} 0.900–0.915; n_D^{20} 1.4670–1.4720; α_D^{20} −25° to −10°; solubility: 1 vol in at least 1 vol of 90% ethanol; solutions may become hazy when >2 vol of 90% ethanol are added; ester content (calculated as bornyl acetate): 37–45% [723].

Pine oil, see Turpentine Oil.

Rose oil and **rose absolute** are used mainly in fine fragrances. Rose oil is also used in small amounts for flavoring purposes.

1.   *Rose oil* is obtained by steam distillation of blossoms of *Rosa × damascena* Mill. that is mainly cultivated in Turkey and Bulgaria (Kazanlik rose). Since a reasonable amount of rose oil is still dissolved in the aqueous phase after steam distillation, the distillation water (*rose water*) is redistilled or extracted.

Rose oil is a yellow liquid with the characteristic odor of rose blossoms and distinct tea and honey notes.

d_{20}^{20} 0.848–0.861; n_D^{20} 1.4530–1.4640; α_D^{20} −2° to −5°; *fp* ca. 20 °C; ester value: 7.5–23.5; ester value after acetylation: min. 204 to max. 231, corresponding to an alcohol content calculated as citronellol: 71–77%; content by GC: citronellol: 34–44%; geraniol 12–18%; phenylethyl alcohol <2% [724].

The major constituents of rose oil are (−)-citronellol, geraniol, and nerol. In contrast to the absolute (see below), the oil contains only a minor amount of phenethyl alcohol, which is extremely soluble in water.

In addition to the major components mentioned above, rose oil also contains a number of components which, although present in low concentrations, contribute to the characteristic fragrance [725–731]. Among these are β-damascenone (see p. 66) and rose oxide (see p. 139).

Since ca. 3 t of blossoms are required to prepare ca. 1 kg of oil, rose oil is one of the most expensive essential oils. Production is just a few tons per year.

2. *Rose absolute* is prepared from the concrete extracted from *Rosa × damascena* in the countries mentioned above, as well as from *Rosa centifolia* L. types in Morocco and the south of France (rose de mai).

The absolute is a reddish liquid with a typical rose odor. The phenethyl alcohol content of its volatile fraction is 60–75%.
FCT 1974 (**12**) p. 979, p. 981, 1975 (**13**), p. 911, p. 913; [*8007-01-0*].

Rosemary oil is obtained by steam distillation of the twigs and flowering tops of *Rosmarinus officinalis* L. (Lamiaceae). It is an almost colorless to pale yellow liquid with a characteristic, refreshing, pleasant odor.

d_{20}^{20} 0.895–0.905/0.908–0.920; n_D^{20} 1.4670–1.4740 (both); α_D^{20} −3° to +10°/−1° to +6°; ester value: 2–20 (both); ester value after acetylation: 30–50/43–72 (Spanish/ North African type) [732].

Major constituents are 1,8-cineole (ca. 20%/45%), α-pinene (ca. 20%/10%) and camphor (ca. 15%/10%) (Spanish/North African oils). Verbenone [*18309-32-5*] is usually a trace constituent in North African oils, but occurs in Spanish oils in quantities up to nearly 3% [733–742].

Verbenone

The main producers of rosemary oil are Tunisia, Morocco and Spain with ca. 150 t annually. Rosemary oil is used widely in perfumery and in large amounts for perfuming bath foams, shampoos, and hair tonics. FCT 1974 (**12**), p. 977; [*8000-25-7*].

Rosewood oil, Brazilian (Bois de rose oil) is obtained by steam distillation of wood from *Aniba rosaeodora* A. Ducke var. *amazonica* A. Ducke and/or *A. parvifora*

Meissner Mez. (Lauraceae). It is an almost colorless to pale yellow liquid with a characteristic, sweet odor, reminiscent of linalool its main constituent [743].

d_{20}^{20} 0.872–0.887; n_D^{20} 1.4620–1.4690; α_D^{20} −2° to +4°; solubility: 1 vol in 9 vol of 60% ethanol at 20 °C; ester number (after acetylation): 247–280, corresponding to an alcohol content of 82–96% (calculated as linalool); linalool content by GC: 75–95% [744].

Brazilian rosewood oil is no longer competitive as a raw material for linalool. Increasing production costs, as well as the development of large-scale processes for fully synthetic linalool (for production of vitamin A), have led to a sharp decline in production. Currently, rosewood oil is, if ever, only produced in very small quantities. FCT 1978 (16) p. 653; [*8015-77-8*].

Sage oils are of three main types:

1.   *Clary sage oil* is obtained by steam distillation of flowering tops and foliage of cultivated *Salvia sclarea* L. (Lamiaceae). It is a pale yellow to yellow liquid with a fresh herbaceous odor and a winelike bouquet.

d_{20}^{20} 0.890–0.908; n_D^{20} 1.4560–1.4660; α_D^{20} −10° to −26°; ester value: 180–235 corresponding to an ester content of 63–82% calculated as linalyl acetate; solubility: 1 vol in max 3 vol 80% ethanol; content by GC: 56–78% linalyl acetate and 6.5–24% linalool [745].

In addition to linalyl acetate, the oil contains linalool and other terpene alcohols, as well as their acetates. When the volatile components are evaporated, a distinct ambergris note develops that is attributed to oxidative degradation products of sclareol [*515-03-7*] [746–750]. Sclareol is the main component in the concrete, obtained by solvent extraction of *S. sclarea* L. leaves.

Sclareol

Sclareol is used as a starting material for a number of ambergris fragrances. Major cultivation areas for *S. sclarea* L. are Russia, the Mediterranean countries, and the United States.

Clary sage oil is used extensively in fine fragrances. FCT 1974 (12) p. 865; 1982 (20) p. 823; [*8016-63-5*].

2.   *Dalmatian sage oil* is obtained by steam distillation of the partially dried leaves of *Salvia officinalis* L. (Lamiaceae). It is a yellowish to greenish-yellow liquid with a warm camphoraceous, thujone-like odor and taste.

d_{20}^{20} 0.910–0.930; n_D^{20} 1.4580–1.4740; α_D^{20} +2° to +30°; solubility 1 vol in max. 8.5 vol 70% ethanol; carbonyl value: 103–288, corresponding to carbonyl compound content of 28–78%, calculated as thujone; content by GC: 18–43% α-thujone, 3–8.5% β-thujone, 5.5–13% 1,8-cineole, 4.5–24.5% camphor [751]. For further constituents see [752–758].

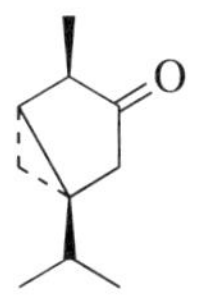

α-Thujone
[546-80-5]
[(–)-thujone]

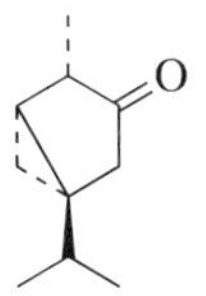

β-Thujone
[471-15-8]
[(+)-isothujone]

The oil is used in pharmaceutical preparations and in perfumery to create dry, spicy-herbaceous notes. FCT 1974 (**12**) p. 987; [*8022-56-8*].

3.   *Spanish sage oil* is produced by steam distillation of leaves and twigs of *Salvia lavandulifolia* Vahl growing in Spain in the provinces Granada, Murcia, Almaria, and Jaén. The oil is an almost colorless to pale yellow liquid with the characteristic camphoraceous odor of the leaves.
d_{20}^{20} 0.913–0.933; n_D^{20} 1.467–1.473; α_D^{20} −12° to 0° for Jaén oils, +16° to +24° for other oils; solubility: 1 vol in 2 vol of 80% ethanol at 20 °C; ester number: min. 14, max. 57 [759].
Unlike Dalmatian sage oil, Spanish sage oil does not contain α- or β-thujone; camphor and 1,8-cineole are the major components [760–762] and are responsible for the odor of the oil, which is used primarily in pharmaceutical preparations and in technical perfumery. FCT 1976 (**14**) p. 857; [*8022-56-8*].

Sandalwood oil, East Indian is obtained by steam distillation of the heartwood of *Santalum album* L. (Santalaceae). It is a slightly viscous, almost colorless to yellow liquid with a characteristic, sweet, woody, long-lasting odor.
d_{20}^{20} 0.968–0.983; n_D^{20} 1.5030–1.5080; α_D^{20} −21° to −15°; solubility: 1 vol in 5 vol of 70% ethanol at 20 °C; ester number: max. 10; free alcohol content (calculated as santalol): min. 90% [763].
East Indian sandalwood oil consists almost exclusively of closely related sesquiterpenoids; the alcohols α-santalol [*115-71-9*] and β-santalol [*77-42-9*] are mainly responsible for the odor and make up >90% of the oil. α-Santalol is more abundant [261, 764–769].

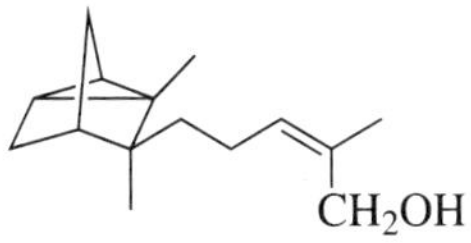

CH₂OH

α-Santalol

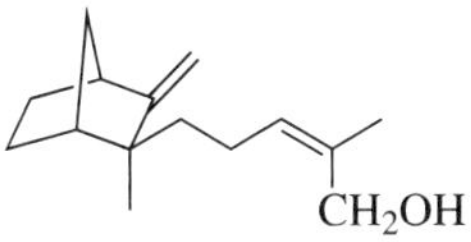

CH₂OH

β-Santalol

The trees cultivated for oil production must be at least 30 years old to make oil production profitable. The oil yield, calculated on the amount of wood used for steam distillation, is 4–6.5%. The production in India and Indonesia amounts to ca. 100 t/a.

East Indian sandalwood oils are used extensively in perfumery as very valuable, stable fixatives. FCT 1974 (**12**) p. 989; [*8006-87-9*].

Sassafras oil

Sassafras oil is used as a generic term for commercial essential oils containing high quantities of safrole. They originate from different botanical species. The main use of these oils is the isolation of safrole as the starting material for the production of heliotropin (see p. 132)

1.   *Brazilian sassafras oil* is obtained by steam distillation of the roots, trunks, and branches of *Ocotea pretiosa* (Nees) Mez. (Lauraceae) a tree growing wild in South America (Brazil). The oil was formerly called *Ocotea cymbarum* oil due to incorrect botanical naming. It is a yellow to brownish liquid with the characteristic odor of safrole.

d_{20}^{20} 1.079–1.098; n_D^{20} 1.5330–1.5370; α_D^{20} −2.2° to −0.8°; solubility: 1 vol in 2 vol of 90% ethanol; *fp* +7.5 to +9.3 °C [*770*].

The main component of the oil is safrole [*771*], which may make up more than 90% of the oil and determines its freezing point.

2.   *Chinese sassafras oils* are oils or fractions of oils, rich in safrole, that are obtained from different species of the camphor tree (see Camphor Oils) [*772*].

Worldwide production of sassafras oils exceeds 1500 t/a. FCT 1978 (**16**) p. 831; [*68917-09-9*].

Savory oil is obtained by steam distillation of the whole dried herb *Satureja hortensis* L. or *S. montana* L. (Lamiaceae). It is a light yellow to dark brown liquid with a spicy odor, reminiscent of thyme or origanum.

d_{25}^{25} 0.875–0.954; n_D^{20} 1.4860–1.5050; α_D^{20} −5° to +4°; phenol content: 20–59%; saponification number: max. 6; solubility: 1 vol in 2 vol of 80% ethanol at 25 °C; solutions in 10 vol of ethanol may be slightly hazy [*773*].

Several qualities of savory oil exist, depending on the *Satureja* species used. The main cultivation areas are France, Spain, some of the Balkan countries, and some midwestern states of the United States. Characteristic of savory oil is its high content of carvacrol [*499-75-2*] [*774–777*].

Carvacrol

Savory oil is used mainly in the food industry, e.g., for flavoring sauces and soups. FCT 1976 (**14**) p. 859; [*8016-68-0*].

Spearmint oil, see Mint Oils.

Spike oil, see Lavandula Products.

Star anise oil is obtained by steam distillation of the star-shaped fruits of *Illicium verum* Hook. f. (Illiciaceae), an evergreen tree growing in Vietnam and China. It is a colorless to pale yellow liquid, which solidifies on cooling.

d_{20}^{20} 0.979–0.985; n_D^{20} 1.5530–1.5560; α_D^{20} −2° to +2°; solubility: 1 vol in max. 3 vol 90% alcohol; *fp* min. 15 °C; content of anethole by GC: 86–93% [778].

The main component of star anise oil, as in anise oil, is *trans*-anethole (80–90%) [779–782]. Pure anethole can be obtained by rectification (see p. 122). Star anise oil has replaced true anise oil derived from *Pimpinella anisum* in the production of natural anethole.

Star anise oil and its product, anethole, are used primarily in the alcoholic beverage industry (anise liquors), but also for flavoring food and toothpaste. FCT 1975 (**13**) p. 715; [*68952-43-2*].

Styrax oil and **styrax resinoid** are obtained from styrax balsam, which is collected from the artificially injured trees, *Liquidamber orientalis* Mill. (Asia Minor) and *L. styraciflua* L. (Central America) (Hamamelidaceae).

Styrax resinoid is a light grey to dark grey-brown viscous liquid that is prepared by solvent extraction. It has a sweet balsamic, slightly grasslike odor and is used in perfumery as a fixative.

Steam distillation of the balsam gives *styrax oil*, a light yellow liquid that contains a relatively large amount of cinnamic acid, which partially crystallizes [783]. Styrax oil has a sweet-balsamic odor with a styrene-like top note. It is used in perfumery in flowery compositions, [*8024-01-9*].

Tagetes oil is produced by steam distillation of the flowering plants *Tagetes minuta* L. (*T. glandulifera* Schrank., Asteraceae). The main producers are in Zimbabwe, South Africa, and India. The oil is a yellow to dark orange liquid with a strong, aromatic-fruity odor. Its main components are *cis*-ocimene, dihydrotagetone, tagetone, and *cis*- and *trans*-ocimenone [784].

It is used in accentuated flowery fragrances and in aroma compositions to achieve fruity effects. FCT 1982 (**20**) 829; [*8016-84-0*].

Tarragon oil (estragon oil) is produced by steam distillation of leaves, stems, and flowers of *Artemisia dracunculus* L. (Asteraceae). It is a pale yellow to amber liquid with a characteristic, spicy, delicate estragon odor reminiscent of liquorice and sweet basil. The following specifications refer to tarragon oil obtained from plants cultivated in southern France and in Piedmont (Italy).

d_{20}^{20} 0.914–0.956; n_D^{20} 1.5040–1.5200; α_D^{20} +2° to +6°; acid number: max. 1; saponification number: max. 18; solubility: 1 vol in max. 4 vol of 90% ethanol; estragole content by GC: 68–80% [785].

Estragole as the main constituent [786–788] of tarragon oil determines primarily the organoleptic properties.

Worldwide production of tarragon oil only amounts to a few tons per year. It is used mainly in aroma compositions, smaller quantities are employed in perfumery. FCT 1974 (**12**) p. 706; [*8016-88-4*].

Tea tree oil is obtained by steam distillation of leaves and twig tips of *Melaleuca alternifolia* L., and other *Melaleuca* species, which are trees growing in Australia (Myrtaceae). It is a pale to light yellow liquid with an earthy, spicy odor.

d_{20}^{20} 0.885–0.906; n_D^{20} 1.4750–1.4820; α_D^{20} +5° to +15°; solubility: 1 vol in max. 2 vol of 85% ethanol; GC content: cineole max. 15%, 1-terpinen-4-ol min. 30% [789].

The main component of the oil is 1-terpinen-4-ol (up to 40%, or occasionally more) [790–792].

Tea tree oil is used in perfumery for creating certain nuances and for earthy notes. FCT 1988 (**26**) 407; [*68647-73-4*].

Thuja oil, see Cedar Leaf Oil.

Thyme oil (*Spanish red thyme oil*) is obtained by steam distillation of flowering plants of *T. zygis* L. var. *gracilis* Boissier (Lamiaceae). Thyme oil is a red or reddish liquid with a strong, characteristic, spicy-phenolic odor and a sharp, lasting taste.

d_{20}^{20} 0.915–0.935; n_D^{20} 1.4960–1.5050; solubility: 1 vol in max. 2 vol of 80% ethanol; total phenol content: 38–56% [793].

The main constituent of thyme oil is thymol (36–55%) which also determines predominantly the organoleptic properties of the oil [794–799].

Thyme shrubs grow in France, Spain, Algeria, and Morocco. Spain is the main producer of the oil. Thyme oil is used mainly for flavoring foods and oral hygiene products, but is also used in perfumery to create spicy, leathery notes. Because of its high phenol content, thyme oil has germicidal and antiseptic properties. FCT 1976 (**12**) p.1003; [*8007-46-3*].

Tolu balsam resinoid is produced by extraction of the balsam of *Myroxylon balsamum* (L.) Harms (Fabaceae). It is a dark orange brown mass with a sweet, resinous, long-lasting odor, reminiscent of hyacinth. An essential oil is also distilled from the balsam.

Tolu balsam resinoid contains a fairly large amount of benzyl and cinnamyl esters of benzoic and cinnamic acid [800].

Both the resinoid and the balsam are used in perfumery, mainly for their fixative properties. FCT 1976 (**14**) p. 689; [*9000-64-0*].

Tonka bean absolute is prepared by solvent extraction either directly from so-called Tonka beans or via the concrete. Tonka beans are the seeds of fruits from *Dipteryx odorata* (Aubl.) Willd. (Fabaceae).

Tonka bean absolute is a solid or crystalline mass with a sweet, caramel-like odor, reminiscent of coumarin. In perfumery tonka bean absolute is used as a fixative and produces a dry sweetness in men's fragrances. FCT 1974 (**12**) p. 1005; [*8046-22-8*].

Tree moss absolute, see Oakmoss Absolute.

Tuberose absolute is obtained by solvent extraction, via the concrete, from the blossoms of *Polianthes tuberosa* L. (Agavaceae). The plants are cultivated in Egypt and India. Tuberose absolute was formerly produced in southern France by enfleurage. It is an orange to brown liquid with a sweet-narcotic blossom odor and is used in modern, flowery perfume compositions. Its main constituents are esters of benzoic acid [801, 802] [*8024-05-3*].

Turpentine oils [803] are used in large quantities by the fragrance industry as starting materials for the manufacture of single fragrance and flavor compounds. Turpentine oils are obtained either from balsams or from the wood of various Pinaceae species. They are less valuable as odor materials than the oils obtained from needles, young twigs, and cones (see Pinaceae Needle Oils). Nevertheless, they are used either as such or indirectly as fragrant solvents for perfuming household products.

Balsam turpentine oil is obtained from the resins of living trees of suitable *Pinus* species by distillation at atmospheric pressure and temperatures up to 180 °C, or by other fractionation methods, which do not change the terpene composition of the resins. *Wood turpentine oils*, on the other hand, are generally obtained by steam distillation of chopped tree trunks, dead wood, or of resin extracted from this wood. *Sulfate turpentine oil* is produced as waste in the manufacture of cellulose by the sulfate process and is also a wood turpentine. *Pine oil* is another wood turpentine oil that is obtained by dry distillation of suitable pine and fir trees, followed by fractionation. However, the term *pine oil* is nowadays used for a product which is manufactured by hydration of turpentine oil (α-pinene). The resulting product is a mixture of monoterpenes containing α-terpineol as the main component. In addition to many other technical purposes, it is used to a large extent in cheap perfumes for technical applications.

Balsam and wood turpentine oils are colorless liquids with a mild, characteristic odor. Oils obtained by dry distillation often also have a phenolic note. The specifications of turpentine oils are listed in Table 6 [804].

Worldwide production of turpentine oils was ca. 280 000 t in 1990, of which approximately one third was produced in the United States and Canada as sulfate turpentine oil. Other major producers are Russia, China, and Scandinavia.

The main components of all turpentine oils are terpene hydrocarbons. The balsam and sulfate turpentine oils produced in the Southeast of the United States contain mainly α- and β-pinene (60–70% and 20–25%, respectively); *P. palustris* Mill. (long leaf pine) gives the (+)-enantiomers and *P. caribea* Morelet (slash pine) gives the (−)-enantiomers. The remaining constituents are *p*-menthadienes,

Table 6. Specifications of turpentine oils

Parameter	Type of turpentine oil			
	Balsam	Wood, steam distilled	Sulfate	Wood, dry distilled
d_4^{20}	0.855–0.870	0.855–0.870	0.860–0.870	0.850–0.865
n_D^{20}	1.465–1.478	1.465–1.478	1.465–1.478	1.463–1.483
Distillate up to *bp* 170 °C, %	90	90	90	60
Evaporation residue, %	2.5	2.5	2.5	2.5
Acid number	1	1	1	1

oxygen-containing terpenoids, and anethole. In contrast with the Scandinavian and Russian turpentine oils, the American oils contain very little 3-carene or camphene. Large amounts of the American oils are separated by fractional distillation into their components, which are used as starting materials in the manufacture of fragrance and flavor compounds. European turpentine oils originate mainly from Portugal, where distillation of *P. pinaster* Soland resin or wood gives a turpentine oil that consists of 72–85% (−)-α-pinene and 12–20% (−)-β-pinene: d_{20}^{20} 0.860–0.872; n_D^{20} 1.4650–1.4750; $α_D^{20}$ −28° to −35°; acid value: max. 1 [805].

Valerian oil is obtained by steam distillation of well-dried ground roots of *Valeriana officinalis* L. (Valerianaceae). It is a yellow-green to yellow-brown liquid with an extremely strong, characteristic, penetrating odor. The oil becomes dark and viscous on aging or on exposure to air.

d_{25}^{25} 0.942–0.984; n_D^{20} 1.4860–1.5025; $α_D^{20}$ −28° to −2°; acid number: 5–50; saponification number: 30–107; solubility: 1 vol in 0.5–2.5 vol of 90% ethanol at 25 °C, solutions are clear to opalescent when up to 10 vol of ethanol is added [806].

The main component of the oil is (−)-bornyl acetate, but it also contains other bornyl esters (e.g., bornyl isovalerate), terpene and sesquiterpene hydrocarbons, as well as free isovaleric acid, which contributes strongly to the odor of the oil [807].

Valerian oil is produced in limited quantities and is used in flavor and fragrance compositions only in very low dosages to create certain effects, [*8008-88-6*].

Vanilla extract (vanilla oleoresin) is produced by extraction of the pods of *Vanilla planifolia* G. Jacks. or *V. tahitensis* Moore (Orchidaceae) with a polar solvent (e.g., methanol, ethanol, or acetone, which may also contain water). The composition of the extract depends on the type and amount of solvent used. Generally, the percentage of vanillin in the extract (yield 25–30%) is 3–4 times higher than that in the pods.

Vanillin and phenol derivatives are primarily responsible for its aroma [808–811]. The main producers of *V. planlifolia* pods are Madagascar, the Comoro Islands, Réunion (Bourbon), Mexico, and some Pacific Islands; *V. tahitensis* pods are grown mainly in Tahiti.

Vanilla extracts are used extensively in chocolate and baked products, but even more so in ice cream. FCT 1982 (**20**) p. 849; [*8023-78-7*].

Vetiver oil is produced by steam distillation of the roots of the grass *Vetiveria zizaniodes* (L.) Nash (Poaceae), which grows wild or is cultivated in many tropical and subtropical countries. The oil is a brown to reddish-brown, viscous liquid with a characteristic precious-wood and rootlike, long-lasting odor. The following specifications are for oil produced in Haiti:

d_{20}^{20} 0.986–0.998; n_D^{20} 1.5210–1.5260; α_D^{20} +22° to +38°; acid number: max. 14; ester number: 5–16; ester number (after acetylation): 160–192; carbonyl number: 23–59; solubility: 1 vol in 2 vol of 80% ethanol at 20 °C [812].

The tenacity of the highly complex vetiver oil is attributed to its high sesquiterpene content. The ketones α-vetivone [*15764-04-2*] and β-vetivone [*18444-79-6*], which usually form more than 10% of the oil, as well as khusimol [*16223-63-5*] (ca. 15%) and its derivatives, contribute significantly to its odor.

α-Vetivone   β-Vetivone   Khusimol

The oil contains a considerable number of bi- and tricyclic primary, secondary, and tertiary sesquiterpene alcohols [174b, 280g–k] called vetiverols or vetivenols [261, 813–817]. These alcohols, as well as their acetates, are valuable fragrance materials. Since several varieties of vetiver grass exist and since fresh as well as air-dried roots are distilled under conditions that vary with the producer, the quality of the commercial oils differs considerably. Oil yields are up to 3%. The main producer is Indonesia (ca. 100 t/a, 'Java type'), followed by Haiti and Réunion.

Vetiver oil is used in fine fragrances for long-lasting precious-wood notes. It is also used as starting material for vetiveryl acetate (see p. 71). FCT 1974 (**12**) p. 1013; [*8016-94-4*].

Violet leaf absolute is obtained by solvent extraction, via the concrete, from the leaves of *Viola odorata* L. (Violaceae), which is grown predominantly in southern France.

The absolute is a dark green to brown liquid with a strong, green odor. The main constituent of the volatile fraction is 2-*trans*-6-*cis*-nonadienal [*557-48-2*] (violet leaf aldehyde) [818, 819].

Violet leaf absolute is used frequently in perfume compositions, but only in very low concentration because of its intense odor. FCT 1976 (**14**) p. 893; [*8024-08-6*].

Ylang-ylang and cananga oils are essential oils that are obtained from two subspecies of the cananga tree.

Table 7. Specifications of ylang-ylang oils from the Comoro Islands (C) and Madagascar (M)

Parameter		Ylang-ylang fractions			
		Extra	I	II	III
d_{20}^{20}	(C)	0.956–0.976	0.940–0.950	0.926–0.936	0.906–0.921
	(M)	0.950–0.965	0.933–0.945	0.923–0.929	0.906–0.921
n_D^{20}	(C)	1.4980–1.5060	1.5000–1.5090	1.5050–1.5100	1.5070–1.5110
	(M)	1.5010–1.5090	1.5000–1.5100	1.5050–1.5110	1.5060–1.5130
α_D^{20}	(C)	$-40°$ to $-25°$	$-46°$ to $-38°$	$-55°$ to $-42°$	$-63°$ to $-49°$
	(M)	$-45°$ to $-36°$	$-44°$ to $-28°$	$-55°$ to $-40°$	$-63°$ to $-49°$
Ester value	(C)	145–185	110–140	75–100	45–70
	(M)	125–160	90–120	65–80	38–58
Acid value	(C/M)	max. 3	max. 3	max. 3	max. 3

1.   *Ylang-ylang oils* are obtained by steam distillation of freshly picked blossoms of *Cananga odorata* (DC.) Hook. f. et Thoms. subsp. *genuina* (Annonaceae). These cananga trees normally grow to a height of 20 m but are pruned to a height of 1.60–1.80 m and flower throughout the year. The oil is produced mainly in Madagascar and the Comoro Islands. Four fractions are collected at progressively longer distillation times and are known as 'Extra,' 'I,' 'II,' and 'III.' Occasionally, a first fraction called 'Extra superior' is collected. They are all pale to dark yellow liquids with a characteristic floral, spicy, balsamic odor. The first fractions are the most valuable; they have a higher density and a higher saponification number. Specifications of fractions obtained from Comoro Islands and Madagascar oils are given in Table 7 [820].

The compositions of the various oil fractions depend on the duration of distillation. The first fraction, ylang-ylang oil Extra, has the highest content of strongly odoriferous constituents such as *p*-cresyl methyl ether (15–20%), methyl benzoate (4–6%), (−)-linalool (10–15%), benzyl acetate (20–25%), and geranyl acetate (5–10%). The other fractions contain increasing amounts of sesquiterpenes (>70% in ylang-ylang III). Components such as *p*-cresol, eugenol, and isoeugenol are important for the odor, although they are present only in low concentration [821–824].

Ylang-ylang Extra and I are used mostly in fine fragrances; ylang-ylang II and III are employed in soap perfumes. FCT 1974 (**12**) p. 1015; [*8006-81-3*].

2.   *Cananga oil* is produced by steam distillation of the flowers of *Cananga odorata* (DC) Hook. f. et Thoms. subsp. *macrophylla*. The yield, based on the weight of the flowers, is ca. 1%: The oil is a light to dark yellow liquid with a characteristic, floral, slightly woody odor.

d_{20}^{20} 0.906–0.923; n_D^{20} 1.4950–1.5030; α_D^{20} −30° to −15°; solubility: partially soluble in 95% ethanol at 20 °C; acid number: max. 2; ester number: 13–35 [825].

The qualitative composition of cananga oil resembles that of ylang-ylang III oil but is distinguished by its higher caryophyllene content.

Cananga oil originates almost exclusively in Java, where the flowers are collected throughout the year; annual production is ca. 50 t. The oil is used mainly in perfuming soaps where it is more stable due to its lower ester content in comparison to ylang-ylang oils. FCT 1973 (**11**) p. 1049; [*68606-83-7*].

4 Quality Control

Quality control of fragrance and flavor substances as well as the products derived from them, comprises the comparison of sensory, analytical and if necessary, microbiological data with standards and specifications. To a large extent these have been established in official specification collections (Pharmacopoeias, ISO, Essential Oil Association).

In the past few decades, a precise analytical methodology has been developed for sensory evaluation and has proved to give reliable results [826]. Analytical determination of identity and purity aids greatly in establishing the acceptability of fragrance and flavor materials. To meet the customer requirements all of these methods should be validated by quality assurance tools [827].

Single fragrance and flavor compounds are identified by generally accepted analytical methods like density, refractive index, optical rotation, and melting point. The advantage of these methods is the short analysis time, which provides assess-ment criteria allowing comparison with other laboratories all over the world. Spectroscopic methods such infrared (IR) and near infrared (NIR) are becoming more important for fast identification checks [828]. NIR techniques may also be used for identification of single and complex fragrance and flavor materials [829, 830].

Content, as well as impurity determinations, are done by chromatographic procedures such as gas chromatography (GC), high pressure liquid chromatography (HPLC) [831], capillary electrophoresis (CE) [832], and by spectroscopic techniques (UV, IR, MS, and NMR) [833, 834]. For complex mixtures modern coupling techniques such as GC-MS, GC-FTIR, HPLC-MS, CE-MS [832, 835, 836, 837] and enzymic procedures [838] are becoming important. Classical sample preparation methods such as distillation, soxhlete extraction [839, 840], but also specific techniques such as supercritical fluid extraction (SFE) [841] are used for isolation, separation, and identification of flavor and fragrance compounds. For administration of products, methods, and analytical data modern analytical quality control labs use powerful information and management systems (LIMS) [842].

Standardization of specifications for complex fragrance and flavor materials, such as essential oils and animal secretions, is far more difficult than for single compounds. In addition to organoleptic and physical properties, the content of certain typical components is determined. Problems concerning the natural, botanical, and geographical origins of these products are also solved by using modern chromatographic methods such as enantiomer separations [843], and spectroscopic analytical techniques such as isotope ratio mass spectrometry (IRMS) [844].

The analysis of trace components (halogens, heavy metals, pesticides, aflatoxins, and restsolvents) in flavor and fragrance materials used in foods and cosmetics is becoming increasingly important.

5 Safety Evaluation and Legal Aspects

5.1 Flavoring Substances

Flavoring substances are used to compose special tastes, e.g., strawberry. When flavor compounds are added to foods, no health hazards should arise from the concentration used. The flavor contains flavoring substances and solvents or carriers; the concentration of a single flavoring substance in the food does not exceed 10–20 ppm. Because of the taste and smell of the substances, high concentrations cannot be used, i.e., flavoring substances are self-limiting.

Normal toxicological testing for food additives is not necessary and would be too expensive in view of the relatively small amounts of substance used. Nevertheless, toxicological testing of flavoring substances has provided very useful data. Those substances which may cause adverse effects on human health (including allergens) have been identified and have been prohibited or limited in use.

The JECFA (Joint FAO/WHO Expert Committee on Food Additives/FAO Food and Nutrition Paper No. 30 Rev. 1 and amendments (Food and Agriculture Organisation of the UN, Via della Terme di Caracalla, Rome/Italy)) has made recommendations for the ADI (acceptable daily intake (in mg/kg body weight)) of food additives based on toxicological tests; these include flavoring substances.

Another international body which has dealt with the safety of flavoring substances for human health is the 'Partial Agreement in the Social and Public Health Field' of the 'Council of Europe.' Lists of flavoring substances have been published in the 'blue book' ('Flavouring substances and natural sources of flavourings,' Vol. 1, 4th Edition (Council of Europe, Strassbourg 1992)), which contains 900 flavoring substances with their structures, use levels in beverages and food, main natural food occurrence, and main toxicological data.

Furthermore the IOFI (International Organisation of the Flavor Industry (8, Rue Charles Humbert, CH-1205 Geneva)) has compiled a list of artificial flavoring substances (approved by experts) in the 'Code of Practice' and a short restrictive list of natural and nature-identical flavoring substances with recommended maximum concentrations for foods and beverages. Natural flavoring substances are obtained by physical, microbiological, or enzymatic processes from a foodstuff or material of vegetable or animal origin. Nature-identical substances are obtained by synthesis or isolated by chemical processes from a raw material and are chemically identical to a substance present in natural products. Artificial flavoring substances are made in the same way, but have not yet been identified in a natural product.

Lists of approved flavoring substances have been compiled in the United States since 1959. According to the American Federal Food, Drug, and Cosmetic Act, every natural and artificial flavoring substance has to be approved by the US FDA or a reliable expert panel. Lists of approved and 'GRAS' (generally recognized as safe) flavor substances are published in the CFR (Code of Federal Regulations of USA (US Government Printing Office, Washington, DC)) and by the FEMA (Flavor and Extract Manufacturers Association of the United States (1620 Eye Street, N.W., Washington, DC)). These lists have been adopted by other countries.

In the European Union a Regulation (EC) No. 2232/96 of the European Parliament and of the Council laying down a Community procedure for flavoring substances used or intended for use in or on foodstuffs has been issued (Official Journal of the European Communities No. L 299, 1–4, 23.11.1996).

This regulation, under Article 3, shall provide a list of flavoring substances by the year 2004 whereby member States shall notify the Commission of all flavoring substances which may be used in products marketed on their territory. These flavoring substances shall be entered onto a register which will be adopted by the Commission.

Amendments to the register will be possible, the intellectual property rights of the flavor manufacturers will be protected. In Article 4 details are put down for a programme for the evaluation of these flavoring substances to be initiated within 10 months of the register being adopted. That programme shall define the order of priorities according to which the flavoring substances are to be examined, the time limits, and the flavoring substances which are to be the subject of the scientific co-operation. According to Article 5 the (positive) list of flavoring substances shall be evaluated and adopted within 5 years. New additions to that list may be authorized following evaluation.

5.2 Fragrance Compounds

Fragrance materials are used in a wide variety of products, e.g., cosmetics, soaps, detergents, etc. As is the case for flavoring substances, the concentrations of fragrances used must not cause any adverse effects on human health. Contrary to flavoring substances, special legislative regulations do not exist for fragrance materials. Safety evaluation and regulation for fragrance materials and compounds is mostly done on the basis of voluntary self-control by the industry. Increasing awareness of possible risks has initiated extensive toxicological testing, which has, in turn, generated useful data on many fragrance ingredients. Hundreds of monographs on fragrance materials and essential oils have been published by the Research Institute for Fragrance Materials (RIFM*) in 'Food and Chemical Toxicology' (formerly 'Food and Cosmetics Toxicology'); they report specifications, data on biological activity, and testing results. Details about the organization of the RIFM, test materials, and methods are published in [845]. Based on RIFM and other data, the International Fragrance Association (IFRA**) has published industrial guidelines for limiting or prohibiting the use of certain fragrance ingredients.

Recently an indicative, nonexhaustive list of fragrance ingredients used in cosmetics has been published by the EC Commission (Resolution 96/335/EG of May 8, 1996 published in OJL 132 of June 1, 1996). This is noteworthy, since it is the first time that fragrance ingredients as a group appear in an official piece of legislation. The sole purpose of this list is to enhance transparency of the generic term 'perfume,' which has to be used for ingredient labeling of cosmetic products as prescribed by the EC Cosmetic Directive.

Some time ago discussions about the environmental fate of fragrance compounds, as used in consumer products, roused some interest after residual amounts of certain fragrance ingredients have been identified in various biota. This has triggered extensive environmental testing of the major, high-volume fragrance ingredients. Based on this data risk assessments have been carried out. These assessments did not indicate any immediate risk, nor did they call for risk management measures. Therefore, the IFRA has not seen any necessity to publish guidelines on environmental grounds. However, the issue is under continued scrutiny by the industry's self-regulatory bodies, so that appropriate measures can be taken without delay if any new evidence requires action.

---

* Two University Plaza, Suite 406 Hackensack, N.J. 07601, United States
** 8, Rue Charles Humbert, CH-1205 Genève, Switzerland

6 References

The abbreviation EOA refers to the Essential Oil Association of the United States.

General References

R. C. Weast (ed.): *Handbook of Chemistry and Physics*, 75th ed., CRC Press, Boca Raton, Ann Arbor, Tokyo 1994/95.

S. Arctander: *Perfume and Flavor Chemicals*, publ. by S. Arctander, Elizabeth, N.J., 1969.

Fenaroli's Handbook of Flavor Ingredients, 2nd ed., CRC Press. Cleveland, Ohio 1975.

D. Merkel: *Riechstoffe*, Akademie Verlag, Berlin 1972.

E. Guenther: *The Essential Oils*, D. van Nostrand Co., Toronto, New York, London 1952.

Gildemeister-Hoffmann: *Die Ätherischen Öle*, Akademie-Verlag, Berlin 1966.

Specific References

[1]   H. Rein, M. Schneider: *Physiologie des Menschen*, 19th ed., Springer Verlag, Berlin–Heidelberg–New York 1977.

[2]   Handbook of Olfaction and Gustation, ed. R. L. Doty, Marcel Dekker, Inc., New York 1995.

[3]   G. Ohloff, *Scent and Fragrances*, Springer Verlag, Berlin, Heidelberg 1994.

[4]   L. B. Buck, *Annu. Rev. Neurosci.* **19** (1996) 517–544.

[5]   G. Ohloff, B. Winter, C. Fehr, in *Perfumes, Art, Science and Technology*, eds. P. M. Müller, D. Lamparsky, Elsevier Science Publishers Ltd., London, New York 1991, p. 287–330.

[6]   M. Chastrette, C. El Aidi, J. F. Peyraud, *Eur. J. Med. Chem.* **30** (1995) 679–686.

[7]   J. A. Bajgrowicz, C. Broger, in *Flavours, Fragrances and Essential Oils, Proc. 13th Int, Congr. Flav., Fragr., Ess. Oils*, Istanbul 1995, ed. K. H. C. Baser, AREP Publ., Istanbul 1995, Vol. 3, p. 1.

[8]   R. R. Calkin, J. S. Jellinek, *Perfumery. Practice and Principles*, John Wiley & Sons, Inc., New York 1994.

[9]   N. Neuner-Jehle, F. Etzweiler, in *Perfumes, Art, Science and Technology*, eds. P. M. Müller, D. Lamparsky, Elsevier Science Publishers Ltd., London, New York, 1991, p. 153–212.

[9a] K.-O. Schnabel, H.-D. Belitz, C. v. Ranson, *Z. Lebensm. Unters. Forsch.* **187** (1988) 215–223.

[10] Firmenich SA, CH 581446, 1973 (F. Naef, G. Ohloff, A. Eschenmoser).

[11] R. G. Berger, *Aroma Biotechnology*, Springer Verlag, Berlin, 1995, p. 60.

[12] Pernod-Ricard SA, FR 2652587, 1989 (P. Brunerie),
       Firmenich SA, WO 93 24644, 1993 (B. Müller, A. Gautier, C. Dean, J. C. Kuhn).

[13] C. M. Wu, S. E. Liou, Y. H. Chang, W. Chiang, *J. Food Sci.* **52(1)** (1987) 132–134.

[14] International Flavors & Fragrances Inc., US 4395370, 1983 (R. M. Boden, T. J. Tyszkiewicz, M. Licciardello).

[15] International Flavors & Fragrances Inc., FR 1 508 854, 1968 (H. J. Blumenthal).

[16] Ruhrchemie AG, EP 7 609, 1978 (H. Bahrmann, B. Cornils, G. Diekhaus, W. Kascha *et al.*).

[17] Union Carbide, US 2 628 257, 1953 (R. I. Hoaglin, D. H. Hirsh).

[18] R. G. Berger, *Aroma Biotechnology*, Springer Verlag, Berlin, 1995, p. 117.

[19] Naarden International N.V., NL 76 08 333, 1976 (P. C. Traas, H. Boelens, H. J. Wille).

[20] T. L. Potter, I. S. Fagerson, *J. Agric. Food. Chem.* **36(11)** (1990) 2054–2056.

[21] Consortium für Elektrochemie DE 3 341 605, 1983 (H. Gebauer, M. Regiert).

[22] Monsanto Chemical Corp., US 2 455 631, 1945 (O. J. Weinkauff).

[23] R. G. Berger, *Aroma Biotechnology*, Springer Verlag, Berlin, 1995, p. 24.

[24] R. G. Berger, *Aroma Biotechnology*, Springer Verlag, Berlin, 1995, p. 61.

[25] Firmenich SA, CH 544 803, 1970 (F. Naef).

[26] Haarmann & Reimer, EP 770 685, 1996 (I. Gatfield, G. Kindel).

[27] Hasegawa, T., Co., Ltd., JP 04 217 642, 1990 (T. Kawanabe, M. Iwamoto, M. Inagaki).

[28] IG Farbenindustrie AG, DE 705 651, 1941 (W. Flemming, H. Buchholz).

[29] Union Camp Corp., EP 132 544, 1984 (P. W. Mitchell, L. T. McElligott, D. E. Sasser).

[30] The Glidden Company, US 3 031 442, 1958 (R. L. Webb).

[31] Kuraray Co., JP 75 58 004, 1973 (O. Yoshiaki, Y. Ninagawa, Y. Jujita, T. Hosogai *et al.*).

[32] SCM Corp., US 4 254 291, 1978 (B. J. Kane).

[33] G. Ohloff, E. Klein, *Tetrahedron* **18** (1962) 37–42.

[34] Rhône-Poulenc Ind., SA, DE-AS 2 025 727, 1970 (P. S. Gradeff, B. Finer).

[35] SCM Corp., US 4 018 842, 1976 (L. A. Canova).

[36] Studiengesellschaft Kohle, US 3 240 821, 1966 (G. Ohloff, E. Klein, G. Schade).

[37] Glidden Co., US 3 076 839, 1958 (R. L. Webb).

[38] Hoffmann-La Roche, CH 342 947, 1959 (W. Kimel, W. N. Sax).

[39] BASF, GB 848 931, 1960 (H. Pasedach, M. Seefelder).

[40] Hoffmann-La Roche, BE 634 738, 1964 (R. Marbet, G. Saucy).

[41] Kuraray, DE 2 356 866, 1974 (Y. Tamai, T. Nishida, F. Mori, Y. Omura *et al.*).

[42] BASF, DE 1 267 682, 1967 (H. Müller, H. Köhl, H. Pommer).

[43] BASF, DE-AS 1 618 098, 1967 (H. Müller, H. Overwien, H. Pommer).

[44] BASF, DE-AS 1 286 020, 1967 (H. Pommer, H. Müller, H. Overwien).

[45] BASF, DE-AS 1 643 710, 1967 (H. Pasedach, W. Hoffmann, W. Himmele).

[46] H. Dorn, N. Tian, H. Qiao, X. Lu, *Sepu* 2(2) (1985) 75–79; *Chem. Abstr.* **103** (1985) 119978k.

[47] Takasago Perfumery, EP-A 112 727, 1984 (S. Mitsuhashi, H. Kumobayashi, S. Akuta-gawa).

[48] Universal Oil Products, US 3 275 696, 1961 (E. Goldstein).

[49] BASF, DE-OS 2 934 250, 1979 (M. Horner, M. Irrgang, A. Nissen).

[50] K. Ziegler, DE-AS 1 118 775, 1960.

[51] Hercules Powder, US 2 388 084, 1945 (A. L. Rummelsburg).

[52] Glidden, US 2 902 510, 1959 (R. L. Webb).

[53] Bush Boake Allen Ltd., DE 2 255 119, 1972 (B. N. Jones, H. R. Ansari, B. G. Jaggers, J. F. Janes).

[54] K. Kogami, J. Kumanotani, *Bull. Chem. Soc. Japan* **41** (1968) 2508–2514; *Chem. Abstr.* **70** (1969) 20230y.

[55] Rhône-Poulenc, DE-AS 1 811 517, 1968 (P. Charbardes).

[56] Hoffmann-La Roche, DE 2 353 145, 1973 (N. Hindley, D. A. Andrews).

[57] Kuraray Co., JP 51 048 608, 1974 (T. Hosogai, T. Nishida, K. Itoi); *Chem. Abstr.* **85** (1976) 142638t.

[58] BASF, DE-AS 1 901 709, 1969 (H. Overwien, H. Müller).

[59] BASF, DE-AS 2 715 209, 1977 (C. Dudeck, H. Diem, F. Brunnmüller, B. Meissner *et al.*).

[60] BASF, DE-AS 2 715 208, 1977 (B. Meissner, W. Fliege, O. Woerz, C. Dudeck *et al.*).

[61] BASF, DE-AS 2 625 074, 1976 (A. Nissen, G. Kaibel).

[62] BASF, EP 21 074, 1980 (A. Nissen, W. Rebafka, W. Aquila).

[63] V. N. Kraseva, A. A. Bag, L. L. Malkina, O. M. Khol'mer *et al.*, SU 118 498, 1958; *Chem. Abstr.* **53** (1959) 22067h.

[64] Glidden, US 3 028 431, 1962 (R. L. Webb).

[65] Rhodia Inc., US 3 971 830, 1976 (P. S. Gradeff).

[66] D.-J. Wang, *K'o Hsueh Fa Chan Yuch K'an* **7** (1979) no. 10, 1036–1048; *Chem. Abstr.* **92** (1980) 124929d.

[67] L'Air liquide, DE-AS 2 045 888, 1970 (M. Vilkas, G. Senechal).

[68] T. Kumano, JP 7 003 366, 1965; *Chem. Abstr.* **72** (1970) 111650t.

[69] Givaudan, DE-AS 2 211 421, 1972 (L. M. Polinski, B. Venugopal, E. Eschinasi, J. Dorsky).

[70] Takasago Perfumery, DE 3 044 760, 1980 (S. Akutagawa, T. Taketomi).

[71] Distillers Company, GB 878 680, 1960 (P. Nayler).

[72] BASF, DE 1 768 980, 1968 (H. Pasedach).

[73] Hoffmann-La Roche, GB 774 621, 1957 (J. A. Birbiglia, G. O. Chase, I Galender).

[74] Haarmann & Reimer GmbH, BE 615 962, 1962 (K. Bauer, J. Pelz).

[75] International Flavors and Fragrances Inc., EP 170 205, 1984 (M. A. Sprecker, S. R. Wilson, L. Steinbach, T. Orourke).

[76] SCM Corp., US 4 000 207, 1976 (G. L. Kaiser).

[77] R. Emberger, R. Hopp in *Topics in Flavour Research*, ed. R. G. Berger, S. Nitz, P. Schreier, H. Eichhorn Marzling-Hangenham 1985, p. 201–218.

[78] Takasago Perfumery, EP 68 506, 1982 (S. Otsuka, K. Tani, T. Yamagata, S. Akutagawa, H. Kumobayashi, M. Yagi).

[79] Bush Boake Allen, FR 1 374 732, 1964 (B. T. D. Sully, P. L. Williams).

[80] Glidden, US 3 031 442, 1958 (R. L. Webb).

[81] Sukh Dev: 'New Industrial Processes Based on Carene,' *IUPAC Int. Symp. Chem. Nat. Prod. 11th* 1978, **4** (Part 1) 433–447 (N. Marekov, I. Ognyanov, A. Arahovates ed.).

[82] Haarmann & Reimer, DE 2 109 456, 1971 (J. Fleischer, K. Bauer, R. Hopp).

[83] Hoechst AG, DE 955 499, 1956 (H. Hoyer, G. Keicher, H. Schubert).

[84] Dragoco, DE 1 235 306, l965 (E. Klein).

[85] Firmenich SA, CH 581 592, 1976 (A. F. Thomas, G. Ohloff).

[86] B. Masumoto, T. Funahashi, JP 5 368, 1952; *Chem. Abstr.* **48** (1954) 8263g.

[87] Dow Chemical Co., US 3 124 614, 1958 (L. J. Dankert, D. A. Pernoda).

[88] Glidden, US 3 293 301, 1964 (J. M. Derfer, B. J. Kane, D. G. Young).

[89] Y.-R. Naves, *J. Soc. Cosmet. Chem.* **22** (1971) 439–456.

[90] Y.-R. Naves, *Riv. Ital. Essenze* **58** (1976) no. 10, 505–514.

[91] Haarmann & Reimer, DE 2 455 761, 1974 (G. Blume, R. Hopp, W. Sturm).

[92] BASF, DE-AS 1 286 019, 1967 (H. Pommer, W. Reif, W. Pasedach, W. Hoffmann).

[93] BASF, DE-AS 1 286 018, 1967 (H. Pommer, W. Reif, W. Pasedach, W. Hoffmann).

[94] Hoffmann-La Roche, DE 1 109 677, 1959 (R. Marbet, G. Saucy).

[95] Polak & Schwarz, GB 812 727, 1959 (M. G. J. Beets, H. van Essen).

[96] Hoffmann-La Roehe, GB 865 478, 1961.

[97] Studiengesellschaft Kohle, FR 1 355 944, 1962 (G. Ohloff, G. Schade).

[98] Firmenich, CH 563 951, 1972 (K. H. Schulte-Elte).

[99] Firmenich SA, EP 260 472, 1987 (C. Fehr, J. Galindo).

[100] Dragoco, DE 2 120 413, 1971 (E. Klein).

[101] J. Kulesza, J. Podlejski, A. Falkowski, PL 57 707, 1966; *Chem. Abstr.* **72** (1970) 12914a.

[102] Japan Terpene Chemical Co., JP 68 27 107, 1966 (Y. Matsubara, K. Takei); *Chem. Abstr.* **70** (1969) 58071k.

[103] Hoechst AG, DE 4 419 686, 1994 (M. Gscheideier, R. Gütmann, J. Wiesmüller, A. Riedel).

[104] Firmenich, DE 2 008 254, 1970 (E. Sundt, G. Ohloff).

[105] E. Demole, P. Enggist, G. Ohloff, *Helv. Chim. Acta* **65(6)** (1982) 1785–1794.

[106] Dragoco, DE 2 827 957, 1978 (E. Klein, E.-J. Brunke).

[107] Givaudan, CH 629 461, 1977 (W. M. Easter, R. E. Naipawer).

[108] Givaudan, L. et Cie. SA, EP 203 528, 1985 (R. E. Naipawer).

[109] Firmenich SA, EP 155 391, 1984 (K. H. Schulte-Elte, B. Mueller, H. Pamingle).

[110] International Flavors & Fragrances, US 4 014 944, 1976 (J. B. Hall, W. J. Wiegers).

[111] Haarmann & Reimer, EP 19 845, 1981 (G. K. Lange, K. Bauer).

[112] K. H. Schulte-Elte, W. Giersch, B. Winter, H. Pamingle, G. Ohloff, *Helv. Chim. Acta* **68(7)** (1985) 1961–1985.

[113] Dragoco, DE 28 07 584, 1978 (E. Klein, W. Rojahn).

[114] Firmenich SA, EP 118 809, 1984 (K. H. Schulte-Elte, G. Ohloff, B. L. Müller, W. Giersch).

[115] Firmenich SA, DE 2 733 928, 1977 (H. Strickler).

[116] Henkel KGaA, DE 2 427 500, 1974 (K. Bruns, P. Meins).

[117] International Flavors & Fragrances, US 4 007 137, 1975 (J. M. Sanders, W. L. Schreiber, J. B. Hall).

[118] International Flavors & Fragrances, US 2 947 780, 1960 (R. W. Teegarden, L. Stein-bach).

[119] Firmenich SA, DE 2 721 002, 1977 (K. P. Dastur).

[120] Bayer AG, DE 2 437 219, 1974 (H. Stetter, H. Kuhlmann).

[121] International Flavors & Fragrances, US 4 045 489, 1976 (W. J. Wiegers, J. B. Hall).

[122] Heraeus, W. C. GmbH, DE 2 909 780, 1979 (H. J. Brockmeyer, H. Winter).

[123] International Flavors & Fragrances Inc., US 3 847 993, 1974 (J. B. Hall, L. K. Lala).

[124] Firmenich, US 4 302 607, 1981 (G. H. Büchi, H. Wüst).

[125] B. D. Mookherjee, R. A. Wilson in *Fragrance Chemistry. The Science of the Sense of Smell*, ed. E. T. Theimer, Academic Press, 1982.

[126] Toray Industries, DE 2 141 309, 1971 (A. Miyake, M. Nishino, H. Kondo).

[127] Firmenich SA, US 4 302 607, 1981 (G. H. Büchi, H. Wüst).

[128] A. F. Morris, F. Naef, R. L. Snowden, *Perfumer & Flavorist* **16** (1991) July/August 33–35.

[129] International Flavors & Fragrances, US 3 929 677, 1975 (J. B. Hall, J. M. Sanders).

[130] International Flavors & Fragrances Inc., US 3 754 036, 1973 (J. H. Blumenthal).

[131] Roure-Bertrand Fils & Justin Dupont, US 3 578 715, 1971 (B. P. Corbier, P. J. Teisseire).

[132] Bayer, US 2 582 743, 1952 (M. Bollmann, E. Kroning).

[133] Van Ameringen-Haebler, US 2 927 127, 1960 (W. T. Somerville, E. T. Theimer).

[134] Givaudan L. et Cie. SA, EP 53 704, 1982 (H. Schenck).

[135] Firmenich et Cie., DE 2 163 868, 1970 (F. Naef).

[136] Firmenich, GB 907 431, 1961 (A. Firmenich, R. Firmenich, G. Firmenich, R. E. Firmenich).

[137] Roure Bertrand Dupont SA, DE 2 732 107, 1977 (P. J. Teisseire, M. Plattier, E. Giraudi).
[138] Firmenich SA, PCT 96/00206, 1996 (V. Rautenstrauch, J.-J. Riedhauser).
[139] Polak's Frutal Works, NL 7 115 065, 1972 (A. M. Cohen).
[140] Toyo Soda, JP 7 831 635, 1976 (K. Kondo, K. Komiya, H. Iwamoto); *Chem. Abstr.* **89** (1978) 75313a.
[141] N. Ya. Zyryanova, A. V. Gurevich, S. A. Voitkevich, *Maslo-Zhir. Promst.* **4** (1978) 35–37. *Chem Abstr.* **89** (1978) 117 528g.
[142] Ruhrchemie, FR 2 015 475, 1970.
[143] Consortium für Elektrochemische Industrie, DE 3 531 585, 1985 (W. Hafner, H. Gebauer, M. Regiert, P. Ritter).
[144] Firmenich SA, CH 655 932, 1983 (C. Tarchini).
[145] Engelhard Minerals and Chemicals Corp., US 3 655 777, 1968 (P. N. Rylander, D. R. Steele).
[146] H. I. Schlesinger, H. C. Brown, US 2 683 721, 1954.
[147] Ajinomoto, JP 7 242 278, 1971 (Y. Matsuzawa, T. Yamashita, S. Ninagawa); *Chem. Abstr.* **78** (1973) 135878y.
[148] Mitsubishi Petrochemical Co., JP 74 25 654, 1970 (T. Hashimoto, M. Tanaka); *Chem. Abstr.* **82** (1975) 170375m.
[149] Lee *et al. Gas Chromatography*, Academic Press, New York 1962, p. 475.
[150] Universal Oil Products, US 3 520 934, 1968 (W. Dunkel, D. J. Eckardt, A. Stern).
[151] Rhône-Poulenc, DE-AS 1 145 161, 1958 (I. Scriabine).
[152] BASF, EP 32 576, 1980 (N. Goetz, W. Hoffmann).
[153] Albright and Wilson Ltd., EP 45 571, 1982 (D. Webb).
[154] United States Rubber Co., US 2 529 186, 1950 (H. R. Richmond).
[155] Polak's Frutal Works, FR 1 392 804, 1963 (J. Stofberg, B. van der Wal, K. N. Nieuwland).
[156] Givaudan, US 3 078 319, 1960 (T. F. Wood).
[157] Naarden Int. N.V., NL-A 7 802 038, 1977.
[158] Polak's Frutal Works, DE-AS 1 045 399, 1954 (E. H. Polak).
[159] Givaudan, US 3 856 875, 1973 (T. F. Wood, E. Heilweil).
[160] Recherche et Industrie Therapeutiques, GB 1 032 879. 1965 (P. Crooy).
[161] Warszawa Fabryka Syntetykow Zapachowych, PL 50 349, 1963 (T. Zardecki); *Chem. Abstr.* **66** (1967) 55210m.
[162] Haarmann & Reimer, DE-AS 2 256 483, 1972 (K. Bauer, U. Harder, W. Sturm).
[163] Council of Scientific and Industrial Research, IN 146 359, 1977 (Y. R. Ramachandra, S. N. Mahapatra); *Chem. Abstr.* **92** (1980) 22281u.
[164] Rhône-Poulenc, DE-AS 1 936 727, 1969 (P. Gandilhon).
[165] Givaudan, DE 2 508 347, 1975 (A. J. Chalk).
[166] S. G. Polyakova, L. L. Malkina, O. M. Khol'mer, I. M. Lebedev *et al.*, SU 114 197, 1958; *Chem. Abstr.* **53** (1959) 14140e.
[167] O. M. Khol'mer, S. G. Polyakova, SU 126 879, 1960; *Chem. Abstr.* **54** (1960) 19596c.
[168] All-Union Scientific Research Institute of Synthetic and Natural Perfumes, SU 407 872, 1973 (G. E. Svadkovskaya, S. G. Polyakova, N. A. Daev, N. N. Zelenetskii *et al.*); *Chem. Abstr.* **80** (1974) 108715c.
[169] Collins Chemical Co., US 3 104 265, 1963 (A. Fiecchi).
[170] All-Union Scientific Research Institute of Synthetic and Natural Perfumes, DE 2 221 116, 1972 (L. K. Andreeva, I. A. Kheifits, T. P. Cherkasova, E. A. Berger *et al.*); *Chem. Abstr.* **78** (1973) 58046w.

[171] J. Imamura, *Yuko Gosei Kagaku Kyokaishi* **37** (1979) 667–677; *Chem. Abstr.* **92** (1980) 6158d.

[172] BASF, DE 2 848 397, 1978 (D. Degner, M. Barl, H. Siegel).

[173] D. W. Goheen in *Lignins*, ed. K. V. Sarkanen, Interscience, New York 1971, pp. 797–831.

[174] Ube Industries, JP 76 128 934, 1975 (S. Nagai, T. Enomiya, T. Nakamura); *Chem. Abstr.* **86** (1977) 189514k.

[175] R. G. Berger, *Aroma Biotechnology*, Springer Verlag, Berlin, 1995, p. 65.

[176] Ube Industries, DE 2 804 063, 1978 (S. Umemura, N. Takamitsu, T. Enomiya, H. Shiraishi *et al.*).

[177] Brichima, DE 2 703 640, 1977 (P. Maggioni).

[178] Haarmann & Reimer, DE 2 754 490, 1977 (K. Bauer, R. Mölleken).

[179] Dragoco Gerberding & Co., GB 876 684, 1958 (F. Porsch).

[180] BASF, DE-AS 2 145 308, 1971 (R. Fischer, W. Körnig).

[181] P. Buil, J. Garnero, D. Joulain, *Parfums, Cosmet., Aromes* **52** (1983) 45–49.

[182] PFW Beheer, DE 2 359 233, 1974 (A. M. Cohen).

[183] Studiengesellsehaft Kohle, DE 1 137 730, 1961 (G. O. Schenk, G. Ohloff, E. Klein).

[184] Consortium für Elektrochemische Industrie GmbH, DE 3 240 054, 1982 (G. Staiger, A. Macri).

[185] Firmenich SA, EP 403 945, 1989 (K.-H. Schulte-Elte, R. L. Snowden, C. Tarchini, B. Baer, C. Vial).

[186] Firmenich, CH 474 500, 1967 (L. Re, G. Ohloff).

[187] Polak's Frutal Works, DE 2 812 713, 1978 (A. M. Cohen).

[188] R. G. Berger, *Aroma Biotechnology*, Springer Verlag, Berlin, 1995, p. 101.

[189] Givaudan Corp., US 4 208 338, 1979 (H. Huber, H. J. Wild).

[190] Pfizer Inc. CA 1 110 254, 1981 (T. M. Brennan, D. P. Brannegan, P. D. Weeks, D. E. Kuhla).

[191] U. G. Ibatullin, Y. V. Pavlov, M. G. Safarov, *Khim. Geterotsikl. Soedin.* **10** (1989) 1326–1328.

[192] International Flavors & Fragrances, US 3 910 964, 1975 (J. M. Sanders, L. H. Michael).

[193] Haarmann & Reimer, EP 120 257, 1984 (H. Finkelmeier, R. Hopp).

[194] NV Polak & Schwarz's Essencefabrieken, GB 714 645, 1954.

[195] May & Baker, FR 1 577 817, 1967.

[196] Ube Industries, JP 7 695 058, 1975 (T. Yamasaki, M. Nakai, Y. Kuroki, M. Oda, Y. Kawaguchi, M. Kobayashi); *Chem. Abstr.* **85** (1976) 192191n.

[197] R. G. Berger, *Aroma Biotechnology*, Springer Verlag, Berlin, 1995, p. 70.

[198] Toyotama Perfumery Co., Ltd., JP 88 41 473, 1988 (H. Tsukasa, Y. Shono), *Chem. Abstr.* **109** (1988) 230790f.

[199] H. Sulser, M. Habegger, W. Büchi, *Z. Lebensm. Untersuch. Forsch.* **148** (1972) 215–221.

[200] H. Stach, W. Huggenberger, M. Hesse, *Helv. Chim. Acta* **70** (1987) 369–374.

[201] K. W. Rosenmund, H. Bach, *Chem. Ber.* **94** (1961) 2934–2405.

[202] Haarmann & Reimer, DE 2 136 496, 1971 (R. Hopp, K. Bauer).

[203] Firmenich, DE 2 026 056, 1970 (J. J. Becker).

[204] Haarmann & Reimer, EP 512 348, 1992 (R. Hopp, H. Finkelmeier, O. Koch, A. Körber).

[205] Soda Koryo, DE 2 818 126, 1978 (K. Suzuki, T. Eto, T. Otsuka, S. Abe).

[206] International Flavors & Fragrances, US 4 014 902, 1977 (C. Y. Tseng).

[207] Chemische Werke Hüls, DE 2 547 267, 1975 (J. Ritter, K. Burzin, K. H. Magosch, R. Feinauer).

[208] K. D. Carlson, V. E. Sohns, R. B. Perkins, E. L. Huffmann, *Ind. Eng. Chem., Prod. Res. Dev.* **16(1)** (1977) 95–101.

[209] Dow Chemical, US 4 036 854, 1975 (K. Y. Chang).

[210] Stamicarbon, NL 7 202 539, 1972 (J. A. Thoma, J. M. Deumens, E. J. Verheijen).

[211] Stamicarbon, DE 2 309 536, 1973 (S. Schaafsma, J Deumens).

[212] Union Carbide, GB 863 446, 1961 (B. Phillips, P. Staacher, D. L. Mac'Peek).

[213] M. Ishihara, K. Yonetani, T. Shibai, Y. Tanaka, *7th International Congress of Essential Oils*, Kyoto 1977, pp. 266–269; *Chem. Abstr.* **92** (1980) 75722v.

[214] H. Maarse, E. A. Visscher, *Volatile Compounds in Foods, Qualitative Data*, TNO-Division for nutrition and food research TNO-CIVO Food Analysis Institute, Zeist (Netherlands) 1989–1994.

[215] R. Tressl, T. Kossa, M. Holzer in *Geruch- und Geschmackstoffe*, ed. F. Drawert, 'Bildung flüchtiger Aromastoffe durch Maillard-Reaktionen,' Verlag Hans Carl, Nürnberg 1975, pp. 33–47.

[216] G. Vernin, 'Mechanism of Formation of Heterocyclic Compounds in Maillard and Pyrolysis Reactions,' *Chem. Heterocycl. Compd. Flavours Aromas* **1982**, 151–207.

[217] G. Ohloff, I. Flament, 'The Role of Heteroaromatie Substances in the Aroma Compounds of Foodstuffs' in *Progress in the Chemistry of Organic Natural Products*, vol. 36, Springer Verlag, Wien 1979, pp. 231–283.

[218] G. Ohloff, I. Flament, W. Pickenhagen, 'Flavor Chemistry,' *Food Reviews International* **1** (1985) no. 1, 99–148.

[219] N. Verlet, *Riv. Ital. EPPOS* (1992) (3, numero speciale), p. 384.

[220] K.-D. Protzen in *Ätherische Öle – Anspruch und Wirklichkeit*, ed. R. Carle, Wissenschaftliche Verlagsgesellschaft, Stuttgart 1993, p. 23–30.

[221] Anon., *Parf. Cosmet. Arom.* **127** (1996) 26.

[222] ITC UNCTAD/GATT, *Essential Oils and Oleoresins: A Study of Selected Producers and Major Markets*, Geneva 1986.

[223] B. Meyer-Warnod, *Perfum. Flavor* **9(2)** (1984) 93.

[224] K. Formazek, K. H. Kubeczka, *Essential Oil Analysis by Capillary Chromatography and Carbon-13 NMR Spectroscopy*, J. Wiley & Sons, New York 1982.

[225] B. M. Lawrence, *Progress in Essential Oils*, bimonthly column in the journal *Perfumer & Flavorist*. The contributions to volumes 1976–1991 were comprehensively reprinted and have been published in book form [226–230].

[226] B. M. Lawrence, *Essential Oils 1976–1977*, Allured Publishing Corp., Wheaton IL 1978.

[227] B. M. Lawrence, *Essential Oils 1978*, Allured Publishing Corp., Wheaton IL 1979.

[228] B. M. Lawrence, *Essential Oils 1979–1980*, Allured Publishing Corp., Wheaton IL 1981.

[229] B. M. Lawrence, *Essential Oils 1981–1987*, Allured Publishing Corp., Wheaton IL 1989.

[230] B. M. Lawrence, *Essential Oils 1988–1991*, Allured Publishing Corp., Carol Stream IL 1993.

[231] D. Zander, (eds. F. Encke, G. Buchheim, S. Seybold), *Handwörterbuch der Pflanzennamen*, 13th ed., Verlag E. Ulmer, Stuttgart 1984.

[232] EOA No. 180.

[233] G. Mazza, S. Ciaravole, G. Chiricosta, S. Celli, *Flav. Fragr. J.* **7** (1992) 111.

[234] EOA No. 183.

[235] H. Kallio, L. Salorinne, *J. Agric. Food Chem.* **38** (1990) 1560.

[236] EOA No. 147.

[237] B. M. Lawrence, *Essential Oils 1978*, Allured Publishing Corp., Wheaton IL 1979 p. 13.

[238] B. M. Lawrence, *Essential Oils 1988–1991*, Allured Publishing Corp., Carol Stream IL 1993 p. 117.

[239] B. M. Lawrence, *Progress in Essential Oils, Perfum. Flavor.* **21(4)** (1996) 58.

[240] ISO 3525-1979.

[241] B. M. Lawrence, *Essential Oils 1979–1980*, Allured Publishing Corp., Wheaton IL 1981 p. 20.

[242] B. M. Lawrence, *Essential Oils 1988–1991*, Allured Publishing Corp., Carol Stream IL 1993 p. 1.

[243] B. M. Lawrence, *Essential Oils 1988–1991*, Allured Publishing Corp., Carol Stream IL 1993, p. 91.

[244] B. M. Lawrence, *Progress in Essential Oils, Perfum. Flavor.* **21(1)** (1996) 38.

[245] EOA No. 96.

[246] EOA No. 97.

[247] B. M. Lawrence, *Essential Oils 1976–1977*, Allured Publishing Corp., Wheaton IL 1978 p. 17.

[248] B. M. Lawrence, *Essential Oils 1981–1987*, Allured Publishing Corp., Wheaton IL 1989 p. 10.

[249] B. M. Lawrence, *Essential Oils 1981–1987*, Allured Publishing Corp., Wheaton IL 1989, p. 34.

[250] B. M. Lawrence, *Essential Oils 1988–1991*, Allured Publishing Corp., Carol Stream IL 1993 p. 60.

[251] B. M. Lawrence, *Progress in Essential Oils, Perfum. Flavor.* **21(5)** (1996) 57.

[252] I. Nykänen, L. Nykänen, M. Alkio, *J. Essent. Oil. Res.* **3** (1991) 229.

[253] J. C. Chalchat, R. Ph. Garry, *J. Essent. Oil. Res.* **5** (1993) 447.

[254] K. Awano, S. Tamogami, T. Houda, T. Honda, O. Takazawa, I. Watanabe in *Flavours, Fragrances and Essential Oils*, Proc. 13th Intern. Congress Flav. Fragr. Essent. Oils Istanbul 1995, ed. K. H. C. Baser, AREP Publ., Istanbul 1995, vol. 2, p. 211.

[255] B. M. Lawrence, *Essential Oils 1981–1987*, Allured Publishing Corp., Wheaton IL 1989, p. 63.

[256] B. M. Lawrence, *Essential Oils 1976–1977*, Allured Publishing Corp., Wheaton IL 1978, p. 17.

[257] J. P. Petitdidier, *Parf. Cosmet. Arom.* **63** (1985) 65.

[258] D. S. Deng, D. Y. Yu, Z. M. Liu, C.-Z. Zhang, in *Dev. Food Sci.* **18**, eds. B. M. Lawrence, B. D. Mookherjee, B. J. Willis, Elsevier, Amsterdam 1988, p. 587.

[259] J. P. Petitdidier, *Parf. Cosmet. Arom.* **88** (1989), 63.

[260] B. D. Mookherjee, R. W. Trenkle in Fedarom, *Technical Data of the 8th Intern. Congress of Essential Oils*, Cannes 1980, Grasse 1982, p. 587.

[261] B. D. Mookherjee, R. W. Trenkle, R. A. Wilson, in *Proc. 12th Int. Congress Flavours, Fragrances and Ess. Oils*, eds. H. Woidich, G. Buchbauer, Fachzeitschriftenverlags GmbH, Vienna 1992, p. 234.

[262] ISO 3475-1975.

[263] B. M. Lawrence, *Essential Oils 1979–1980*, Allured Publishing Corp., Wheaton IL 1981, p. 24.

[264] B. M. Lawrence, *Essential Oils 1981–1987*, Allured Publishing Corp., Wheaton IL 1989, p. 72.

[265] B. M. Lawrence, *Essential Oils 1981–1987*, Allured Publishing Corp., Wheaton IL 1989, p. 186.

[266] B. M. Lawrence, *Essential Oils 1981–1987*, Allured Publishing Corp., Wheaton IL 1989, p. 214.

[267] B. M. Lawrence, *Essential Oils 1981–1987*, Allured Publishing Corp., Wheaton IL 1989, p. 1.

[268] B. M. Lawrence, *Essential Oils 1981–1987*, Allured Publishing Corp., Wheaton IL 1989, p. 53.

[269] B. M. Lawrence, *Essential Oils 1988–1991*, Allured Publishing Corp., Carol Stream IL 1993, p. 52.

[270] B. M. Lawrence, *Progress in Essential Oils, Perfum. Flavor.* **19(5)** (1994) 94.

[271] AFNOR NF T 75-357: 1991.

[272] AFNOR NF T 75-244: 1992.

[273] B. M. Lawrence, *Essential Oils 1978*, Allured Publishing Corp., Wheaton IL 1979, p. 18.

[274] B. M. Lawrence, *Essential Oils 1979–1980*, Allured Publishing Corp., Wheaton IL 1981, p. 20.

[275] B. M. Lawrence, *Essential Oils 1981–1987*, Allured Publishing Corp., Wheaton IL 1989, p. 181.

[276] B. M. Lawrence, *Essential Oils 1988–1991*, Allured Publishing Corp., Carol Stream IL 1993, p. 70.

[277] B. M. Lawrence, *Essential Oils 1988–1991*, Allured Publishing Corp., Carol Stream IL 1993, p. 114.

[278] B. M. Lawrence, *Progress in Essential Oils, Perfum. Flavor.* **17(4)** (1992) 47.

[279] B. M. Lawrence, *Progress in Essential Oils, Perfum. Flavor.* **20(4)** (1995) 35.

[280] ISO 3045-1974.

[281] B. M. Lawrence, *Essential Oils 1976–1977*, Allured Publishing Corp., Wheaton IL 1978, p. 36.

[282] B. M. Lawrence, *Essential Oils 1979–1980*, Allured Publishing Corp., Wheaton IL 1981, p. 28.

[283] B. M. Lawrence, *Essential Oils 1981–1987*, Allured Publishing Corp., Wheaton IL 1989, p. 46.

[284] H. M. Boelens, D. de Rijke, H. G. Haring, *Perfum. Flavor.* **6(6)** (1981) 15.

[285] N. A. Shaath, B. Benveniste, *Perfum. Flavor.* **16(6)** (1991) 17.

[286] B. M. Lawrence, *Essential Oils 1981–1987*, Allured Publishing Corp., Wheaton IL 1989, p. 234.

[287] B. M. Lawrence, *Essential Oils 1988–1991*, Allured Publishing Corp., Carol Stream IL 1993, p. 180.

[288] Institut National de la Recherche Agronomique, EP 217719, 1986 (J. Rigaud, P. Etievant, R. Henry, A. Latrasse).

[289] N. F. Collins, E. H. Graven, T. A. v. Beek, G. P. Lelyveld, *J. Essent. Oil. Res.* **8** (1996) 229.

[290] B. M. Lawrence, *Essential Oils 1976–1977*, Allured Publishing Corp., Wheaton IL 1978, p. 5.

[291] EOA No. 101.

[292] B. M. Lawrence, *Essential Oils 1979–1980*, Allured Publishing Corp., Wheaton IL 1981, p. 47.

[293] B. M. Lawrence, *Essential Oils 1981–1987*, Allured Publishing Corp., Wheaton IL 1989, p. 77.

[294] B. M. Lawrence, *Essential Oils 1981–1987*, Allured Publishing Corp., Wheaton IL 1989, p. 183.

Flavouring substances and natural sources of flavourings Vol 1, 4th Edition, Council of Strassbourg 1992

[295] ISO 7357-1985.

[296] EOA No. 98.

[297] EOA No. 69.

[298] B. M. Lawrence, *Progress in Essential Oils, Perfum. Flavor.* **20(4)** (1995) 30.

[299] AFNOR NF T 75-347: 1986.

[300] ISO 4733-1981.

[301] B. M. Lawrence, *Essential Oils 1976–1977*, Allured Publishing Corp., Wheaton IL 1978, p. 13.

[302] B. M. Lawrence, *Essential Oils 1979–1980*, Allured Publishing Corp., Wheaton IL 1981, p. 24.

[303] B. M. Lawrence, *Essential Oils 1981–1987*, Allured Publishing Corp., Wheaton IL 1989, p. 79.

[304] B. M. Lawrence, *Essential Oils 1981–1987*, Allured Publishing Corp., Wheaton IL 1989, p. 170.

[305] B. M. Lawrence, *Essential Oils 1988–1991*, Allured Publishing Corp., Carol Stream IL 1993, p. 78.

[306] B. M. Lawrence, *Essential Oils 1988–1991*, Allured Publishing Corp., Carol Stream IL 1993, p. 130.

[307] B. M. Lawrence, *Progress in Essential Oils, Perfum. Flavor.* **17(6)** (1992) 51.

[308] AFNOR NF T 75-352: 1989.

[309] B. M. Lawrence, *Essential Oils 1976–1977*, Allured Publishing Corp., Wheaton IL 1978, p. 2.

[310] B. M. Lawrence, *Essential Oils 1979–1980*, Allured Publishing Corp., Wheaton IL 1981, p. 37.

[311] B. M. Lawrence, *Essential Oils 1988–1991*, Allured Publishing Corp., Carol Stream IL 1993, p. 27.

[312] B. M. Lawrence, *Essential Oils 1988–1991*, Allured Publishing Corp., Carol Stream IL 1993, p. 104.

[313] B. M. Lawrence, *Progress in Essential Oils, Perfum. Flavor.* **17(3)** (1992) 71.

[314] B. M. Lawrence, *Essential Oils 1979–1980*, Allured Publishing Corp., Wheaton IL 1981, p. 1.

[315] B. M. Lawrence, *Essential Oils 1981–1987*, Allured Publishing Corp., Wheaton IL 1989, p. 181.

[316] B. M. Lawrence, *Essential Oils 1988–1991*, Allured Publishing Corp., Carol Stream IL 1993, p. 29.

[317] B. M. Lawrence, *Essential Oils 1988–1991*, Allured Publishing Corp., Carol Stream IL 1993, p. 121.

[318] B. M. Lawrence, *Progress in Essential Oils, Perfum. Flavor.* **18(6)** (1993) 54.

[319] B. M. Lawrence, *Progress in Essential Oils, Perfum. Flavor.* **20(5)** (1995) 103.

[320] EOA No. 86.

[321] ISO 9843-1991.

[322] ISO 4725-1986.

[323] ISO 4724-1984.

[324] G. Ohloff, *Scent and Fragrances*, Springer Verlag, Berlin, 1994, p. 173.

[325] B. M. Lawrence, *Essential Oils 1979–1980*, Allured Publishing Corp., Wheaton IL 1981, p. 33.

[326] B. M. Lawrence, *Essential Oils 1988–1991*, Allured Publishing Corp., Cold Stream IL 1993, p. 176.

[327] ISO 3760-1979.

[328] B. M. Lawrence, *Essential Oils 1979–1980*, Allured Publishing Corp., Wheaton IL 1981, p. 24.

[329] B. M. Lawrence, *Essential Oils 1981–1987*, Allured Publishing Corp., Wheaton IL 1989, p. 74.

[330] B. M. Lawrence, *Essential Oils 1981–1987*, Allured Publishing Corp., Wheaton IL 1989, p. 157.

[331] B. M. Lawrence, *Essential Oils 1988–1991*, Allured Publishing Corp., Carol Stream IL 1993, p. 74.

[332] B. M. Lawrence, *Essential Oils 1988–1991*, Allured Publishing Corp., Carol Stream IL 1993, p. 109.

[333] B. M. Lawrence, *Progress in Essential Oils, Perfum. Flavor.* **20(1)** (1995) 52.

[334] G. Vernin, C. Parkanyi in *Dev. Food Sci 34*, ed. G. Charalambous, Elsevier, Amsterdam 1994, p. 329.

[335] B. M. Lawrence, *Essential Oils 1988–1991*, Allured Publishing Corp., Carol Stream IL 1993, p. 205.

[336] B. M. Lawrence, *Progress in Essential Oils, Perfum. Flavor.* **21(3)** (1996) 55.

[337] EOA No. 156.

[338] AFNOR NF T 75-253: 1992.

[339] B. M. Lawrence, *Essential Oils 1979–1980*, Allured Publishing Corp., Wheaton IL 1981, p. 15.

[340] B. M. Lawrence, *Essential Oils 1981–1987*, Allured Publishing Corp., Wheaton IL 1989, p. 9.

[341] B. M. Lawrence, *Essential Oils 1981–1987*, Allured Publishing Corp., Wheaton IL 1989, p. 112.

[342] B. M. Lawrence, *Essential Oils 1988–1991*, Allured Publishing Corp., Carol Stream IL 1993, p. 101.

[343] B. M. Lawrence, *Progress in Essential Oils, Perfum. Flavor.* **17(5)** (1992) 145.

[344] B. M. Lawrence, *Progress in Essential Oils, Perfum. Flavor.* **18(4)** (1993) 71.

[345] ISO/DIS 3216-1994.

[346] B. M. Lawrence, *Essential Oils 1978*, Allured Publishing Corp., Wheaton IL 1979, p. 1.

[347] B. M. Lawrence, *Essential Oils 1978*, Allured Publishing Corp., Wheaton IL 1979, p. 13.

[348] B. M. Lawrence, *Essential Oils 1981–1987*, Allured Publishing Corp., Wheaton IL 1989, p. 6.

[349] B. M. Lawrence, *Progress in Essential Oils, Perfum. Flavor.* **19(4)** (1994) 33.

[350] G. Vernin, C. Vernin, J. Metzger, L. Pujol, C. Parkanyi in *Dev. Food Sci. 34*, ed. G. Charalambous, Elsevier, Amsterdam 1994, p. 411.

[351] ISO 3524-1977.

[352] B. M. Lawrence, *Essential Oils 1976–1977*, Allured Publishing Corp., Wheaton IL 1978, p. 29.

[353] B. M. Lawrence, *Essential Oils 1978*, Allured Publishing Corp., Wheaton IL 1979, p. 14.

[354] B. M. Lawrence, *Progress in Essential Oils, Perfum. Flavor* **19(2)** (1994) 69.

[355] B. M. Lawrence, *Essential Oils 1978*, Allured Publishing Corp., Wheaton IL 1979, p. 14.

[356] B. M. Lawrence, *Progress in Essential Oils, Perfum. Flavor* **19(3)** (1994) 59.

[357] EOA No. 87.

[358] G. Ohloff, *Scent and Fragrances*, Springer Verlag, Berlin, 1994, p. 127.

[359] ISO/DIS 3520-1996.

[360] G. Ohloff, *Scent and Fragrances*, Springer Verlag, Berlin, 1994, p. 138.

[361] B. M. Lawrence, *Essential Oils 1979–1980*, Allured Publishing Corp., Wheaton IL 1981, p. 9.

[362] B. M. Lawrence, *Essential Oils 1981–1987*, Allured Publishing Corp., Wheaton IL 1989, p. 39.

[363] B. M. Lawrence, *Essential Oils 1981–1987*, Allured Publishing Corp., Wheaton IL 1989, p. 55.

[364] B. M. Lawrence, *Essential Oils 1981–1987*, Allured Publishing Corp., Wheaton IL 1989, p. 73.

[365] B. M. Lawrence, *Essential Oils 1981–1987*, Allured Publishing Corp., Wheaton IL 1989, p. 214.

[366] B. M. Lawrence, *Essential Oils 1988–1991*, Allured Publishing Corp., Carol Stream IL 1993, p. 7.

[367] B. M. Lawrence, *Essential Oils 1988–1991*, Allured Publishing Corp., Carol Stream IL 1993, p. 173.

[368] B. M. Lawrence, *Progress in Essential Oils, Perfum. Flavor.* **19(6)** (1994) 57.

[369] G. Dugo. *Perfum. Flavor.* **19(6)** (1994) 29.

[370] L. Mondello, P. Dugo, K. D. Bartle, G. Dugo, A. Cotroneo, *Flav. Frag. J.* **10** (1995) 33.

[371] ISO 3053-1975.

[372] G. Ohloff, *Scent and Fragrances*, Springer Verlag, Berlin, 1994, p. 135.

[373] B. M. Lawrence, *Essential Oils 1979–1980*, Allured Publishing Corp., Wheaton IL 1981, p. 3.

[374] B. M. Lawrence, *Essential Oils 1981–1987*, Allured Publishing Corp., Wheaton IL 1989, p. 41.

[375] B. M. Lawrence, *Essential Oils 1981–1987*, Allured Publishing Corp., Wheaton IL 1989, p. 91.

[376] B. M. Lawrence, *Essential Oils 1981–1987*, Allured Publishing Corp., Wheaton IL 1989, p. 157.

[377] B. M. Lawrence, *Essential Oils 1988–1991*, Allured Publishing Corp., Carol Stream IL 1993, p. 25.

[378] B. M. Lawrence, *Essential Oils 1988–1991*, Allured Publishing Corp., Carol Stream IL 1993, p. 167.

[379] B. M. Lawrence, *Progress in Essential Oils, Perfum. Flavor.* **19(3)** (1994) 61.

[380] ISO 855-1981.

[381] G. Ohloff, *Scent and Fragrances*, Springer Verlag, Berlin, 1994, p. 134.

[382] B. M. Lawrence, *Essential Oils 1978*, Allured Publishing Corp., Wheaton IL 1979, p. 24.

[383] B. M. Lawrence, *Essential Oils 1979–1980*, Allured Publishing Corp., Wheaton IL 1981, p. 30.

[384] B. M. Lawrence, *Essential Oils 1981–1987*, Allured Publishing Corp., Wheaton IL 1989, p. 41.

[385] B. M. Lawrence, *Essential Oils 1981–1987*, Allured Publishing Corp., Wheaton IL 1989, p. 46.

[386] B. M. Lawrence, *Essential Oils 1981–1987*, Allured Publishing Corp., Wheaton IL 1989, p. 113.

[387] B. M. Lawrence, *Essential Oils 1981–1987*, Allured Publishing Corp., Wheaton IL 1989, p. 149.

[388] B. M. Lawrence, *Essential Oils 1988–1991*, Allured Publishing Corp., Carol Stream IL 1993, p. 61.

[389] B. M. Lawrence, *Progress in Essential Oils, Perfum. Flavor.* **17(1)** (1992) 45.

[390] B. M. Lawrence, *Progress in Essential Oils, Perfum. Flavor.* **19(3)** (1994) 64.

[391] B. M. Lawrence, *Progress in Essential Oils, Perfum. Flavor.* **21(1)** (1996) 41.

[392] D. Chouchi, D. Barth, *J. Chrom. A*, **672** (1994) 177.

[393] O. Philipp, H.-D. Isengaard, *Z. Lebensm. Unters. Forsch.*, **201** (1995) 551.

[394] M. H. Boelens, *Perfum. Flavor.*, **16(2)** (1991) 17.

[395] L. Haro, W. E. Faas, *Perfum. Flavor.*, **10(5)** (1985) 67.

[396] L. Haro-Guzman in Fedarom, *Technical Data of the 8th Intern. Congress of Essential Oils*, Cannes 1980, Grasse 1982, p. 181.

[397] D. McHale in Fedarom, *Technical Data of the 8th Intern. Congress of Essential Oils*, Cannes 1980, Grasse 1982, p. 177.

[398] ISO 3809-1987.

[399] B. M. Lawrence, *Essential Oils 1976–1977*, Allured Publishing Corp., Wheaton IL 1978, p. 26.

[400] B. M. Lawrence, *Essential Oils 1979–1980*, Allured Publishing Corp., Wheaton IL 1981, p. 35.

[401] B. M. Lawrence, *Essential Oils 1981–1987*, Allured Publishing Corp., Wheaton IL 1989, p. 25.

[402] B. M. Lawrence, *Essential Oils 1981–1987*, Allured Publishing Corp., Wheaton IL 1989, p. 42.

[403] B. M. Lawrence, *Essential Oils 1981–1987*, Allured Publishing Corp., Wheaton IL 1989, p. 59.

[404] B. M. Lawrence, *Essential Oils 1981–1987*, Allured Publishing Corp., Wheaton IL 1989, p. 152.

[405] B. M. Lawrence, *Essential Oils 1988–1991*, Allured Publishing Corp., Carol Stream IL 1993, p. 137.

[406] B. M. Lawrence, *Progress in Essential Oils, Perfum., Flavor.* **21(4)** (1996) 62.

[407] ISO 3519-1976.

[408] G. Ohloff, *Scent and Fragrances*, Springer Verlag, Berlin, 1994, p. 139.

[409] B. M. Lawrence, *Essential Oils 1981–1987*, Allured Publishing Corp., Wheaton IL 1989, p. 25.

[410] B. M. Lawrence, *Essential Oils 1981–1987*, Allured Publishing Corp., Wheaton IL 1989, p. 42.

[411] B. M. Lawrence, *Essential Oils 1981–1987*, Allured Publishing Corp., Wheaton IL 1989, p. 142.

[412] B. M. Lawrence, *Essential Oils 1988–1991*, Allured Publishing Corp., Carol Spring IL 1993, p. 137.

[413] ISO/DIS 3528-1994.

[414] B. M. Lawrence, *Essential Oils 1979–1980*, Allured Publishing Corp., Wheaton IL 1981, p. 22.

[415] B. M. Lawrence, *Essential Oils 1979–1980*, Allured Publishing Corp., Wheaton IL 1981, p. 45.

[416] B. M. Lawrence, *Essential Oils 1981–1987*, Allured Publishing Corp., Wheaton IL 1989, p. 152.

[417] B. M. Lawrence, *Essential Oils 1981–1987*, Allured Publishing Corp., Wheaton IL 1989, p. 230.

[418] B. M. Lawrence, *Essential Oils 1988–1991*, Allured Publishing Corp., Carol Stream IL 1993, p. 132.

[419] B. M. Lawrence, *Progress in Essential Oils, Perfum. Flavor.* **17(4)** (1992) 53.

[420] B. M. Lawrence, *Progress in Essential Oils, Perfum. Flavor.* **21(2)** (1996) 25.

[421] P. Dugo, L. Mondello, E. Cogliandro, I. Stagno d'Alcontres, A. Cotroneo, *Flav. Fragr. J.* **9** (1994) 105.

[422] ISO 9844-1991.

[423] B. M. Lawrence, *Essential Oils 1981–1987*, Allured Publishing Corp., Wheaton IL 1989, p. 41.

[424] B. M. Lawrence, *Essential Oils 1981–1987*, Allured Publishing Corp., Wheaton IL 1989, p. 55.

[425] B. M. Lawrence, *Essential Oils 1981–1987*, Allured Publishing Corp., Wheaton IL 1989, p. 125.

[426] B. M. Lawrence, *Essential Oils 1988–1991*, Allured Publishing Corp., Carol Stream IL 1993, p. 170.

[427] B. M. Lawrence, *Progress in Essential Oils, Perfum. Flavor.* **19(5)** (1994) 83.

[428] M. H. Boelens, A. Oporto, *Perfum. Flavor.* **16(6)** (1991) 1.

[429] ISO 3140-1990.

[430] G. Ohloff, *Scent and Fragrances*, Springer Verlag, Berlin, 1994, p. 132.

[431] B. M. Lawrence, *Essential Oils 1979–1980*, Allured Publishing Corp., Wheaton IL 1981, p. 22.

[432] B. M. Lawrence, *Essential Oils 1981–1987*, Allured Publishing Corp., Wheaton IL 1989, p. 113.

[433] B. M. Lawrence, *Essential Oils 1981–1987*, Allured Publishing Corp., Wheaton IL 1989, p. 126.

[434] B. M. Lawrence, *Essential Oils 1981–1987*, Allured Publishing Corp., Wheaton IL 1989, p. 222.

[435] B. M. Lawrence, *Essential Oils 1988–1991*, Allured Publishing Corp., Carol Stream IL 1993, p. 116.

[436] B. M. Lawrence, *Progress in Essential Oils, Perfum. Flavor.* **17(5)** (1992) 133.

[437] B. M. Lawrence, *Progress in Essential Oils, Perfum. Flavor.* **19(4)** (1994) 35.

[438] A. Verzera, A. Trozzi, I. Stagno d'Alcontres, A. Cotroneo, *J. Essent. Oil. Res.*, **8** (1996) 159.

[439] ISO/DIS 3142-1994.

[440] ISO/DIS 3141-1994.

[441] ISO/DIS 3143-1994.

[442] B. M. Lawrence, *Essential Oils 1979–1980*, Allured Publishing Corp., Wheaton IL 1981, p. 33.

[443] B. M. Lawrence, *Essential Oils 1981–1987*, Allured Publishing Corp., Wheaton IL 1989, p. 87.

[444] B. M. Lawrence, *Essential Oils 1981–1987*, Allured Publishing Corp., Wheaton IL 1989, p. 227.

[445] B. M. Lawrence, *Essential Oils 1988–1991*, Allured Publishing Corp., Carol Stream IL 1993, p. 36.

[446] B. M. Lawrence, *Essential Oils 1988–1991*, Allured Publishing Corp., Carol Stream IL 1993, p. 166.

[447] B. M. Lawrence, *Progress in Essential Oils, Perfum. Flavor.* **19(5)** (1994) 92.

[448] B. M. Lawrence, *Progress in Essential Oils, Perfum. Flavor.* **20(5)** (1995) 102.

[449] G. Vernin, E. Vernin,, J. Metzger, L. Pujol, C. Parkanyi in *Dev. Food Sci. 34*, ed. G. Charalambous, Elsevier, Amsterdam 1994, p. 483.

[450] EOA No. 10.

[451] ISO/DIS 3516-1994.

[452] B. M. Lawrence, *Essential Oils 1976–1977*, Allured Publishing Corp., Wheaton IL 1978, p. 29.

[453] B. M. Lawrence, *Essential Oils 1979–1980*, Allured Publishing Corp., Wheaton IL 1981, p. 26.

[454] B. M. Lawrence, *Essential Oils 1981–1987*, Allured Publishing Corp., Wheaton IL 1989, p. 24.

[455] B. M. Lawrence, *Essential Oils 1981–1987*, Allured Publishing Corp., Wheaton IL 1989, p. 100.

[456] B. M. Lawrence, *Essential Oils 1981–1987*, Allured Publishing Corp., Wheaton IL 1989, p. 228.

[457] B. M. Lawrence, *Essential Oils 1988–1991*, Allured Publishing Corp., Carol Stream IL 1993, p. 14.

[458] B. M. Lawrence, *Essential Oils 1988–1991*, Allured Publishing Corp., Carol Stream IL 1993, p. 128.

[459] B. M. Lawrence, *Essential Oils 1988–1991*, Allured Publishing Corp., Carol Stream IL 1993, p. 182.

[460] B. M. Lawrence, *Progress in Essential Oils, Perfum. Flavor.* **19(1)** (1994) 42.

[461] ISO 3849-1981.

[462] ISO 3848-1976.

[463] M. H. Boelens, *Perfum. Flavor.* **19(2)** (1994) 29.

[464] B. M. Lawrence, *Progress in Essential Oils, Perfum. Flavor.* **21(1)** (1996) 38.

[465] ISO 3217-1974.

[466] ISO 4718-1981.

[467] ISO 4727-1988.

[468] R. Soman, *Chem. Eng. World* **30** (1995) 34.

[469] AFNOR NF T 75-254-1992.

[470] B. M. Lawrence, *Essential Oils 1976–1977*, Allured Publishing Corp., Wheaton IL 1978, p. 18.

[471] B. M. Lawrence, *Essential Oils 1981–1987*, Allured Publishing Corp., Wheaton IL 1989, p. 29.

[472] B. M. Lawrence, *Essential Oils 1981–1987*, Allured Publishing Corp., Wheaton IL 1989, p. 128.

[473] B. M. Lawrence, *Progress in Essential Oils, Perfum. Flavor.* **20(4)** (1995) 34.

[474] B. M. Lawrence, *Essential Oils 1978*, Allured Publishing Corp., Wheaton IL 1979, p. 4.

[475] B. M. Lawrence, *Essential Oils 1988–1991*, Allured Publishing Corp., Carol Spring IL 1993, p. 15.

[476] B. M. Lawrence, *Progress in Essential Oils, Perfum. Flavor.* **20(1)** (1995) 54.

[477] B. M. Lawrence, *Essential Oils 1976–1977*, Allured Publishing Corp., Wheaton IL 1978, p. 25.

[478] B. M. Lawrence, *Essential Oils 1981–1987*, Allured Publishing Corp., Wheaton IL 1989, p. 2.

[479] B. M. Lawrence, *Essential Oils 1981–1987*, Allured Publishing Corp., Wheaton IL 1989, p. 88.

[480] B. M. Lawrence, *Essential Oils 1981–1987*, Allured Publishing Corp., Wheaton IL 1989, p. 164.

[481] B. M. Lawrence, *Essential Oils 1988–1991*, Allured Publishing Corp., Carol Stream IL 1993, p. 131.

[482] B. M. Lawrence, *Progress in Essential Oils, Perfum. Flavor.* **19(5)** (1994) 90.

[483] B. M. Lawrence, *Progress in Essential Oils, Perfum. Flavor.* **21(3)** (1996) 64.

[484] S. Nitz, M. H. Spraul, F. Drawert, *Chem. Mikrobiol. Technol. Lebensm.* **13** (1991) 183.

[485] I. Blank, A. Sen, W. Grosch, *Food Chem.*, **43** (1992) 337.

[486] B. M. Lawrence, *Perfum. Flavor.* **5(4)** (1980) 6.

[487] ISO/DIS 10624-1995.

[488] B. M. Lawrence, *Essential Oils 1979–1980*, Allured Publishing Corp., Wheaton IL 1981, p. 26.

[489] B. M. Lawrence, *Essential Oils 1981–1987*, Allured Publishing Corp., Wheaton IL 1989, p. 109.

[490] C. H. Brieskorn, G. Krauß, *Planta Med.*, (1986) 305.

[491] M. A. Villanueva, R. C. Torres, K. H. C. Baser, T. Özek, M. Kürkcüoglu, *Flav. Fragr. J.*, **8** (1993) 35.

[492] ISO 3065-1974.

[493] B. M. Lawrence, *Essential Oils 1979–1980*, Allured Publishing Corp., Wheaton IL 1981, p. 13.

[494] B. M. Lawrence, *Essential Oils 1981–1987*, Allured Publishing Corp., Wheaton IL 1989, p. 18.

[495] B. M. Lawrence, *Essential Oils 1981–1987*, Allured Publishing Corp., Wheaton IL 1989, p. 199.

[496] B. M. Lawrence, *Essential Oils 1988–1991*, Allured Publishing Corp., Carol Stream IL 1993, p. 122.

[497] B. M. Lawrence, *Progress in Essential Oils, Perfum. Flavor.* **18(3)** (1993) 72.

[498] B. M. Lawrence, *Progress in Essential Oils, Perfum. Flavor.* **19(6)** (1994) 61.

[499] M. H. Boelens, *Perfum. Flavor.* **9(6)** (1984) 1.

[500] ISO/DIS 770.2-1995.

[501] ISO/DIS 3044-1994.

[502] B. M. Lawrence, *Essential Oils 1979–1980*, Allured Publishing Corp., Wheaton IL 1981, p. 2.

[503] B. M. Lawrence, *Essential Oils 1981–1987*, Allured Publishing Corp., Wheaton IL 1989, p. 187.

[504] B. M. Lawrence, *Essential Oils 1988–1991*, Allured Publishing Corp., Carol Stream IL 1993, p. 2.

[505] D. J. Boland, J. J. Brophy, A. P. N. House, *Eucalyptus Leaf Oils: Use, Chemistry, Distillation and Marketing*, Inkata Pres, Melbourne and Sidney 1991, p. 71.

[506] B. M. Lawrence, *Essential Oils 1979–1980*, Allured Publishing Corp., Wheaton IL 1981, p. 6.

[507] B. M. Lawrence, *Essential Oils 1979–1980*, Allured Publishing Corp., Wheaton IL 1981, p. 17.

[508] B. M. Lawrence, *Essential Oils 1981–1987*, Allured Publishing Corp., Wheaton IL 1989, p. 6.

[509] B. M. Lawrence, *Essential Oils 1981–1987*, Allured Publishing Corp., Wheaton IL 1989, p. 97.

[510] B. M. Lawrence, *Essential Oils 1988–1991*, Allured Publishing Corp., Carol Stream IL 1993, p. 48.

[511] B. M. Lawrence, *Progress in Essential Oils, Perfum. Flavor.* **17(2)** (1992) 44.

[512] B. M. Lawrence, *Progress in Essential Oils, Perfum. Flavor.* **19(1)** (1994) 31.

[513] B. M. Lawrence, *Progress in Essential Oils, Perfum. Flavor.* **21(2)** (1996) 28.

[514] AFNOR NF T 75-350: 1992.

[515] G. Ohloff, *Scent and Fragrances*, Springer Verlag, Berlin, 1994, p. 178.

[516] B. M. Lawrence, *Essential Oils 1978*, Allured Publishing Corp., Wheaton IL 1979, p. 15.

[517] B. M. Lawrence, *Essential Oils 1979–1980*, Allured Publishing Corp., Wheaton IL 1981, p. 14.

[518] B. M. Lawrence, *Essential Oils 1981–1987*, Allured Publishing Corp., Wheaton IL 1989, p. 59.

[519] B. M. Lawrence, *Essential Oils 1988–1991*, Allured Publishing Corp., Carol Stream IL 1993, p. 82.

[520] AFNOR NF T 75-212: 1987.

[521] G. Ohloff, *Scent and Fragrances*, Springer Verlag, Berlin, 1994, p. 158.

[522] B. M. Lawrence, *Essential Oils 1976–1977*, Allured Publishing Corp., Wheaton IL 1978, p. 8.

[523] B. M. Lawrence, *Essential Oils 1976–1977*, Allured Publishing Corp., Wheaton IL 1978, p. 33.

[524] B. M. Lawrence, *Essential Oils 1978*, Allured Publishing Corp., Wheaton IL 1979, p. 1.

[525] B. M. Lawrence, *Essential Oils 1981–1987*, Allured Publishing Corp., Wheaton IL 1989, p. 119.

[526] B. M. Lawrence, *Essential Oils 1981–1987*, Allured Publishing Corp., Wheaton IL 1989, p. 158.

[527] B. M. Lawrence, *Essential Oils 1981–1987*, Allured Publishing Corp., Wheaton IL 1989, p. 184.

[528] B. M. Lawrence, *Essential Oils 1988–1991*, Allured Publishing Corp., Carol Stream IL 1993, p. 30.

[529] B. M. Lawrence, *Progress in Essential Oils, Perfum. Flavor.* **17(2)** (1992) 46.

[530] B. M. Lawrence, *Progress in Essential Oils, Perfum. Flavor.* **17(6)** (1992) 59.

[531] B. M. Lawrence, *Progress in Essential Oils, Perfum. Flavor.* **19(1)** (1994) 40.

[532] EOA No. 13.

[533] B. M. Lawrence, *Essential Oils 1976–1977*, Allured Publishing Corp., Wheaton IL 1978, p. 1.

[534] B. M. Lawrence, *Essential Oils 1976–1977*, Allured Publishing Corp., Wheaton IL 1978, p. 40.

[535] B. M. Lawrence, *Essential Oils 1981–1987*, Allured Publishing Corp., Wheaton IL 1989, p. 30.

[536] B. M. Lawrence, *Essential Oils 1988–1991*, Allured Publishing Corp., Carol Stream IL 1993, p. 20.

[537] B. M. Lawrence, *Essential Oils 1988–1991*, Allured Publishing Corp., Carol Stream IL 1993, p. 87.

[538] B. M. Lawrence, *Essential Oils 1988–1991*, Allured Publishing Corp., Carol Stream IL 1993, p. 183.

[539] B. M. Lawrence, *Progress in Essential Oils, Perfum. Flavor.* **20(2)** (1995) 54.

[540] O. Nishimura, *J. Agric. Food Chem.*, **43** (1995) 2941.

[541] EOA No. 63.

[542] B. M. Lawrence, *Progress in Essential Oils, Perfum. Flavor.* **17(6)** (1992) 52.

[543] B. M. Lawrence, *Essential Oils 1979–1980*, Allured Publishing Corp., Wheaton IL 1981, p. 34.

[544] G. Ohloff, *Scent and Fragrances*, Springer Verlag, Berlin, 1994, p. 151.

[545] B. M. Lawrence, *Essential Oils 1976–1977*, Allured Publishing Corp., Wheaton IL 1978, p. 36.

[546] B. M. Lawrence, *Essential Oils 1988–1991*, Allured Publishing Corp., Carol Stream IL 1993, p. 8.

[547] B. M. Lawrence, *Progress in Essential Oils, Perfum. Flavor.* **17(3)** (1992) 68.

[548] B. M. Lawrence, *Progress in Essential Oils, Perfum. Flavor.* **19(2)** (1994) 64.

[549] B. M. Lawrence, *Progress in Essential Oils, Perfum. Flavor.* **20(4)** (1995) 39.

[550] ISO 8897-1991.

[551] B. M. Lawrence, *Essential Oils 1976–1977*, Allured Publishing Corp., Wheaton IL 1978, p. 37.

[552] B. M. Lawrence, *Essential Oils 1979–1980*, Allured Publishing Corp., Wheaton IL 1981, p. 21.

[553] B. M. Lawrence, *Essential Oils 1981–1987*, Allured Publishing Corp., Wheaton IL 1989, p. 110.

[554] B. M. Lawrence, *Essential Oils 1981–1987*, Allured Publishing Corp., Wheaton IL 1989, p. 240.

[555] B. M. Lawrence, *Essential Oils 1988–1991*, Allured Publishing Corp., Carol Stream IL 1993, p. 104.

[556] B. M. Lawrence, *Progress in Essential Oils, Perfum. Flavor.* **21(1)** (1996) 40.

[557] EOA No. 181.

[558] G. Ohloff, *Scent and Fragrances*, Springer Verlag, Berlin, 1994, p. 181.

[559] B. M. Lawrence, *Essential Oils 1976–1977*, Allured Publishing Corp., Wheaton IL 1978, p. 9.

[560] B. M. Lawrence, *Essential Oils 1979–1980*, Allured Publishing Corp., Wheaton IL 1981, p. 5.

[561] B. M. Lawrence, *Essential Oils 1981–1987*, Allured Publishing Corp., Wheaton IL 1989, p. 24.

[562] B. M. Lawrence, *Essential Oils 1981–1987*, Allured Publishing Corp., Wheaton IL 1989, p. 93.

[563] B. M. Lawrence, *Essential Oils 1981–1987*, Allured Publishing Corp., Wheaton IL 1989, p. 186.

[564] B. M. Lawrence, *Essential Oils 1988–1991*, Allured Publishing Corp., Carol Stream IL 1993, p. 91.

[565] K. Tajima, J. Yamamoto, N. Toi, S. Tanaka, J. Koshino, Y. Fujikura, in *Flavours, Fragrances and Essential Oils, Proc. 13th Intern. Congess Flav. Fragr. Essent. Oils*, Istanbul 1995, ed. K. H. C. Baser, AREP Publ. Istanbul, 1995, vol. 2, p. 217.

[566] EOA No. 119.

[567] B. M. Lawrence, *Essential Oils 1976–1977*, Allured Publishing Corp., Wheaton IL 1978, p. 41.

[568] B. M. Lawrence, *Essential Oils 1981–1987*, Allured Publishing Corp., Wheaton IL 1989, p. 7.

[569] B. M. Lawrence, *Essential Oils 1981–1987*, Allured Publishing Corp., Wheaton IL 1989, p. 64.

[570] B. M. Lawrence, *Essential Oils 1981–1987*, Allured Publishing Corp., Wheaton IL 1989, p. 183.

[571] B. M. Lawrence, *Essential Oils 1981–1987*, Allured Publishing Corp., Wheaton IL 1989, p. 228.

[572] B. M. Lawrence, *Essential Oils 1988–1991*, Allured Publishing Corp., Carol Stream IL 1993, p. 97.

[573] AFNOR NF T 75-301: 1992.

[574] G. Ohloff, *Scent and Fragrances*, Springer Verlag, Berlin, 1994, p. 143.

[575] B. M. Lawrence, *Essential Oils 1976–1977*, Allured Publishing Corp., Wheaton IL 1978, p. 5.

[576] B. M. Lawrence, *Essential Oils 1978*, Allured Publishing Corp., Wheaton IL 1979, p. 3.

[577] B. M. Lawrence, *Essential Oils 1978*, Allured Publishing Corp., Wheaton IL 1979, p. 10.

[578] B. M. Lawrence, *Essential Oils 1979–1980*, Allured Publishing Corp., Wheaton IL 1981, p. 29.

[579] B. M. Lawrence, *Essential Oils 1981–1987*, Allured Publishing Corp., Wheaton IL 1989, p. 135.

[580] B. M. Lawrence, *Essential Oils 1981–1987*, Allured Publishing Corp., Wheaton IL 1989, p. 243.

[581] B. M. Lawrence, *Progress in Essential Oils, Perfum. Flavor.* **18(1)** (1993) 58.

[582] B. M. Lawrence, *Progress in Essential Oils, Perfum. Flavor.* **19(4)** (1994) 38.

[583] B. M. Lawrence, *Progress in Essential Oils, Perfum. Flavor.* **21(3)** (1996) 67.

[584] AFNOR NF T 75-201: 1992

[585] B. M. Lawrence, *Essential Oils 1978*, Allured Publishing Corp., Wheaton IL 1979, p. 11.

[586] B. M. Lawrence, *Essential Oils 1979–1980*, Allured Publishing Corp., Wheaton IL 1981, p. 27.

[587] B. M. Lawrence, *Essential Oils 1979–1980*, Allured Publishing Corp., Wheaton IL 1981, p. 31.

[588] B. M. Lawrence, *Essential Oils 1981–1987*, Allured Publishing Corp., Wheaton IL 1989, p. 82.

[589] B. M. Lawrence, *Essential Oils 1981–1987*, Allured Publishing Corp., Wheaton IL 1989, p. 97.

[590] B. M. Lawrence, *Essential Oils 1981–1987*, Allured Publishing Corp., Wheaton IL 1989, p. 239.

[591] B. M. Lawrence, *Essential Oils 1988–1991*, Allured Publishing Corp., Carol Stream IL 1993, p. 109.

[592] ISO 3054-1987.

[593] AFNOR NF T 75-304: 1992.

[594] B. M. Lawrence, *Essential Oils 1976–1977*, Allured Publishing Corp., Wheaton IL 1978, p. 33.

[595] B. M. Lawrence, *Essential Oils 1979–1980*, Allured Publishing Corp., Wheaton IL 1981, p. 14.

[596] B. M. Lawrence, *Essential Oils 1979–1980*, Allured Publishing Corp., Wheaton IL 1981, p. 17.

[597] B. M. Lawrence, *Essential Oils 1979–1980*, Allured Publishing Corp., Wheaton IL 1981, p. 29.

[598] B. M. Lawrence, *Essential Oils 1981–1987*, Allured Publishing Corp., Wheaton IL 1989, p. 241.

[599] B. M. Lawrence, *Progress in Essential Oils, Perfum. Flavor.* **18(2)** (1993) 43.

[600] B. M. Lawrence, *Progress in Essential Oils, Perfum. Flavor.* **19(5)** (1994) 84.

[601] B. M. Lawrence, *Progress in Essential Oils, Perfum. Flavor.* **20(3)** (1995) 23.

[602] EOA No. 157.

[603] B. M. Lawrence, *Essential Oils 1978*, Allured Publishing Corp., Wheaton IL 1979, p. 24.

[604] ISO 3214-1974.

[605] B. M. Lawrence, *Essential Oils 1981–1987*, Allured Publishing Corp., Wheaton IL 1989, p. 11.

[606] B. M. Lawrence, *Progress in Essential Oils, Perfum. Flavor.* **18(1)** (1993) 55.

[607] B. M. Lawrence, *Progress in Essential Oils, Perfum. Flavor.* **21(5)** (1996) 62.

[608] ISO/DIS 11019-1995.

[609] B. M. Lawrence, *Essential Oils 1979–1980*, Allured Publishing Corp., Wheaton IL 1981, p. 40.

[610] B. M. Lawrence, *Essential Oils 1981–1987*, Allured Publishing Corp., Wheaton IL 1989, p. 35.

[611] B. M. Lawrence, *Essential Oils 1981–1987*, Allured Publishing Corp., Wheaton IL 1989, p. 201.

[612] B. M. Lawrence, *Essential Oils 1988–1991*, Allured Publishing Corp., Carol Stream IL 1993, p. 113.

[613] EOA No. 76.

[614] B. M. Lawrence, *Essential Oils 1976–1977*, Allured Publishing Corp., Wheaton IL 1978, p. 19.

[615] B. M. Lawrence, *Essential Oils 1981–1987*, Allured Publishing Corp., Wheaton IL 1989, p. 18.

[616] B. M. Lawrence, *Essential Oils 1981–1987*, Allured Publishing Corp., Wheaton IL 1989, p. 71.

[617] B. M. Lawrence, *Essential Oils 1981–1987*, Allured Publishing Corp., Wheaton IL 1989, p. 96.

[618] B. M. Lawrence, *Essential Oils 1981–1987*, Allured Publishing Corp., Wheaton IL 1989, p. 230.

[619] B. M. Lawrence, *Essential Oils 1988–1991*, Allured Publishing Corp., Carol Stream IL 1993, p. 42.

[620] B. M. Lawrence, *Progress in Essential Oils, Perfum. Flavor.* **19(4)** (1994) 39.

[621] M. Petrzilka, Ch. Ehret, in *Perfumes – Art, Science and Technology*, eds. P. M. Müller, D. Lamparsky, Elsevier Science Publ. Ltd., London/New York, 1991, p. 515.

[622] ISO/DIS 9776-1996.

[623] B. M. Lawrence, *Essential Oils 1981–1987*, Allured Publishing Corp., Wheaton IL 1989, p. 10.

[624] B. M. Lawrence, *Essential Oils 1981–1987*, Allured Publishing Corp., Wheaton IL 1989, p. 14.

[625] B. M. Lawrence, *Essential Oils 1981–1987*, Allured Publishing Corp., Wheaton IL 1989, p. 67.

[626] B. M. Lawrence, *Essential Oils 1988–1991*, Allured Publishing Corp., Carol Stream IL 1993, p. 41.

[627] ISO 856-1981.

[628] B. M. Lawrence, *Essential Oils 1981–1987*, Allured Publishing Corp., Wheaton IL 1989, p. 8.

[629] B. M. Lawrence, *Essential Oils 1981–1987*, Allured Publishing Corp., Wheaton IL 1989, p. 122.

[630] B. M. Lawrence, *Essential Oils 1981–1987*, Allured Publishing Corp., Wheaton IL 1989, p. 172.

[631] B. M. Lawrence, *Essential Oils 1988–1991*, Allured Publishing Corp., Carol Stream IL 1993, p. 31.

[632] B. M. Lawrence, *Progress in Essential Oils, Perfum. Flavor.* **18(4)** (1993) 60.

[633] B. M. Lawrence, C.-K. Shu, W. R. Harris, *Perfum. Flavor.* **14(6)** (1989) 21.

[634] ISO 3033-1988.

[635] B. M. Lawrence, *Essential Oils 1976–1977*, Allured Publishing Corp., Wheaton IL 1978, p. 20.

[636] B. M. Lawrence, *Progress in Essential Oils, Perfum. Flavor.* **18(2)** (1993) 46.

[637] B. M. Lawrence, *Progress in Essential Oils, Perfum. Flavor.* **18(3)** (1993) 61.

[638] EOA No. 212.

[639] EOA No. 141.

[640] G. Ohloff, *Scent and Fragrances*, Springer Verlag, Berlin, 1994, p. 189.

[641] B. M. Lawrence, *Essential Oils 1981–1987*, Allured Publishing Corp., Wheaton IL 1989, p. 84.

[642] ISO 3517-1975.

[643] G. Ohloff, *Scent and Fragrances*, Springer Verlag, Berlin, 1994, p. 143.

[644] L. Mondello, P. Dugo, K. D. Bartle, B. Frere, G. Dugo, *Chromatographia*, **39** (1994) 529.

[645] ISO/DIS 3215-1994.

[646] B. M. Lawrence, *Essential Oils 1976–1977*, Allured Publishing Corp., Wheaton IL 1978, p. 12.

[647] B. M. Lawrence, *Essential Oils 1981–1987*, Allured Publishing Corp., Wheaton IL 1989, p. 12.

[648] B. M. Lawrence, *Essential Oils 1981–1987*, Allured Publishing Corp., Wheaton IL 1989, p. 153.

[649] B. M. Lawrence, *Essential Oils 1988–1991*, Allured Publishing Corp., Carol Stream IL 1993, p. 125.

[650] B. M. Lawrence, *Progress in Essential Oils, Perfum. Flavor.* **17(4)** (1992) 146.

[651] G. Ohloff, *Scent and Fragrances*, Springer Verlag, Berlin, 1994, p. 185.

[652] B. M. Lawrence, *Essential Oils 1976–1977*, Allured Publishing Corp., Wheaton IL 1978, p. 14.

[653] R. Tabacchi, *Parf. Cosmet. Arom.*, **16** (1977) 37.

[654] H. Schultz, G. Albroscheit, *J. Chrom.*, **466** (1989) 301.

[655] T. Terajima, H. Ichikawa, K. Tokuda, S. Nakamura in *Dev. Food Sci. 18*, eds. B. M. Lawrence, B. D. Mookherjee, B. J. Willis, Elsevier, Amsterdam 1988, p. 685.

[656] T. H. Moxham in *Progress in Essential Oil Research*, ed. E. Brunke, De Gruyter, Berlin 1986, p. 491.

[657] EOA No. 68.

[658] G. Ohloff, *Scent and Fragrances*, Springer Verlag, 1994, p. 188.

[659] B. M. Lawrence, *Essential Oils 1976–1977*, Allured Publishing Corp., Wheaton IL 1978, p. 34.

[660] B. M. Lawrence, *Essential Oils 1981–1987*, Allured Publishing Corp., Wheaton IL 1989, p. 31.

[661] B. M. Lawrence, *Progress in Essential Oils, Perfum. Flavor.* **17(3)** (1992) 62.

[662] EOA No. 67.

[669] B. M. Lawrence, *Essential Oils 1981–1987*, Allured Publishing Corp., Wheaton IL 1989, p. 86.

[664] A. O. Tucker, M. J. Maciarello in *Dev. Food. Sci. 34*, ed. G. Charalambous, Elsevier, Amsterdam 1994, p. 439.

[665] B. M. Lawrence, *Essential Oils 1976–1977*, Allured Publishing Corp., Wheaton IL 1978, p. 42.

[666] B. M. Lawrence, *Essential Oils 1979–1980*, Allured Publishing Corp., Wheaton IL 1981, p. 31.

[667] B. M. Lawrence, *Essential Oils 1981–1987*, Allured Publishing Corp., Wheaton IL 1989, p. 35.

[668] B. M. Lawrence, *Essential Oils 1988–1991*, Allured Publishing Corp., Carol Stream IL 1993, p. 24.

[669] B. M. Lawrence, *Essential Oils 1988–1991*, Allured Publishing Corp., Carol Stream IL 1993, p. 44.

[670] B. M. Lawrence, *Progress in Essential Oils, Perfum. Flavor.* **18(1)** (1993) 53.

[671] B. M. Lawrence, *Progress in Essential Oils, Perfum. Flavor.* **20(4)** (1995) 29.

[672] B. M. Lawrence, *Progress in Essential Oils, Perfum. Flavor.* **20(5)** (1995) 98.

[673] EOA No. 142.

[674] EOA No. 116.

[675] G. Ohloff, *Scent and Fragrances*, Springer Verlag, Berlin, 1994, p. 164.

[676] B. M. Lawrence, *Progress in Essential Oils, Perfum. Flavor.* **20(6)** (1995) 35.

[677] B. Belcour, D. Courtois, Ch. Ehret, V. Petiard, *Phytochemistry*, **34** (1993) 1313.

[678] EOA No. 293.

[679] B. M. Lawrence, *Essential Oils 1981–1987*, Allured Publishing Corp., Wheaton IL 1989, p. 26.

[680] B. M. Lawrence, *Essential Oils 1981–1987*, Allured Publishing Corp., Wheaton IL 1989, p. 27.

[681] B. M. Lawrence, *Essential Oils 1981–1987*, Allured Publishing Corp., Wheaton IL 1989, p. 178.

[682] B. M. Lawrence, *Essential Oils 1988–1991*, Allured Publishing Corp., Carol Stream IL 1993, p. 75.

[683] B. M. Lawrence, *Essential Oils 1988–1991*, Allured Publishing Corp., Carol Stream IL 1993, p. 178.

[684] B. M. Lawrence, *Essential Oils 1988–1991*, Allured Publishing Corp., Carol Stream IL 1993, p. 179.

[685] B. M. Lawrence, *Progress in Essential Oils, Perfum. Flavor.* **16(5)** (1991) 81.

[686] B. M. Lawrence, *Progress in Essential Oils, Perfum. Flavor.* **16(5)** (1991) 82.

[687] B. M. Lawrence, *Progress in Essential Oils, Perfum. Flavor.* **19(2)** (1994) 72.

[688] B. M. Lawrence, *Progress in Essential Oils, Perfum. Flavor.* **21(4)** (1996) 65.

[689] ISO 3527-1975.

[690] ISO 3757-1978.

[691] G. Ohloff, *Scent and Fragrances*, Springer Verlag, Berlin, 1994, p. 149.

[692] B. M. Lawrence, *Essential Oils 1976–1977*, Allured Publishing Corp., Wheaton IL 1978, p. 7.

[693] B. M. Lawrence, *Essential Oils 1981–1987*, Allured Publishing Corp., Wheaton IL 1989, p. 14.

[694] B. M. Lawrence, *Essential Oils 1988–1991*, Allured Publishing Corp., Carol Stream IL 1993, p. 90.

[695] B. M. Lawrence, *Progress in Essential Oils, Perfum. Flavor.* **20(3)** (1995) 72.

[696] ISO 3061-1979.

[697] B. M. Lawrence, *Essential Oils 1976–1977*, Allured Publishing Corp., Wheaton IL 1978, p. 15.

[698] B. M. Lawrence, *Essential Oils 1981–1987*, Allured Publishing Corp., Wheaton IL 1989, p. 139.

[699] B. M. Lawrence, *Progress in Essential Oils, Perfum. Flavor.* **17(3)** (1992) 67.

[700] B. M. Lawrence, *Progress in Essential Oils, Perfum. Flavor.* **20(2)** (1995) 49.

[701] EOA No. 65.

[702] ISO 3064-1977.

[703] G. Ohloff, *Scent and Fragrance*, Springer Verlag, Berlin, 1994, p. 141.

[704] B. M. Lawrence, *Essential Oils 1976–1977*, Allured Publishing Corp., Wheaton IL 1978, p. 19.

[705] B. M. Lawrence, *Essential Oils 1979–1980*, Allured Publishing Corp., Wheaton IL 1981, p. 44.

[706] B. M. Lawrence, *Progress in Essential Oils, Perfum. Flavor.* **18(5)** (1993) 43.

[707] ISO 3043-1975.

[708] B. M. Lawrence, *Essential Oils 1979–1980*, Allured Publishing Corp., Wheaton IL 1981, p. 41.

[709] B. M. Lawrence, *Essential Oils 1988–1991*, Allured Publishing Corp., Carol Stream IL 1993, p. 86.

[710] B. M. Lawrence, *Progress in Essential Oils, Perfum. Flavor.* **19(6)** (1994) 60.

[711] H. Kollmannsberger, S. Nitz, *Chem Mikrobiol. Technol. Lebensm.*, **15** (1993) 116.

[712] A. Bello, M. L. Rodriguez, A. Rosado, J. A. Pino, *J. Essent. Oil. Res.*, **7** (1995) 423.

[713] B. M. Lawrence, *Essential Oils 1978*, Allured Publishing Corp., Wheaton IL 1979, p. 15.

[714] B. M. Lawrence, *Essential Oils 1978*, Allured Publishing Corp., Wheaton IL 1979, p. 23.

[715] B. M. Lawrence, *Essential Oils 1981–1987*, Allured Publishing Corp., Wheaton IL 1989, p. 140.

[716] B. M. Lawrence, *Essential Oils 1988–1991*, Allured Publishing Corp., Carol Stream IL 1993, p. 138.

[717] B. M. Lawrence, *Essential Oils 1988–1991*, Allured Publishing Corp., Carol Stream IL 1993, p. 139.

[718] B. M. Lawrence, *Progress in Essential Oils, Perfum. Flavor.* **20(5)** (1995) 95.

[719] EOA No. 126.

[720] EOA No. 50.

[721] EOA No. 112.

[722] EOA No. 133.

[723] EOA No. 32.

[724] ISO 9842-1991.

[725] G. Ohloff, *Scent and Fragrances*, Springer Verlag, Berlin, 1994, p. 154.

[726] B. M. Lawrence, *Essential Oils 1976–1977*, Allured Publishing Corp., Wheaton IL 1978, p. 4.

[727] B. M. Lawrence, *Essential Oils 1976–1977*, Allured Publishing Corp., Wheaton IL 1978, p. 19.

[728] B. M. Lawrence, *Essential Oils 1978*, Allured Publishing Corp., Wheaton IL 1979, p. 6.

[729] B. M. Lawrence, *Essential Oils 1979–1980*, Allured Publishing Corp., Wheaton IL 1981, p. 7.

[730] B. M. Lawrence, *Essential Oils 1988–1991*, Allured Publishing Corp., Carol Stream IL 1993, p. 144.

[731] B. M. Lawrence, *Progress in Essential Oils, Perfum. Flavor.* **17(1)** (1992) 55.

[732] ISO 1342-1988.

[733] B. M. Lawrence, *Essential Oils 1976–1977*, Allured Publishing Corp., Wheaton IL 1978, p. 34.

[734] B. M. Lawrence, *Essential Oils 1979–1980*, Allured Publishing Corp., Wheaton IL 1981, p. 15.

[735] B. M. Lawrence, *Essential Oils 1979–1980*, Allured Publishing Corp., Wheaton IL 1981, p. 50.

[736] B. M. Lawrence, *Essential Oils 1981–1987*, Allured Publishing Corp., Wheaton IL 1989, p. 60.

[737] B. M. Lawrence, *Essential Oils 1981–1987*, Allured Publishing Corp., Wheaton IL 1989, p. 115.

[738] B. M. Lawrence, *Essential Oils 1981–1987*, Allured Publishing Corp., Wheaton IL 1989, p. 179.

[739] B. M. Lawrence, *Essential Oils 1988–1991*, Allured Publishing Corp., Carol Stream IL 1993, p. 49.

[740] B. M. Lawrence, *Essential Oils 1988–1991*, Allured Publishing Corp., Carol Stream IL 1993, p. 136.

[741] B. M. Lawrence, *Progress in Essential Oils, Perfum. Flavor.* **20(1)** (1995) 47.

[742] M. H. Boelens, *Perfum. Flavor.* **10(5)** (1985) 21.

[743] B. M. Lawrence, *Essential Oils 1981–1987*, Allured Publishing Corp., Wheaton IL 1989, p. 118.

[744] ISO/DIS 3761-1990.

[745] AFNOR NF T 75-255: 1992.

[746] G. Ohloff, *Scent and Fragrances*, Springer Verlag, Berlin, 1994, p. 147.

[747] B. M. Lawrence, *Essential Oils 1976–1977*, Allured Publishing Corp., Wheaton IL 1978, p. 11.

[748] B. M. Lawrence, *Essential Oils 1981–1987*, Allured Publishing Corp., Wheaton IL 1989, p. 192.

[749] B. M. Lawrence, *Essential Oils 1988–1991*, Allured Publishing Corp., Carol Stream IL 1993, p. 106.

[750] B. M. Lawrence, *Progress in Essential Oils, Perfum. Flavor.* **21(5)** (1996) 61.

[751] ISO/DIS 9909-1991.

[752] B. M. Lawrence, *Essential Oils 1976–1977*, Allured Publishing Corp., Wheaton IL 1978, p. 38.

[753] B. M. Lawrence, *Essential Oils 1981–1987*, Allured Publishing Corp., Wheaton IL 1989, p. 66.

[754] B. M. Lawrence, *Essential Oils 1981–1987*, Allured Publishing Corp., Wheaton IL 1989, p. 127.

[755] B. M. Lawrence, *Essential Oils 1988–1991*, Allured Publishing Corp., Carol Stream IL 1993, p. 16.

[756] B. M. Lawrence, *Essential Oils 1988–1991*, Allured Publishing Corp., Carol Stream IL 1993, p. 80.

[757] B. M. Lawrence, *Essential Oils 1988–1991*, Allured Publishing Corp., Carol Stream IL 1993, p. 168.

[758] B. M. Lawrence, *Progress in Essential Oils, Perfum. Flavor.* **19(6)** (1994) 61.

[759] ISO 3526-1976.

[760] B. M. Lawrence, *Essential Oils 1981–1987*, Allured Publishing Corp., Wheaton IL 1989, p. 129.

[761] B. M. Lawrence, *Essential Oils 1988–1991*, Allured Publishing Corp., Carol Stream IL 1993, p. 4.

[762] B. M. Lawrence, *Essential Oils 1988–1991*, Allured Publishing Corp., Carol Stream IL 1993, p. 84.

[763] ISO 3518-1979.

[764] G. Ohloff, *Scent and Fragrances*, Springer Verlag, Berlin, 1994, p. 175.

[765] B. M. Lawrence, *Essential Oils 1976–1977*, Allured Publishing Corp., Wheaton IL 1978, p. 3.

[766] B. M. Lawrence, *Essential Oils 1976–1977*, Allured Publishing Corp., Wheaton IL 1978, p. 15.

[767] B. M. Lawrence, *Essential Oils 1981–1987*, Allured Publishing Corp., Wheaton IL 1989, p. 22.

[768] B. M. Lawrence, *Essential Oils 1988–1991*, Allured Publishing Corp., Wheaton IL 1993, p. 180.

[769] E. J. Brunke, J. Volhardt, G. Schmaus, *Flav. Fragr. J.*, **10** (1995) 211.

[770] ISO 590-1981.

[771] B. M. Lawrence, *Essential Oils 1981–1987*, Allured Publishing Corp., Wheaton IL 1989, p. 154.

[772] L.-F. Zhu, D.-S. Ding, B. M. Lawrence, *Perfum. Flavor.* **19(4)** (1992) 17.

[773] EOA No. 197.

[774] B. M. Lawrence, *Essential Oils 1978*, Allured Publishing Corp., Wheaton IL 1979, p. 26.

[775] B. M. Lawrence, *Essential Oils 1981–1987*, Allured Publishing Corp., Wheaton IL 1989, p. 16.

[776] B. M. Lawrence, *Essential Oils 1988–1991*, Allured Publishing Corp., Carol Stream IL 1993, p. 2.

[777] B. M. Lawrence, *Progress in Essential Oils, Perfum. Flavor.* **17(1)** (1992) 50.

[778] ISO/DIS 11016-1996.

[779] B. M. Lawrence, *Essential Oils 1976–1977*, Allured Publishing Corp., Wheaton IL 1978, p. 31.

[780] B. M. Lawrence, *Essential Oils 1981–1987*, Allured Publishing Corp., Wheaton IL 1989, p. 102.

[781] B. M. Lawrence, *Essential Oils 1981–1987*, Allured Publishing Corp., Wheaton IL 1989, p. 169.

[782] B. M. Lawrence, *Essential Oils 1981–1987*, Allured Publishing Corp., Wheaton IL 1989, p. 202.

[783] J. C. Chalchat, R.-P. Garry, J.-P. Mathieu, *J. Essent. Oil. Res.*, **6** (1994) 73.

[784] B. M. Lawrence, *Progress in Essential Oils, Perfum. Flavor.* **17(5)** (1992) 131.

[785] ISO/DIS 10115-1989.

[786] B. M. Lawrence, *Essential Oils 1988–1991*, Allured Publishing Corp., Carol Stream IL 1993, p. 5.

[787] B. M. Lawrence, *Essential Oils 1988–1991*, Allured Publishing Corp., Carol Stream IL 1993, p. 89.

[788] B. M. Lawrence, *Progress in Essential Oils, Perfum. Flavor.* **20(4)** (1995) 38.

[789] ISO/DIS 4730-1992.

[790] B. M. Lawrence, *Essential Oils 1978*, Allured Publishing Corp., Wheaton IL 1979, p. 21.

[791] B. M. Lawrence, *Essential Oils 1988–1991*, Allured Publishing Corp., Carol Stream IL 1993, p. 57.

[792] B. M. Lawrence, *Essential Oils 1988–1991*, Allured Publishing Corp., Carol Stream IL 1993, p. 99.

[793] AFNOR Pr T 75-349: 1992.

[794] B. M. Lawrence, *Essential Oils 1979–1980*, Allured Publishing Corp., Wheaton IL 1981, p. 37.

[795] B. M. Lawrence, *Essential Oils 1979–1980*, Allured Publishing Corp., Wheaton IL 1981, p. 104.

[796] B. M. Lawrence, *Essential Oils 1981–1987*, Allured Publishing Corp., Wheaton IL 1989, p. 38.

[797] B. M. Lawrence, *Essential Oils 1981–1987*, Allured Publishing Corp., Wheaton IL 1989, p. 111.

[798] B. M. Lawrence, *Progress in Essential Oils, Perfum. Flavor.* **17(5)** (1992) 140.

[799] B. M. Lawrence, *Progress in Essential Oils, Perfum. Flavor.* **20(3)** (1995) 67.

[800] I. Wahlberg, M. B. Heltje, K. Karlsson, C. R. Enzell, *Acta Chem. Scand.*, **25** (1971) 3285.

[801] M. Petrzilka, Ch. Ehret, in *Perfumers – Art, Science and Technology*, eds. P. M. Müller, D. Lamparsky, Elsevier Science Publ. Ltd., London/New York, 1991, p. 511.

[802] B. M. Lawrence, *Progress in Essential Oils, Perfum. Flavor.* **20(6)** (1995) 42.

[803] M. Gscheidmeier, H. Fleig, in *Ullmann's Encyclopedia of Industrial Chemistry*, Vol. A27, VCH Verlagsgesellschaft, Weinheim 1996, p. 267.

[804] ISO 412-1965.

[805] AFNOR NF T 75-359: 1991.

[806] EOA No. 165.

[807] B. M. Lawrence, *Essential Oils 1981–1987*, Allured Publishing Corp., Wheaton IL 1989, p. 160.

[808] B. M. Lawrence, *Essential Oils 1976–1977*, Allured Publishing Corp., Wheaton IL 1978, p. 23.

[809] B. M. Lawrence, *Essential Oils 1976–1977*, Allured Publishing Corp., Wheaton IL 1978, p. 27.

[810] B. M. Lawrence, *Progress in Essential Oils, Perfum. Flavor.* **18(2)** (1993) 25.

[811] D. Ehlers, M. S. Bartholomae, *Z. Lebensm Unters. Forsch.*, **199** (1994) 38.

[812] ISO 4716-1989.

[813] G. Ohloff, *Scent and Fragrances*, Springer Verlag, Berlin, 1994, p. 172.

[814] J. Garnero, *Parf. Cosm. Sav. France*, **1** (1971) 569.

[815] B. Maurer, M. Fracheboud, A. Grieder, G. Ohloff, *Helv. Chim. Acta*, **55** (1972) 2371.

[816] J. Samadja, E. M. Gaydou, G. Lamarty, J.-Y. Conan, *Parf. Cosm. Arom.*, **84** (1988) 61.

[817] N. Sellier, A. Cazaussus, H. Budzinski, M. Lebon, *J. Chrom.*, **577** (1991) 451.

[818] G. Ohloff, *Scent and Fragrances*, Springer Verlag, Berlin, 1994, p. 161.

[819] B. M. Lawrence, *Progress in Essential Oils, Perfum. Flavor.* **19(1)** (1994) 33.

[820] ISO 3063-1983.

[821] B. M. Lawrence, *Essential Oils 1976–1977*, Allured Publishing Corp., Wheaton IL 1978, p. 3.

[822] B. M. Lawrence, *Essential Oils 1981–1987*, Allured Publishing Corp., Wheaton IL 1989, p. 195.

[823] B. M. Lawrence, *Essential Oils 1988–1991*, Allured Publishing Corp., Carol Stream IL 1993, p. 57.

[824] B. M. Lawrence, *Progress in Essential Oils, Perfum. Flavor.* **20(2)** (1995) 57.

[825] ISO 3523-1976.

[826] G. Jellinek, *Sensory Evaluation of Food*, VCH Weinheim/Ellis Horwood, Ltd., Chichester 1985.

[827] W. Funk, V. Dammann, G. Donnevert, *Qualitätssicherung in der Analytischen Chemie*, VCH, Weinheim 1992.

[828] K. Molt, Principles and applications of quality control by near infrared spectroscopy using the example of polymer additives. *Fresenius J. Anal. Chem.*, **348**, 523 (1994).

[829] S. Kling, A. Vogt, A. Fehrmann, A. Hoffmann, L. Rudzik, E. Wüst, Qualitative Analyse pulverförmiger Lebensmittelzusatzstoffe mittels NIR Spektroskopie, *Deutsche Milchwirtschaft* 25/1994, 45. Jg.

[830] K. Molt, NIR Spektroskopie für der Identitätsprüfung, *Nachr. Chem. Tech. Lab.* **43(3)** (1995) 330–336.

[831] *Chromatographic methods.* 5th ed. A. Braithwaite, F. J. Smith, Blackie, Glasgow, UK 1996.

[832] R. L. St. Claire III, Capillary electrophoresis. *Anal Chem.* **68(12)** (1996) 569R–586R.

[833] *Modern Spectroscopy.* 3rd ed., J. M. Hollas, John Wiley and Sons Ltd., Chichester, Sussex, UK 1996.

[834] A. A. Swigar, R. M. Silverstein, 'Monoterpenes' Infrared, Mass, ^{1}H NMR and ^{13}C NMR Spectra, and Kováts Indices, Aldrich Chemical Company Inc., Milwaukee 1981.

[835] H. Surburg, M. Guentert, H. Harder, Volatile compounds from flowers. Analytical and olfactory aspects. In *Bioactive Volatile Compounds from Plants.* eds. R. Teranishi, R. G. Buttery, H. Sugisawa, pp. 168–186, *ACS Symposium Series No. 525, Amer. Chem. Soc.*, Washington 1993.

[836] S. Maeno, P. A. Roderiguez, Simple and versatile injection system for capillary gas chromatographic columns. Performance evaluation of a system including mass spectrometric and light-pipe Fourier-Transform infra-red detection. *J. Chromatogr.* **731(1–2)** (1996) 201–215.

[837] L. D. Rothmann, Column liquid chromatography: equipment and instrumentation. *Anal. Chem.* **68(12)** (1996) 587R–598R.

[838] Boehringer Mannheim, Biochemica, Methoden der enyzmatischen Bioanalytik und Lebensmittelanalytik.

[839] H. Maarse, R. Belz, *Isolation, Separation and Identification of Volatile Compounds in Aroma Research.* Akademie-Verlag, Berlin 1981.

[840] R. H. McCormick, *Food Analysis, Vol. 4: Separation Techniques*, eds. D. W. Gruenwedel and J. R. Whitacker, Marcel Dekker Inc., New York 1987, p. 1.

[841] T. L. Chester, J. D. Pinkston, D. E. Raynie, Supercritical-fluid chromatography and extractions. *Anal. Chem.* **68(12)** (1996) 487R–514R.

[842] User is a four-letter word. R. D. McDowall, *LC–GC Int.* **9(7)** (1996) 404–407.

[843] H. J. Bouwmeester, J. A. R. Davies, H. Toxopeus, Enantiomeric composition of carvone, limonene and carveols in seeds of dill and annual and biennial caraway varieties. *J. Agric. Food Chem.* **43** (1995) 3057–3064.

[844] R. Carle, I. Fleischhauer, J. Beyer, E. Reinhard, Studies on the origin of (−)-α-bisabolol and chamazulene in chamomile preparations. Part I. Investigations by isotope ratio mass spectrometry (IRMS). *Plant Med.* **56** (1990) 456–460.
[845] D. L. J. Opdyke, *Food Cosm. Toxicol.* **11** (1973) 95–96.

Formula Index;
CAS Registry Number Index

Formula	Molecular weight	Name	CAS registry number	Page
$C_3H_6O_2$	74.08	Ethyl formate	[109-94-4]	18
C_4H_5NS	99.14	Allyl isothiocyanate	[57-06-7]	22
$C_4H_6O_2$	86.09	Diacetyl	[431-03-8]	16
C_4H_8OS	104.17	3-Mercapto-2-butanone	[40789-98-8]	22
$C_4H_8O_2$	88.11	(±)-Acetoin	[52217-02-4]	16
		Ethyl acetate	[141-78-6]	18
$C_4H_{10}S_2$	122.24	Methyl propyl disulfide	[2179-60-4]	166
$C_5H_4O_2$	96.09	2-Furaldehyde	[98-01-1]	137
C_5H_6OS	114.16	2-Furylmethanethiol	[98-02-2]	156
		2-Methylfuran-3-thiol	[28588-74-1]	138
$C_5H_6O_3$	114.10	4-Hydroxy-5-methyl-3(2H)-furanone	[19332-27-1]	138
C_5H_7NOS	129.18	2-Acetyl-2-thiazoline	[29926-41-8]	158
C_5H_7NS	113.18	2,5-Dimethylthiazole	[4175-66-0]	158
$C_5H_{10}O_2$	102.13	Ethyl propionate	[105-37-3]	19
$C_6H_6N_2O$	122.13	2-Acetylpyrazine	[22047-25-2]	158
$C_6H_6O_2$	110.11	2-Acetylfuran	[1192-62-7]	137
$C_6H_6O_3$	126.11	Maltol	[118-71-8]	143
		Methyl 2-furoate	[611-13-2]	138
C_6H_7NO	109.13	2-Acetylpyrrole	[1072-83-9]	157
$C_6H_8N_2$	108.14	2,3-Dimethylpyrazine	[5910-89-4]	158
		2,5-Dimethylpyrazine	[123-32-0]	158
		2,6-Dimethylpyrazine	[108-50-9]	158
$C_6H_8O_2$	112.13	3-Methyl-2-cyclopenten-2-ol-1-one	[80-71-7]	86
		2,5-Dimethyl-3(2H)-furanone	[14400-67-0]	138
$C_6H_8O_3$	128.13	2,5-Dimethyl-4-hydroxy-2H-furan-3-one	[3658-77-3]	142
		4,5-Dimethyl-2(5H)-furanone	[28664-35-9]	149
C_6H_9N	95.14	2-Acetyl-3,4-dihydro-5H-pyrrole	[85213-22-5]	157
C_6H_9NOS	143.21	4-Methyl-5-thiazolethanol	[137-00-8]	158
$C_6H_{10}O$	98.14	trans-2-Hexenal	[6728-26-3]	14
$C_6H_{10}O_2$	114.15	2-Methyl-2-pentenoic acid	[3142-72-1]	17
$C_6H_{10}S_2$	146.27	Diallyl disulfide	[2179-57-9]	166
$C_6H_{12}O$	100.16	trans-2-Hexenol	[928-95-0]	9
		cis-3-Hexenol	[928-96-1]	9
		Hexanal	[66-25-1]	12
$C_6H_{12}OS$	132.22	4-Mercapto-4-methyl-2-pentanone	[19872-52-7]	22

Formula	Molecular weight	Name	CAS registry number	Page
		2-Ethyl-2,5-dimethylpyrazine	[13925-07-0]	158
$C_8H_{12}N_2O$	152.20	2-Methoxy-3-isopropylpyrazine	[25773-40-4]	158
$C_8H_{14}O$	126.20	6-Methyl-5-hepten-2-one	[110-93-0]	27, 41
$C_8H_{14}O_2$	142.20	*trans*-2-Hexenyl acetate	[2497-18-9]	19
		cis-3-Hexenyl acetate	[3681-71-8]	19
		γ-Octalactone	[104-50-7]	148
$C_8H_{14}O_4$	174.20	Ethyl 2-methyl-1,3-dioxolane-2- acetate	[6413-10-1]	147
$C_8H_{16}O$	128.21	1-Octen-3-ol	[3391-86-4]	10
		n-Octanal	[124-13-0]	12
		3-Octanone	[106-68-3]	16
		2,5-Diethyltetrahydrofuran	[41239-48-9]	138
$C_8H_{16}OS$	160.27	2-Methyl-4-propyl-1,3-oxathiane	[67715-80-4]	138
$C_8H_{16}O_2$	144.21	Hexyl acetate	[142-92-7]	18
		Butyl butyrate	[109-21-7]	19
		Ethyl octanoate	[123-66-0]	21
$C_8H_{17}NO$	143.23	5-Methyl-3-heptanone oxime	[22457-23-4]	22
$C_8H_{17}NS_2$	191.35	2,4-Dimethyl-6-isopropyldihydro-4H-1,3,5- dithiazine	[104691-41-0]	159
		4,6-Dimethyl-2-isopropyldihydro-4H-1,3,5- dithiazine	[104691-40-9]	159
$C_8H_{18}O$	130.23	3-Octanol	[589-98-0]	9
$C_9H_6O_2$	146.15	Coumarin	[91-64-5]	153
C_9H_7N	129.16	Cinnamonitrile	[4360-47-8]	118
C_9H_8O	132.16	*trans*-Cinnamaldehyde	[14371-10-9]	102
		cis-Cinnamaldehyde	[57194-69-1]	102
$C_9H_8O_2$	148.16	Dihydrocoumarin	[119-84-6]	154
C_9H_9N	131.18	3-Methylindole	[83-34-1]	157
$C_9H_{10}O$	134.18	*trans*-Cinnamic alcohol	[4407-36-7]	98
		cis-Cinnamic alcohol	[4510-34-3]	98
		Dihydrocinnamaldehyde	[104-53-0]	101
		Hydratropaldehyde	[93-53-8]	101
		4-Methylphenylacetaldehyde	[104-09-6]	102
		4-Methylacetophenone	[122-00-9]	106
$C_9H_{10}O_2$	150.18	Benzyl acetate	[140-11-4]	111, 193, 218
		4-Methoxyacetophenone	[100-06-1]	134
$C_9H_{10}O_3$	166.18	Veratraldehyde	[120-14-9]	131
		Ethylvanillin	[121-32-4]	132
$C_9H_{10}O_4$	182.18	Methyl 3-methylresorcylate	[33662-58-7]	136
$C_9H_{11}NO_2$	165.19	Methyl *N*-methylanthranilate	[85-91-6]	119, 182, 206
$C_9H_{12}O$	136.19	Phenethyl methyl ether	[3558-60-9]	95
		Dihydrocinnamic alcohol	[122-97-4]	96
$C_9H_{12}O_2$	152.19	β-Orcinol monomethyl ether	[56426-87-5]	202

Formula	Molecular weight	Name	CAS registry number	Page
C$_9$H$_{14}$N$_2$O	166.22	2-Methoxy-3-isobutylpyrazine	[24683-00-9]	158, 190
C$_9$H$_{14}$O	138.21	2-*trans*-6-*cis*-Nonadienal	[557-48-2]	11
		2,4-Dimethyl-3-cyclohexene carboxaldehyde	[68039-49-6]	78
C$_9$H$_{14}$O$_2$	154.21	Methyl 2-octynoate	[111-12-6]	21
C$_9$H$_{16}$O	140.22	2-*trans*-6-*cis*-Nonadien-1-ol	[28069-72-9]	11
		2,6-Dimethyl-5-hepten-1-al	[106-72-9]	15
C$_9$H$_{16}$OS	172.28	S-*sec*.Butyl 3-methyl-2-butenethioate	[34322-09-3]	190
C$_9$H$_{16}$O$_2$	156.22	2-Propenyl hexanoate	[123-68-2]	20
		γ-Nonalactone	[104-61-0]	148
C$_9$H$_{16}$O$_2$	156.23	β-Methyl-γ-octalactone	[39212-23-2]	148
C$_9$H$_{18}$O	142.24	*n*-Nonanal	[124-19-6]	12
C$_9$H$_{18}$O$_2$	158.24	Isoamyl butyrate	[106-27-4]	19
		Ethyl heptanoate	[106-30-9]	20
C$_9$H$_{19}$NS$_2$	205.37	2-(2-Butyl)-4,6-dimethyldihydro-4H-1,3,5-dithiazine	[104691-36-1]	159
		6-(2-Butyl)-2,4-dimethyldihydro-4H-1,3,5-dithiazine	[104691-37-4]	159
		2,4-Dimethyl-6-isobutyldihydro-4H-1,3,5-dithiazine	[101517-86-6]	159
		4,6-Dimethyl-2-isobutyldihydro-4H-1,3,5-dithiazine	[101517-87-7]	159
C$_9$H$_{20}$O	144.26	2,6-Dimethyl-2-heptanol	[13254-34-7]	9
C$_9$H$_{20}$O$_2$	160.26	Heptanal dimethyl acetal	[10032-05-0]	11
C$_{10}$H$_9$Cl$_3$O$_2$	267.54	Trichloromethyl phenyl carbinyl acetate	[90-17-5]	112
C$_{10}$H$_9$N	143.19	6-Methylquinoline	[91-62-3]	157
C$_{10}$H$_{10}$O$_2$	162.19	Methyl *trans*-cinnamate	[1754-62-7]	116
		Isosafrole	[120-58-1]	127
		Safrole	[94-59-7]	127, 174, 212
C$_{10}$H$_{12}$O	148.20	Cuminaldehyde	[122-03-2]	99
		Benzylacetone	[255-26-7]	107
		Anethole	[104-46-1]	122, 170, 189, 213
		Estragole	[140-67-0]	122, 170, 214
C$_{10}$H$_{12}$O$_2$	164.20	Benzyl propionate	[122-63-4]	111
		Phenethyl acetate	[103-45-7]	112
		(±)-1-Phenylethyl acetate	[50373-55-2]	112
		Ethyl phenylacetate	[101-97-3]	116
		trans-Isoeugenol	[5932-68-3]	124
		cis-Isoeugenol	[5912-86-7]	124
		Eugenol	[97-53-0]	124, 125, 171, 178, 183, 207

Formula	Molecular weight	Name	CAS registry number	Page
		β-Asarone	[5273-86-9]	173
C₁₂H₁₈O	178.28	2,2-Dimethyl-3-(3-methylphenyl)propanol	[103694-68-4]	96
		3-Methyl-5-phenylpentanol	[55066-48-3]	98
		2-Methyl-5-phenylpentanol	[25634-93-9]	97
		1-Phenyl-3-methyl-3-pentanol	[10415-87-9]	98
C₁₂H₂₀O₂	196.29	Geranyl acetate	[105-87-3]	42
		Neryl acetate	[141-12-8]	43
		(±)-Linalyl acetate	[40135-38-4]	43
		Lavandulyl acetate	[25905-14-0]	44, 195
		Ethyl 2-*trans*-4-*cis*-decadienoate	[3025-30-7]	21
		Decahydro-β-naphthyl acetate	[10519-11-6]	88
		α-Terpinyl acetate	[80-26-2]	69, 175
		(+)-Bornyl acetate	[20347-65-3]	70
		(−)-Bornyl acetate	[5655-61-8]	70
		(±)-Bornyl acetate	[36386-52-4]	70
		Isobornyl acetate	[125-12-2]	70
		Allyl 3-cyclohexylpropionate	[2705-87-5]	88
		Acetaldehyde ethyl phenethyl acetal	[2556-10-7]	95
C₁₂H₂₀O₂S	212.35	(+)-*p*-Menthan-8-thiol-3-one *S*-acetate	[57074-23-7]	173
		(−)-*p*-Menthan-8-thiol-3-one *S*-acetate	[57129-12-1]	173
C₁₂H₂₂O	182.30	2-*n*-Heptylcyclopentanone	[137-03-1]	80
C₁₂H₂₂O₂	198.30	Dihydromyrcenyl acetate	[53767-93-4]	45
		(±)-Citronellyl acetate	[67650-82-2]	45
		(−)-Menthyl acetate	[2623-23-6]	68
		(±)-Menthyl acetate	[29066-34-0]	68
		1-*p*-Menthanyl acetate	[26252-11-9]	69
		8-*p*-Menthanyl acetate	[80-25-1]	69
		trans-2-*tert*.Butylcyclohexyl acetate	[20298-70-8]	87
		cis-2-*tert*.Butylcyclohexyl acetate	[20298-69-5]	87
		trans-4-*tert*.Butylcyclohexyl acetate	[1900-69-2]	87
		cis-4-*tert*.Butylcyclohexyl acetate	[10411-92-4]	87
C₁₂H₂₂O₃	214.30	3-Pentyltetrahydro-2H-pyran-4-ol acetate	[18871-14-2]	144
C₁₂H₂₄O	184.32	3,4,5,6,6-Pentamethyl-3-hepten-2-ol	[81787-06-6]	10
		3,4,5,6,6-Pentamethyl-4-hepten-2-ol	[81787-07-7]	10
		3,5,6,6-Tetramethyl-4-methyleneheptan-2-ol	[81787-05-5]	10
		n-Dodecanal	[112-54-9]	13
		2-Methylundecanal	[110-41-8]	13
C₁₂H₂₆O₃	218.34	Hydroxydihydrocitronellal dimethyl acetal	[141-92-4]	40
C₁₃H₁₀O	182.22	Benzophenone	[119-61-9]	107
C₁₃H₁₂	168.24	Diphenylmethane	[101-81-5]	92
C₁₃H₁₅N	185.27	6-Isobutylquinoline	[68141-26-4]	157
C₁₃H₁₆O₂	204.27	2,4-Dimethyl-4,4a,5,9b-tetrahydroindeno-[1,2:d]-*m*-dioxin	[27606-09-3]	146

Formula	Molecular weight	Name	CAS registry number	Page
		3,3-Dimethyl-5-(2,2,3-trimethyl-3-cyclopenten-1-yl)-4-penten-2-ol	[107898-54-4]	75
		(−)-α-Bisabolol	[23089-26-1]	177
		β-Eudesmol	[473-15-4]	167
		epi-γ-Eudesmol	[15051-81-7]	167
		(−)-Guaiol	[489-86-1]	192
		Elemol	[639-99-6]	167, 187
		Cedrol	[77-53-2]	176
		(+)-Carotol	[465-28-1]	175
		Bulnesol	[22451-73-6]	192
		Patchoulol	[5986-55-0]	205
$C_{15}H_{26}O_2$	238.37	Geranyl isovalerate	[109-20-6]	43
		Citronellyl tiglate	[24717-85-9]	45
		Bornyl isovalerate	[76-50-6]	216
$C_{15}H_{26}O_4$	270.37	1,13-Tridecanedioic acid ethylene ester	[105-95-3]	153
$C_{15}H_{28}O$	224.39	Cyclopentadecanone	[502-72-7]	82
$C_{15}H_{28}O_2$	240.39	Citronellyl isovalerate	[68922-10-1]	45
		15-Pentadecanolide	[106-02-5]	150
$C_{15}H_{28}O_3$	256.39	12-Oxa-16-hexadecanolide	[6707-60-4]	152
		11-Oxa-16-hexadecanolide	[3391-83-1]	152
		10-Oxa-16-hexadecanolide	[1725-01-5]	152
$C_{15}H_{30}O$	226.41	1-(2,2,6-Trimethylcyclohexyl)-hexan-3-ol	[70788-30-6]	77
$C_{15}H_{30}O_2$	242.41	(Ethoxymethoxy)cyclododecane	[58567-11-6]	78
$C_{16}H_{14}O_2$	238.29	Benzyl cinnamate	[103-41-3]	116
$C_{16}H_{16}O_2$	240.30	Phenethyl phenylacetate	[102-20-5]	116
$C_{16}H_{24}O$	232.37	Allylionone	[79-78-7]	63
$C_{16}H_{26}O$	234.38	2,3,8,8-Tetramethyl-1,2,3,4,5,6,7,8-octahydro-2-naphthalenyl methyl ketone	[54464-57-2]	85
$C_{16}H_{28}O$	236.40	Cedryl methyl ether	[19870-74-7] [67874-81-1]	58
		3-*trans*-Isocamphylcyclohexanol	[4105-12-8]	76
		5-Cyclohexadecen-1-one	[37609-25-9]	84
		3a,6,6,9a-Tetramethyldodecahydro-naphtho[2,1-b]furan	[6790-58-5]	141, 168
$C_{16}H_{28}O_2$	252.40	7-Hexadecen-16-olide	[123-69-3]	166
		9-Hexadecen-16-olide	[28645-51-4]	152
$C_{16}H_{30}O$	238.41	Cyclohexadecanone	[2550-52-9]	169
		Muscone	[541-91-3]	83, 170
$C_{17}H_{16}O_2$	252.31	Phenethyl cinnamate	[103-53-7]	117
$C_{17}H_{24}O$	244.38	5-Acetyl-1,1,2,3,3,6-hexamethylindane	[15323-35-0]	108
		4-Acetyl-1,1-dimethyl-6-*tert*.butylindane	[13171-00-1]	108
$C_{17}H_{26}O$	246.39	4-Acetyl-1,1,6-trimethyl-6,8a-ethano-1,2,3,5,6,7,8,8a-octahydronaphthalene	[32388-56-0]	68

Formula	Molecular weight	Name	CAS registry number	Page
		Methyl 2,6,10-trimethyl-2,5,9-cyclododecatrien-1-yl ketone	[28371-99-5]	86
$C_{17}H_{26}O_2$	262.39	Khusimyl acetate	[61474-33-7]	71
$C_{17}H_{28}O_2$	264.41	Guaiyl acetate		71
		Cedryl acetate	[77-54-3]	71
$C_{17}H_{30}O$	250.42	Civetone	[542-46-1]	84, 169
		9-Cycloheptadecen-1-one	[542-46-1]	84
$C_{17}H_{32}O$	252.44	Cycloheptadecanone	[3661-77-6]	169
$C_{18}H_{24}O_2$	272.39	Geranyl phenylacetate	[102-22-7]	116
$C_{18}H_{26}O$	258.40	6-Acetyl-1,1,2,4,4,7-hexamethyltetralin	[1506-02-1]	110
		5-Acetyl-1,1,2,6-tetramethyl-3-isopropylindane	[68140-48-7]	109
		4,6,6,7,8,8a-Hexamethyl-1,3,4,6,7,8-hexahydrocyclopenta[g]benzopyran	[1222-05-5]	145
$C_{20}H_{36}O_2$	308.50	Sclareol	[515-03-7]	210

Subject Index